AF334572

Materials in Electronics

BY THE SAME AUTHOR:

Reliable Electronic Assembly Production

Materials in Electronics

C. E. JOWETT

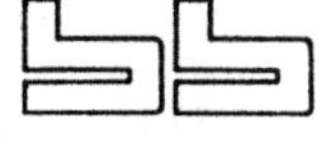

London

BUSINESS BOOKS LIMITED

First published 1971

© CHARLES ERIC JOWETT 1971

All rights reserved.
Except for normal review purposes, no part of this book
may be reproduced or utilised in any form or by any means,
electronic or mechanical, including photocopying, recording,
or by any information storage and retrieval system, without
permission of the publishers

ISBN 0 220 66796 9

*This book has been set 10 on 11 pt Baskerville
and printed by Robert MacLehose, Glasgow
for the publishers, Business Books Limited
(registered office: 180 Fleet Street, London EC4)
publishing offices: Mercury House, Waterloo Road, London SE1*

MADE AND PRINTED IN GREAT BRITAIN

Contents

Preface

This book can neither aspire to be a comprehensive treatment of materials nor an encyclopedia of data on present-day materials practice. In the field of electronic materials much has been written on their behaviour as a function of atomic structure; on the comparative characteristics of gases, liquids and solids; on the causes of absorption, losses and breakdown, or the inconsistencies of material that remain unexplained.

Other books cover the properties of materials employed in electronics under a variety of conditions, approaching the subject from an empirical angle since it is difficult to extrapolate data or predict performance under untried conditions. The technical publications contain a wealth of information on special problems of application, but articles that are specialised are limited in scope. Still further technical writings discuss the long-term performance of material structures, observing the changes, chiefly for the worst, that take place in electronic materials subjected to service conditions for extended periods of time.

The purpose of the present book is to give some of each of these viewpoints, a glimpse at several phases of the problems of materials used in electronics. The author has been associated for many years with several branches of the subject and has had to straddle, mentally, the fences between the divisions of research, materials, equipment design and testing. Knowledge of all these helps to give an understanding of the reason for observed facts.

This book, then, might be termed a survey type of information, introducing an important subject and offering stimulation to the inquiring mind to pursue details of special interest still further. In plan, the book opens with a resumé of the essential concepts of the nature and behaviour of materials. Next are presented a description and comparison of materials used in electronics and microelectronics, sufficiently specific to aid design engineers and production engineers in making a choice.

Following this, the practices in application of materials to the major types of components are described. In the usual sources on the design and construction of equipment, the subject of materials is very sketchily covered, if at all, and yet some of the most difficult problems in design and reliability often lie in the use of materials.

The final chapter covers the purpose of testing materials.

The author is aware that he has not extended the boundary of scientific knowledge. From a large volume of material collected for the purpose, he has selected portions most useful to the Design Engineer, the Production Engineer, the Inspector and the Student who seeks a working knowledge of materials in electronics.

Grateful acknowledgement for invaluable help is given to the author's many associates; to the manufacturers who so generously furnished data and suggestions; and to authors and publishers who permitted inclusion of important parts of the text.

Harpenden   CHARLES E. JOWETT
May 1971

Terminology

With the advance of new technology, so new terms and phases are coined. To establish precise definitions through national and international standards is slow. Below then, is a guide to the terms and expressions according to their usage in this book.

Accelerators   Materials used to speed up curing process.

Automatic gain control (AGC)   A control circuit that automatically changes circuit gain so that the output signal remains essentially constant despite variations in input-signal strength.

Arc resistance   An empirical rating of the relative ability of materials to withstand deterioration of an electric arc under specified conditions.

Active device   A circuit element, e.g. a transistor, which is capable of amplifying.

AND   Logic function — a gate giving an output of 1 only when all its inputs are 1.

Binary counter   A system of flip-flop circuits and gates that counts the number of times the single input changes state in a given direction, for example, from logic 0 to logic 1 but not from 1 to 0.

Basic insulation level (BIL)   Impulse or surge voltage — the withstand voltage of an insulating system to a short time electrical impulse with a fast rise time and specified shape.

Bi-polar   Containing *npn* or *pnp* junctions as distinct from field-effect devices.

Breadboarding   An early trial of a circuit using discrete components and conventional wiring.

Buried N+   A process for making integrated circuits using a particular method of isolation.

Casting   A process of embedding where a removable mould is used, which is separated when the resin has solidified, the cast resin itself acting as a case and holding the parts in the correct positions.

Curing   The solidifying or hardening process by which the material becomes stiff or resilient.

Chip   Unit of semiconducting material containing a circuit or a device.

Cordwood   A method of assembling discrete components with their principal ones perpendicular to the interconnection pattern.

Crossover   A point at which two isolated conductors cross.

Child board   A printed circuit board interconnecting SIC's and so on and attached to interconnections on a mother board — sometimes referred to as daughter board.

Clocked flip-flop   A flip-flop circuit that is set and reset at specific times by adding clock pulses to the inputs so that the circuit is triggered only if both triggers and clock pulses are present simultaneously.

Combined-gate IC   A single IC chip with several gate circuits combined by interconnection on the chip to form a more complex circuit.

Cascade   A tandem or series-series connection for two or more similar circuits or amplifying stages in which the output of one circuit is connected to the input of the next.

Conductor   A material (usually a metal) that conducts electricity by transfer of orbital electrons. Resistivity ranges from 10^{-7} to 1 Ω.cm.

Corona   Any electrically detectable, field-intensified ionisation in an insulating system that does not result immediately in catastrophic breakdown.

Complementary transfer logic (CTL)   A logic system using-emitter circuits with a combination of *pnp* and *npn* transistors.

Current hogging   A condition where one of several parallel logic circuits has a lower resistance and takes most of the available current, resulting in unequal current sharing.

Closed-looped gain   The over-all gain of an amplifier that has an external negative feedback loop.

Common-mode rejection   The ability of a differential amplifier stage to reject signals common to both of its inputs.

Class B transistor amplifier   An amplifier biased so that with no input signal, output current is nearly zero. When an alternating input signal is applied, output current flows for approximately half a cycle.

Dilutents   Used to reduce viscosity of uncured resin systems in order to make the mix easier to pour.

Direct-coupled-transistor logic (DCTL)   A form of saturating logic.

Diffusion area   That area of a semiconductor which is being doped by diffusion in a particular operation.

DTL   Diode-transistor logic — a form of saturating logic circuit, where logic decisions are made by a group of diodes and the output signal is coupled through a transistor.

Dual in-line package   Integrated circuit package with leads in two lines normal to the package for easy insertion on printed circuit boards.

Dielectric   A medium in which it is possible to maintain an electric field with little supply of energy from outside sources. The energy required to produce the field is recoverable in whole or part.

Dielectric absorption   The molecular and macrascopic changes which occur to reduce stress in a dielectric caused by an applied field. Dielectric absorption is indicated by an increase in resistivity and later recovery of voltage.

Dielectric breakdown voltage   The voltage at which a catastrophic failure occurs in an insulating system.

Permittivity (dielectric constant), relative permittivity, specific inductive capacity (Sic, K', ϵ')   The ratio of the total capacitance of a dielectric system to the geometric capacitance, i.e. the capacitance of the identical system with vacuum as the dielectric of that system.

Dielectric loss angle (δ)   The difference between $90°$ and the dielectric phase angle.

Dielectric phase angle (θ)   The angular phase difference between the sinusoidal alternating voltage and the component of the resulting alternating current in the dielectric.

Dielectric strength, electric strength, disruptive gradient   The ratio (V/cm or kV/cm) of the dielectric breakdown voltage to the thickness of the material.

Dielectric withstand voltage, proof voltage   The maximum voltage a system withstands before breakdown.

Dissipation factor, loss tangent, loss (D, $\tan \delta$)   The ratio of energy absorbed by a dielectric to the energy passed back to the electrical system; the ratio of the current absorbed by a dielectric to the capacitive current which passes through the dielectric.

Dust and fog tracking resistance   Empirical tests that rate materials for their ability to withstand the action of electric arcs produced by conduction through surface films of a specified contaminant containing moisture.

EECL   Emitter-emitter-coupled logic — a form of non-saturating logic circuit.

Epitaxy   The growth of a single crystal layer with its lattice aligned with that of a single crystal substrate.

Embedding and encapsulation   In a general sense consists of completely enclosing the component in a resin or other material and filling voids and spaces between the parts.

Elastomer   Material that can be stretched repeatedly to twice its original length at room temperature, and when released returns to its approximate original length with force.

Exotherm   Rise in temperature resulting from the liberation of heat by polymerisation. Heat produced by chemical reaction as in a curing process.

Extender   Material, usually having some adhesive action, added to an adhesive to reduce the amount of primary adhesive required.

Emitter-coupled logic   A logic system using emitter-coupled transistor circuits.

Filler   A non-adhesive substance added to an adhesive to improve its properties.

Fan-out   The number of logic circuit inputs driven from a logic circuit output.

Flat-pack   Integrated circuit package of medium height with co-planar leads.

Flip chip   A chip that can be mounted circuit side downwards on a thin or thick film circuit.

Flip-flop   A circuit with two stable states that can be triggered from one state to the other by an input trigger pulse. In the absence of a trigger pulse, the circuit remains permanently in its prior state.

Heat distortion temperature (*HDT*)   The temperature at which under test conditions a bar of material of a certain size bends.

Hybrid (integrated circuits)   Contains components formed by two or more microelectronic techniques, e.g. SIC and thin film.

Hardeners   Or curing agents — are materials, usually liquids, which are added to uncured resins to make them set. Hardeners are sometimes called catalysts, possibly incorrectly.

Inverting input   An input terminal of a differential amplifier that produces an output signal of opposite phase.

Insulation resistance   The ratio, in ohms, of applied d.c. voltage to resulting current in an insulating system. It includes both the volume and surface resistance components and usually varies with potential gradient, time of electrification and environment.

Insulator   A material in which a voltage produces a small current. Resistivity is above 10^5 Ω.cm, and generally decreases with temperature rise.

Impregnating   A process of applying a resin or other material to closely constructed devices, such as coils and of eliminating internal voids by penetration of the resin.

IC   Integrated circuit.

Linear amplifier   An amplifier that operates on the linear portion of its forward transfer characteristic so that its output signal is always an amplified replica of the input signal.

Loss index   The product of the dielectric and the loss tangent. A measure of power loss per cycle.

Linear integrated circuit   One that fulfils a continuous amplifying or modulating, not a switching function.

Logic circuit   One that fulfils a switching function with a specified relation between inputs and outputs.

Modulus of elasticity   Ratio of stress corresponding strain below the proportional limit.

Monomer   Molecule that can react with like or unlike molecules to form polymers.

Micromodules   Assemblies of miniature components.

Monolithic   Made out of a single piece of material.

MOST   Metal-oxide-silicon transistor array of field-effect transistors.

Negative feedback   The external feeding back of a portion of an amplifiers' output signal to its input with a phase opposite to that of the input signal.

Non-inverting input   An input terminal of a differential amplifier that produces an output signal of the same phase.

Noise margin   The difference between the operating voltage of a binary-logic circuit and a threshold voltage.

NAND   Logic function, not AND — a gate giving an output of 0 when all its inputs are 1.

NOR   Logic function, not OR — a gate giving an output of 1 when all its inputs are 0.

Operational amplifier   A high-gain d.c. amplifier as used for analogue computing and signal processing.

OR   Logic function — a gate giving an output of 1 when any of its inputs are 1.

Open-loop gain   The gain of an amplifier with no external feedback.

Plasticiser   Material added to resins to make them softer and more flexible when cured.

Polymerisation   Chemical reaction in which the molecules of a monomer are linked to form large molecules. The resultant molecular weight is a multiple of that of original substance.

Pot life   Period of time after mixing during which an adhesive can be used.

Power dissipation In a logic circuit the supply power consumed when the circuit is operating with a 50 per cent duty cycle.

Propagation delay The time delay between the application of a signal to the input of a logic circuit and the change of state at the output.

Power factor The ratio of the power in watts in a material to the product of the effective sinusoidal voltage and current in volt-amperes. It is the ratio of the energy absorbed by the dielectric to the total energy entering the dielectric.

Package pins Connector terminals on an IC container.

Photoengraving Process for transferring a pattern to a solid surface.

Potting A can or similar container filled and remaining an integral part of a mould unit to hold the assembly.

Resistor-capacitor transistor logic (RCTL) A variation of the resistor-transistor logic system in which a capacitor is connected across the series resistor to allow fast switching.

Resistor-transistor logic (RTL) A system of transistor logic in which a resistor is included in series with the base of each transistor to reduce current hogging.

Reset terminal The flip-flop input terminal that triggers the circuit back from its second state to its original state.

Set terminal The flip-flop input terminal that triggers the circuit from its first state to its second state.

Surface resistance The ratio (in ohms) of the direct voltage applied to an insulation system to that portion of the current that flows across the surface of the system. (The thin layer of moisture at the gas-solid interface usually has the greatest effect on the surface resistance.)

Surface resistivity The ratio (in ohms) of the potential gradient parallel to the current along a surface of unit length to the current per unit width of the surface.

SIC Silicon integrated circuit formed entirely in a silicon chip.

Slice Thin disc cut from a crystal and used in the fabrication of a multiplicity of circuits. Finally separated into chips.

Treeing A name given to tracking, erosion or corona cutting failure on or in a material and descriptive of the branched configuration of failure.

Transistor-transistor logic (TTL) A logic system similar to DTL in which the logic diodes are replaced by a multi-emitter transistor.

Threshold voltage The input voltage level at which a binary-logic circuit changes from one state to another.

Thixotropy Property of an adhesive upon isothermal agitation to thin and to thicken afterwards.

Thermal SiO_2 Protective layer of silica grown by the thermal oxidation of silicon.

Thick-film circuit Silk-screened circuit; cermet circuit; circuit made by depositing paint on an insulating substrate and burning off the organic matter to leave a purely mineral component.

Thin-film circuit Evaporated circuit; vacuum-deposited circuit; circuit made by vacuum-deposition or sputtering of materials on an insulating substrate.

Transistor base diffusion   Diffusion giving the base region of a transistor, e.g. boron into n-type silicon.

Video amplifier   An amplifier designed for linear amplification over a wide range of frequencies.

Volume resistance   The ratio (in ohms) of the direct voltage applied to an insulation system to that portion of the current that flows through the volume of the insulation system.

Volume resistivity   The ratio (in ohms.cm) of the potential gradient parallel to the current through a material to the current density.

Wired-OR   OR logic function obtained by wiring together the inputs to be operated upon.

Wetting   Ability to adhere to a surface immediately upon contact.

Webbing   Formation of wrinkles on vacuum-formed parts.

Wideband   Capable of passing a broad range of frequencies.

Conversion factors, symbols and units

Conversion factors — Imperial to SI units

In most instances the conversion factors in this part have been kept in a range not greater than 1,000 and not less than 0·1.

To convert		To		Multiply by
Unit name	Symbol	Unit name	Symbol	
acre	—	square metres	m²	4046·86
		hectare	ha	0·404 686
atmosphere (standard)	atm	kilonewton/square metre	kN/m²	101·325
bar†	bar	kilonewton/square metre	kN/m²	100·0
British thermal unit	Btu	kilojoule	J	1·055 06
British thermal unit/pound	Btu/lb	kilojoule/kilogramme	kJ/kg	2·326
calorie (international)	cal$_{IT}$	joule	J	4·186 8
cubic foot	ft³	litre*	l	28·316 8
cubic inch	in³	millilitre*	ml	16·387 1
cubic yard	yd³	cubic metre	m³	0·764 555
		kilolitre*	kl	0·764 555
cubic foot of water	ft³H₂O	kilogramme	kg	28·313 2
cubic foot/minute	ft³/min	litre/second*	l/s	0·471 947
cycle per second	c/s	hertz	Hz	1·0
degree (angle)	. . .°	milliradian	mrad	17·453 3
dyne	dyn	micronewton	μN	10·0
dyne centimetre	dyn cm	micronewton metre	μN m	0·1
erg	erg	microjoule	μJ	0·1
fluid ounce (UK)	UKfl oz	millilitre*	ml	28·413 1
foot	ft	metre	m	0·304 8
foot/minute	ft/min	millimetre/second	mm/s	5·08
foot of water	ft H₂O	kilonewton/square metre	kN/m²	2·989 07
foot per second per second	ft/s²	metre/second squared	m/s²	0·304 8
foot pound — force	ft lbf	joule	J	1·355 82
foot pound — force/minute	ft lbf/min	watt	W	81·349 2
foot pound — force/second	ft lbf/s	watt	W	1·355 82
foot poundal	ft pdl	millijoule	mJ	42·140 1
foot/second	ft/s	metre/second	m/s	0·304 8
foot squared/second	ft²/s	metre squared/second	m²/s	0·092 903 0
gallon (UK)	gal	litre*	l	4·456 09
gallon (US)	USgal	litre*	l	3·785 41
gramme force	gf	millinewton	mN	9·806 65
gravity (standard)	gn	metre/second squared	m/s²	9·806 65
hectobar†	hbar	meganewton/square metre	MN/m²	10·0

To convert		To		Multiply by
Unit name	Symbol	Unit name	Symbol	
horsepower	hp	kilowatt	kW	0·745 700
horsepower hour	hp h	megajoule	MJ	2·684 52
hundredweight	cwt	kilogramme	kg	50·802 3
inch	in	millimetre	mm	25·4
inch of mercury	inHg	kilonewton/square metre	kN/m²	3·386 39
inch of water	inH₂O	newton/square metre	N/m²	249·089
inch to the fourth	in⁴	millimetre to the fourth	mm⁴	416·231 × 10³
inch/second	in/s	millimetre/second	mm/s	25·4
kilogramme — force	kgf	newton	N	9·806 65
kilogramme — force metre	kgf m	joule	J	9·806 65
kilogramme-force/square metre	kgf/m²	newton/square metre	N/m²	9·806 65
kilogramme — force/square millimetre	kgf/mm²	meganewton/square millimetre	MN/m²	9·806 65
kilometre/hour	km/h	metre/second	m/s	0·277 778
kilovolt/inch	kV/in	kilovolt/metre	kV/m	39·370 1
kilowatt hour	kW h	megajoule	MJ	3·60
knot (UK)	UKkn	metre/second	m/s	0·514 773
		kilometre/hour	km/h	1·853 18
knot (international)	kn	metre/second	m/s	0·514 444
		kilometre/hour	km/h	1·852
metric technical unit of mass	kgf s²/m	kilogramme	kg	9·806 65
mile (statute)	mile	kilometre	km	1·609 344
mile/hour	mile/h	metre/second	m/s	0·447 04
		kilometre/hour	km/h	1·609 344
microinch	μin	micrometre§	μm	0·025 4
milli-inch (thou)	m inch	micrometre§	μm	25·4
millibar†	mbar	newton/square metre	N/m²	100
millimetre of mercury	mmHg	newton/square metre	N/m²	133·322
millimetre of water	mmH₂O	newton/square metre	N/m²	9·806 65
minute (angle)	...′	milliradian	mrad	0·290 888
ounce	oz	gramme	g	28·349 5
ounce/cubic inch	oz/in³	megagramme/cubic metre	Mg/m³	1·729 99
pint	pt	litre*	l	0·568 261
poise	P	newton second/square metre	Ns/m²	0·1
pound (mass)	lb	kilogramme	kg	0·453 592 37
poundal	pdl	newton	N	0·138 255
pound-force	lbf	newton	N	4·448 22
pound-force foot	lbf ft	newton metre	N m	1·355 82
pound-force inch	lbf in	newton metre	N m	0·112 985
pound-force/square inch	lbf/in²	kilonewton/square metre	kN/m²	6·894 76
		bar†	bar	0·068 947 6
pound-force/square foot	lbf/ft²	newton/square metre	N/m²	47·880 3
pound/cubic foot	lb/ft³	kilogramme/cubic metre	kg/m³	16·018 5
pound/cubic inch	lb/in³	megagramme/cubic metre	Mg/m³	27·679 9
pound/foot	lb/ft	kilogramme/metre	kg/m	1·488 16
pound/second	lb/s	kilogramme/second	kg/s	0·453 592 37

To convert		To		Multiply by
Unit name	Symbol	Unit name	Symbol	
pound/foot second	lb/ft s	kilogramme/metre second	kg/m s	1·488 16
revolution	rev	radian	rad	6·283 19
revolution/minute	rev/min	radian/second	rad/s	0·104 720
second (angle)	...″	microradian	μrad	4·848 14
slug	—	kilogramme	kg	14·593 9
square foot	ft^2	square metre	m^2	0·092 903 0
square inch	in^2	square millimetre	mm^2	645·16
square mile	sq mile	square kilometre	km^2	2·589 99
		hectare	ha	258·999
square yard	yd^2	square metre	m^2	0·836 127
stokes	St	millimetre squared/second	mm^2/s	100·0
therm	therm	megajoule	MJ	105·506
ton (UK)	ton	kilogramme	kg	1016·05
		tonne*	t	1·016 05
ton-force	tonf	kilonewton	kN	9·964 02
ton-force/square inch	tonf/in^2	meganewton/square metre	MN/m^2	15·444 3
		hectobar†	hbar	1·544 43
yard	yd	metre	m	0·914 4

* The kilolitre referred to in these tables is equal to the SI unit of volume, namely 1 cubic metre; the litre is 1/1,000 part of this, i.e. the volume of a 100-mm cube.

The SI litre differs slightly from the measurement of volume made in the terms of the 1901 litre which is 1·000 028 of the SI litre. The name litre is not strictly an SI unit name, but it is considered that it will continue in use by general acceptance.

† The kilonewton/square metre (kN/m^2) and the meganewton/square metre (MN/m^2) are preferred to bar (10^5 N/m^2) and hectobar (10^7 N/m^2), but the latter may be encountered in certain documents as the units of pressure and stress.

‡ The megagramme (Mg) is preferred to the tonne (1 t = 1,000 kg) but the latter may be encountered in certain documents.

§ Micrometre sometimes called by the obsolescent term 'micron'.

Symbols for units

Unit	Symbol	Unit	Symbol
metre	m	cubic inch (similarly for cubic foot, etc.)	in^3
micrometre (or micron)	μm (or μ)	litre	l
Ångstrom	Å	millilitre	ml
inch	in	gallon	gal
foot	ft	second (time)	sec
yard	yd	minute (time)	min
mile	mile	hour	h
degree: minute: second (angle)	° ′ ″	day	d
radian	rad	year	a
square centimetre (similarly for square metre, etc.)	cm^2	radian per second	rad/s
square inch (similarly for square foot etc.)	in^2	cycle per second	c/s
		revolution per minute	rev/min
cubic centimetre (similarly for cubic metre etc.)	cm^3	mile per hour	mile/h
		knot	kn

Symbols for chemical elements

Name	Symbol	Atomic weight	Name	Symbol	Atomic weight
Actinium	Ac	89	Lead	Pb	82
Aluminium	Al	13	Lithium	Li	3
Americium	Am	95	Lutetium	Lu	71
Antimony	Sb	51			
Argon	A	18	Magnesium	Mg	12
Arsenic	As	33	Manganese	Mn	25
Astatine	At	85	Mercury	Hg	80
			Molybdenum	Mo	42
Barium	Ba	56			
Berkelium	Bk	97	Neodymium	Nd	60
Beryllium	Be	4	Neon	Ne	10
Bismuth	Bi	83	Neptunium	Np	93
Boron	B	5	Nickel	Ni	28
Bromine	Br	35	Niobium*	Nb	41
			Nitrogen	N	7
Cadmium	Cd	48			
Caesium	Cs	55	Osmium	Os	76
Calcium	Ca	20	Oxygen	O	8
Californium	Cf	98			
Carbon	C	6	Palladium	Pd	46
Cerium	Ce	58	Phosphorus	P	15
Chlorine	Cl	17	Platinum	Pt	78
Chromium	Cr	24	Plutonium	Pu	94
Cobalt	Co	27	Polonium	Po	84
Copper	Cu	29	Potassium	K	19
Curium	Cm	96	Praseodymium	Pr	59
			Promethium	Pm	61
Dysprosium	Dy	66	Protoactinium	Pa	91
Erbium	Er	68	Radium	Ra	88
Europium	Eu	63	Radon	Rn	86
			Rhenium	Re	75
Fluorine	F	9	Rhodium	Rh	45
Francium	Fr	87	Rubidium	Rb	37
			Ruthenium	Ru	44
Gadolinium	Gd	64			
Gallium	Ga	31	Samarium	Sm	62
Germanium	Ge	32	Scandium	Sc	21
Gold	Au	79	Selenium	Se	34
			Silicon	Si	14
Hafnium	Hf	72	Silver	Ag	47
Helium	He	2	Sodium	Na	11
Holmium	Ho	67	Strontium	Sr	38
Hydrogen	H	1	Sulphur	S	16
Indium	In	49			
Iodine	I	53	Tantalum	Ta	73
Iridium	Ir	77	Technetium†	Tc	43
Iron	Fe	26	Tellurium	Te	52
			Terbium	Tb	65
Krypton	Kr	36	Thallium	Tl	81
			Thorium	Th	90
Lanthanum	La	57	Thulium	Tm	69

Name	Symbol	Atomic weight	Name	Symbol	Atomic weight
Tin	Sn	50	Xenon	Xe	54
Titanium	Ti	22			
Tungsten	W	74	Ytterbium	Yb	70
			Yttrium	Yt	39
Uranium	U	92			
			Zinc	Zn	30
Vanadium	V	23	Zirconium	Zr	40

* Known in the USA as 'columbium'.
† Also known as 'masurium'.

Abbreviation of words and units

Word	Abbreviation	Word	Abbreviation
absolute	abs	pound	lb
alternating current	a.c.	slug	slug
anhydrous	anhyd.	hundredweight	cwt
aqueous	a.q.	ton	ton
atmospheric	atm.		
atomic weight	at.wt.	dyne	dyn
		newton	N
boiling point	b.p.	poundal	pdl
		pound-force (similarly for kilogramme-force, etc.)	lbf
calculated	calc.		
centre of gravity	c.g.		
coefficient	coeff.	pound-force per square inch	lbf/in²
compound	cpd.		
concentrated	conc.	bar (10^6 dyn/cm²)	bar
concentration	concn.	millibar (10^3 dyn/cm²)	mb
constant	const.	atmosphere, standard	atm
corrected	corr.	millimetre of mercury (conventional)	mmHg
critical	crit.		
crystalline, crystallised	cryst.	inch of mercury (conventional)	inHg
current density	c.d.		
decomposition	decomp.	poise	P
degree	deg.	centipoise	cP
dilute	dil.	stokes	S
direct current	d.c.	centistokes	cS
distilled	dist.		
		joule	J
electromotive force	e.m.f.	erg	erg
equation	eqn.	kilowatt hour	kWh
equivalent	equiv.	horsepower hour	hp h
experiment(-al)	expt.	foot pound-force	ft lbf
freezing point	f.p.	calorie	cal
		kilocalorie	kcal
gramme	g	British Thermal Unit	Btu
kilogramme	kg		
tonne (1,000 kg)	t	watt	W
grain	gr	kilowatt	kW
ounce (avoirdupois)	oz	horsepower	

Word	Abbreviation	Word	Abbreviation
decibel (acoustics)	dB	specific gravity	sp.qr.
decibel (telecommunica-tions)	db	specific heat	sp.ht.
		specific volume	sp.vol.
neper (general)	Np	standard temperature and pressure	s.t.p.
neper (telecommunications except where can be con-fused with Newton)	N	temperature	temp.
degree (temperature value)		ultra-violet	u.v.
degree Celsius (Centi-grade)	°C	vacuum	vac.
degree Kelvin*	°K	vapour density	v.d.
degree Fahrenheit	°F	vapour pressure	v.p.
degree Rankine	°R	volume	vol.
gramme-molecule	mole.	weight	wt.
horse-power: brake	b.h.p.	degree (temperature interval)	
effective	e.h.p.	degree Celsius (Centi-grade)	degC
indicated	i.h.p.	degree Kelvin*	degK
nominal	n.h.p.	degree Fahrenheit	degF
shaft	s.h.p.	degree Rankine	degR
infra-red	i.r.	candela	cd
insoluble	insol.	lumen	lm
		lux	lx
latent heat	lat.ht.	nit	nt
liquid	liq.		
magnetomotive force	m.m.f.	electrostatic unit	e.s.u.
maximum	max.	electromagnetic unit	e.m.u.
melting point	m.p.	coulomb	c
minimum	min.	ampere	A, amp
molecule, molecular	mol.	volt	V
molecular weight	mol.wt.	ohm	Ω
observed	obs.	electron-volt	eV
potential difference	p.d.	volt-coulomb	VC
precipitate (noun)	ppt.	volt-ampere	VA
preparation	prep.	ampere-turn	AT
recrystallised	recryst.	farad	F
relative humidity	r.h.	henry	H
root mean square	r.m.s.	gauss	G
		weber	Wb
soluble	sol.		
solution	soln.	curie	c
specific	sp.	röntgen	r

* Now internationally standardised as the kelvin (K)

Introduction

In the garment and decorating industries 'materials' are referred to as 'fabrics' and they use them in many ingenious ways. For instance, man-made fibres and wool are blended into a plaid to make trousers that keep their creases and are cool for summer wear; polyester and cotton are used for sport shirts in colours; glass-fibre curtains are designed to withstand sun and soot; carpets woven from acrylic are stainless and waterproof.

In electronic engineering, we call them 'materials' or more specifically insulation, magnetic and conductive materials. With the materials available to the design engineer, is sufficient use being made of their potential? Are basic problems such as wear, heat resistance and dielectric strength considered in the original stages of design, when the choice of materials should be a controlling factor?

Are the outstanding properties of new materials fully investigated before an old 'tried-and-true' standby is specified and are we making full use of the materials at hand?

Basic research into almost all materials used in the electronics industry has made available an overwhelming number of new superior, materials for components, equipments and systems; these developments are allowing engineers to design more effectively to meet the needs of customers rather than around the limitations of traditional materials. But, in the process of adapting either new or old materials to new uses, compromise becomes necessary. Performance may be sacrificed for cost, reliability for size or weight or operational limits for stability in varying environments.

Because of the increasing significance materials can have in influencing total design, engineers should give greater attention to the impact of materials at the critical point — the beginning of the design process.

For many years it was adequate in the classification of materials to specify thermal capabilities. The available materials were natural and simple. With the advent of synthetic materials and tailoring of properties by molecular structuring by compounding or adding stabilisers or antioxidants, the old classification system is no longer sufficient.

The most recent method of classification that has been evolved is 'life by test'; this permits the assignment of temperature indices to a material. Of course, these indices depend on the material and the deterioration in properties that govern its suitability for a given application.

The key position of materials in the design of electronic devices and equipment is being increasingly recognised but the knowledge of fundamentals of materials and technology is still patchy among engineers.

The approach to the selection of materials is often too artificial and based on commercial literature. This failure in knowledge becomes a critical problem since the reliability of devices and equipment — and beyond these, of systems — depends essentially on the reliability of the materials used.

The problem is primarily one of communication between the various segments of engineering and material specialists, the managements of materials producers and equipment manufacturers. It takes in many approaches and it has to overcome many misconceptions, a great deal of inertia, and too much indifference born of professional parochialism. It needs all the best efforts and co-operation of the various professional disciplines concerned, as well as their managements, to be successfully solved.

The author in the following chapters neither hopes nor plans to cover all the areas of materials, but rather to examine the basic problems from the practical viewpoints of both design and material selection.

In almost any electrical or electronic engineering there exists the need for continuous up-dating and education of engineers because of new developments. This need is markedly true with respect to materials technology, with its revolutionary consequences in terms of new devices, new uses and new fields of application.

A recent dramatic instance of changes brought about by new materials is in the application of solid-state devices such as diodes, transistors and integrated circuits; so rapid has been the advance that no one can predict where it will lead in ten years hence.

Material selection

At the outset I would like to warn my readers that I am a materials reliability engineer and therefore my viewpoint might be biased. However, I hope that this bias is significantly tempered by two factors. One, unfortunately, is age; the other is the time span of the past twelve years of my career during which I have supported design engineering in development.

After this warning I am decidedly of the opinion that materials selection is one of the elements in successful engineering design because such selection vitally affects three factors which make or break any engineering accomplishment:

1 Performance.
2 Production.
3 Price.

Performance, of course, refers to functional specifications. The most beautiful structure is worthless if it collapses due to materials with inadequate strength. Production means ease of fabrication; the strongest material for the structure is also worthless if it cannot be economically fabricated into proper components. Finally, price stands for economic considerations. While such considerations might be less important in such applications such as military or aerospace projects, most have to design for a highly competitive commercial market in which suitable selection of materials and related processes can frequently mean the difference between a profitable product or a losing proposition.

One of the major difficulties in selecting materials is the ubiquitous and frequently severe limitations of available materials. In today's increasing complex and sophisticated technology, limitations in material properties frequently present the major stumbling block to successful engineering design.

Ironically, however, even if an ideal material for a specific application exists, finding that material can be an equally difficult problem. The sheer number of available materials is truly staggering and the complexity of their characteristics can be downright discouraging.

Figure I.1 presents a quick superficial survey of engineering materials technology. Such a complexity overtaxes the capacity of even the materials engineers, who are usually familiar only with certain categories of materials such as metals, plastics or ceramics. Confronted with this enormous number of materials and the practically limitless range of properties of such materials, the design and materials engineers have to make their selection, considering functional performance, manufacturability, etc.

The list of subjective difficulties is long and frustrating. Information on desired properties of certain materials may not be available or be difficult to obtain and the lack of proper retrieval tools to get the information is becoming increasingly evident. Last, but not least, is the lack of ability or willingness to communicate. This may very well be the most important subjective difficulty. The communication problem can be traced back to the attitudes of the designers and the materials engineers.

Ground rules

Where does this confusing situation leave the design engineer whose training in the field of materials is necessarily scant? Indeed, the question arises, should he be called upon to select the material for the components, structures and mechanisms he has designed or should he simply hand the design over to the materials specialists who will then come up with the proper material? Obviously, neither of these extreme solutions is practical. As a matter of fact, only through close co-operation between the designer and the materials engineer can optimum material selection be obtained, otherwise the selection process will degenerate into chaos.

To avoid this complication I would suggest that the design engineer should follow a three-phase materials selection process:

1 Conduct an initial survey (parallel with the early design concept) to find the most suitable type of material, then mould his design to the characteristics of that material and/or the process dictated by the material.

2 Discuss his selection with the materials engineer, generally at a more advanced stage of the design, when shape, functional requirements and production quantities have sufficiently crystallised to allow a detailed and refined selection.

3 Evaluate the materials engineer's suggestions, consider the alternatives and make the final decision.

The purpose of this book is not to discourage the designer, but to help him

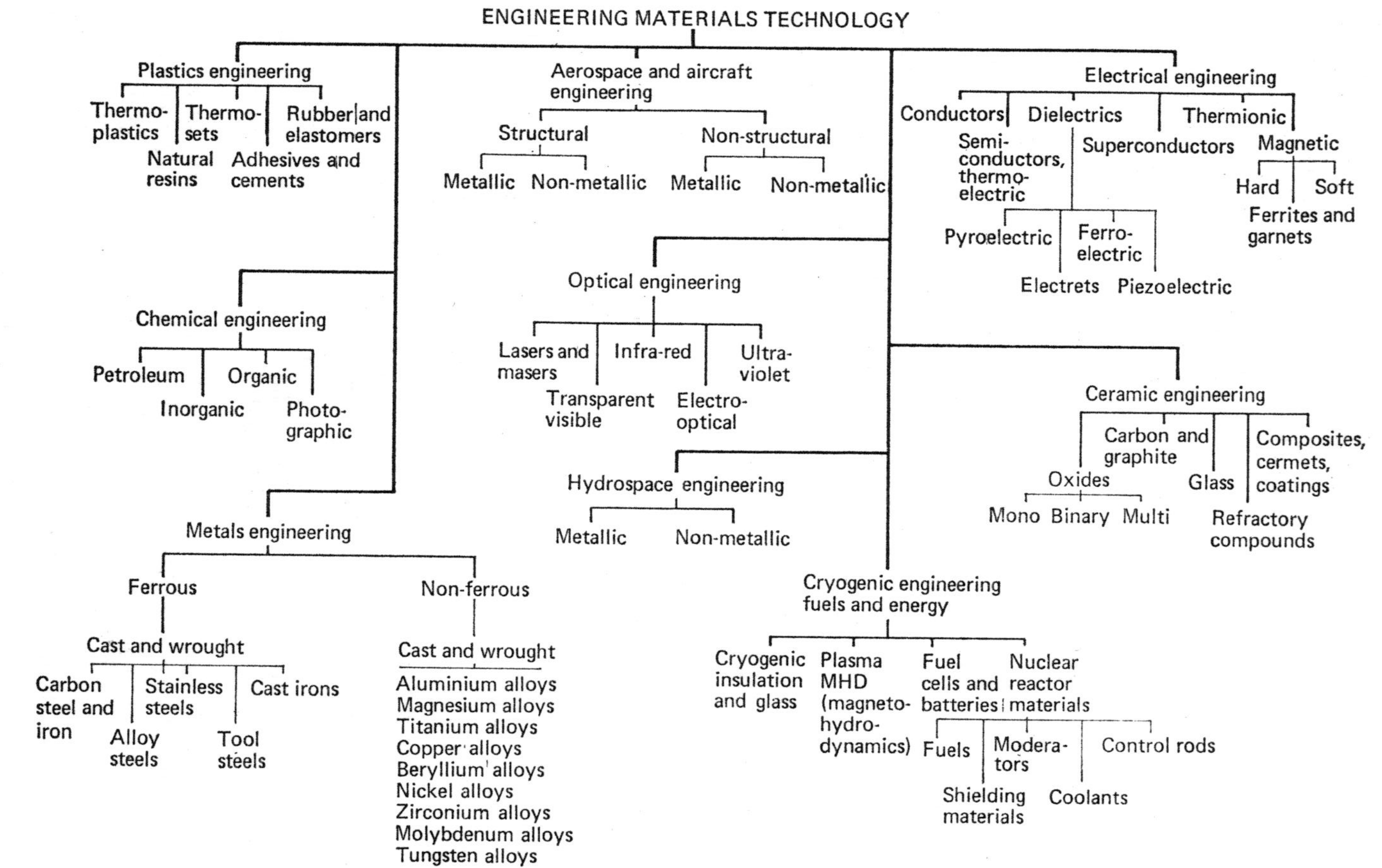

Fig. I.1   Overview of engineering materials technology

around a great many of the pitfalls that can be in his path of selecting and evaluating materials.

Communication

The most vital link in the communications chain lies within the firm that engineers and manufactures electronics systems. Specifically, it is the communication link between the materials and the design engineering, both in the electrical and electromechanical groups. Most large electronic firms have within their organisation the technical staff necessary to optimise most of their designs, providing the communications at the materials and design interface have been sufficiently good to give each group a working knowledge and appreciation of the technology of the other group.

It is unfortunate that many materials engineers have strictly chemical leanings with little perception of mechanical or electronic technology. Hence, the technical advice of both a mechanical and electrical engineer in a materials group can pay rich dividends. When the magnitude and complexity of the materials spectrum is to be considered it might be useful to consult Fig. I.1 again.

In response to a request concerning high-temperature dielectric materials, for example, the engineer would be questioned to determine if his information needs were concerned with radomes, space vehicle or missile surfaces, printed circuit boards, aerials, or cable and wire insulation, all of which may involve high-temperature materials, further it may be necessary to determine which materials the designer was considering, plastic laminates, ceramics, etc., or other factors, such as manufacturing processes and fabrication techniques to be used and the possible problems associated with compatibility of the material with its environment and other materials need to be discussed.

The magnitude of the possible considerations to be made in this regard can be appreciated by referring to Fig. 1.2 in which properties important to plastics design are listed.

What has emerged from the laboratories and factories of the electronics industry during the past few years is the realisation that it is possible to eliminate much of the material formerly utilised in the construction of circuitry.

Since an electron is a rather small thing, if it is a thing at all, and billions of electrons are small, too, we may ask how large a wire is really needed to act as a passageway for the movement of these electrons; how large a resistor is really needed to conduct the electrons somewhat 'less well' than the wire; how large need a capacitor really be — do we need all those rolls of paper — and so on. The answer is that electrically very little material is necessary, in fact several orders of magnitude less than what we normally use. What savings in copper and solder we could effect if the needs of electronic circuits were exclusively electronic?

But there are other factors. Size contributes a measure of mechanical ruggedness, although there are arguments for and against this. Size also contributes thermal paths and sinks, but there is another reason for this being equally important. If you think in terms of resistors, capacitors, inductors

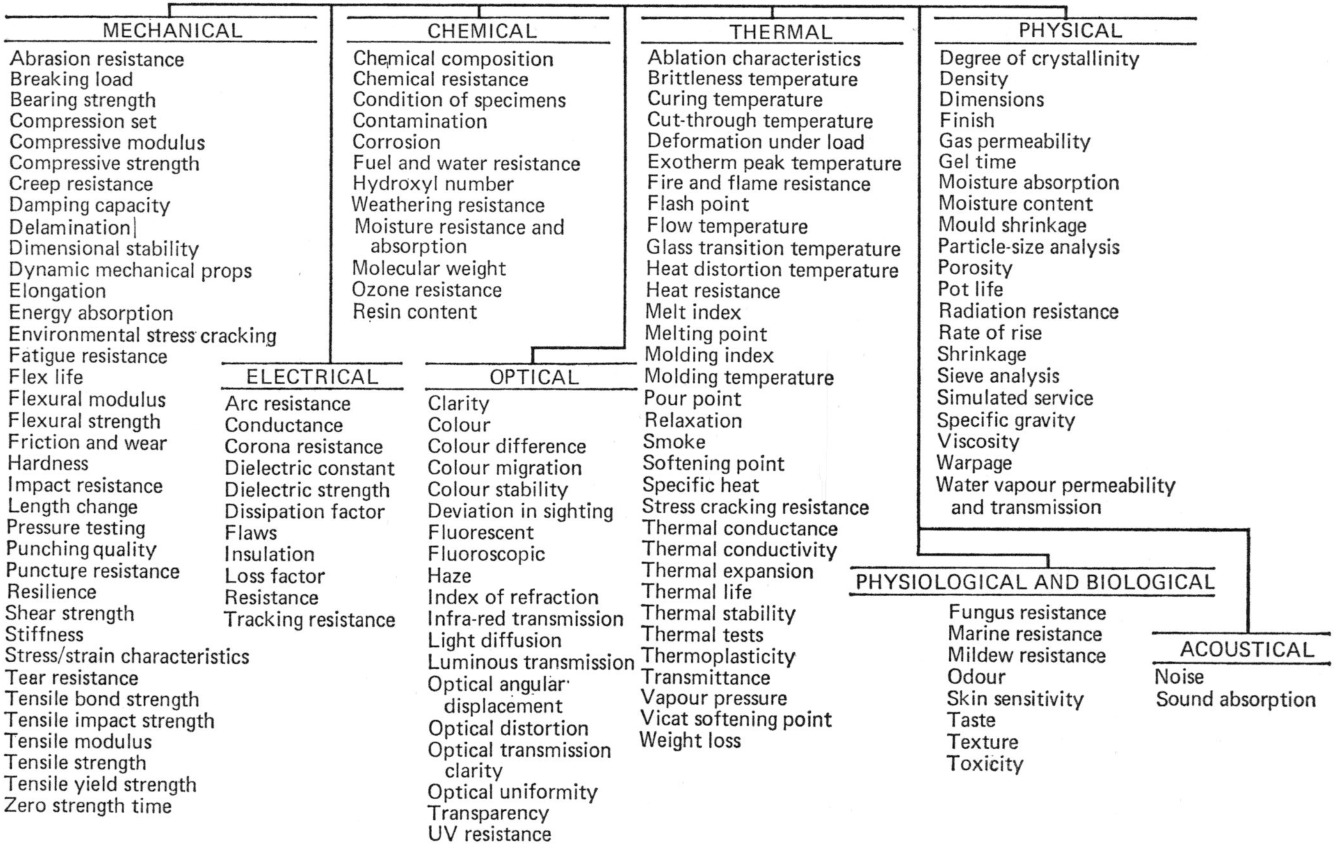

Fig. I.2 Overview of properties important to plastics design

and devices, there is a certain measure of assurance in being able to hold one of each of them in your hands as a discrete thing. If you want to make them do things to electrons you may have to consider them as building blocks which can be built up into a system.

Printed circuits led the way and the semiconductor leap-frogged into what our imaginations must now consider as very close to the ultimate in using materials efficiently. As matters now stand, the spectrum runs from discrete elements soldered together at one extreme to integrated circuits at the other extreme. Each technique presents advantages and disadvantages of size, power, frequency, flexibility and cost.

We have seen a routinely manufactured circuit that is as hybrid as anything can be. It was built on a dielectric substrate using discrete passivated semiconductor diodes and transistors, bonded to thick-film 'pads', interconnected with thick-film 'leads' to thick-film resistors and thin-film capacitors, culminating in multistage integrated circuits; then on the reverse side of the dielectric 'massive' wire connections were made by the printed circuit method of molten-solder flushing, the entire structure was potted to make a rather prosaic looking assembly with about a dozen wires emerging to provide paths for electronic in and out.

The boundaries between discrete components and circuits have been blurred in the past. By the same token, the boundaries delineating the different approaches to putting together commercial electronic circuits are also in the process of consolidation and amalgamation, leading to a new generation of circuit-making and assembly procedures, far advanced in scope and with greatly improved efficiency in the employment of materials and labour over past practice. Until this positive process has been evaluated and proved it is necessary to produce equipment that is reliable.

Adhesive materials

Adhesives used for bonding, sealing, soldering and other applications have become increasingly popular in the electronic industry. Some adhesives have replaced welding, soldering, brazing and other methods of mechanical joining with definite advantages.

Corrosion that results from galvanic action can be minimised; high temperatures required for welding, soldering and brazing are eliminated, as are the stress-concentrating, space-wasting protrusions of mechanical fasteners.

Because of more uniform stress distribution, fatigue strength can be greatly increased. Perfect airtight chambers, which were at one time difficult to achieve, are now easily obtainable through the utilisation of special sealing adhesives.

In electronic and electrical applications, the use of adhesives has helped to reduce the size and weight of microminiature assemblies. Adhesives are insulators, but conductive metal powder or metallised fibre-filled adhesives have replaced solder in applications where heat-sensitive components are involved. These materials can also be used for RF caulking applications.

So far, the greatest use of adhesives has been in structural bonding. The basic type of adhesives for this purpose are the fluid types, which are dispersed in a carrier (such as solvents) that must be dried off before the adhesive is used, and the types that consist of 100 per cent functional ingredients; these latter can be divided into pastes and films.

In the fluid categories there are single-component and two-component systems. The latter require the mixing of the basic resins with a catalyst or hardener to promote curing; films are sub-divided into heat and electrical curing types. Film adhesives are available in sheet or strip form in various sizes and thicknesses and are readily shaped or cut. They offer the advantages of uniform thickness and ease in handling. Some types liberate fumes during curing with heat and during the heat-curing cycle the film becomes a liquid which helps to form a good bond between the two surfaces. Other types require no heat and can be cured by contact pressure.

A relatively new film adhesive is an 'electrically' cured type. Here a resistive heating element is bonded between two layers of film, thus forming a laminate. The laminate is applied between the surfaces to be bonded, current is fed to the heating element and the heat generated cures the adhesive. This type finds use in applications where the assembly to be bonded is too large to be heat-cured by conventional methods.

Basic materials

Most adhesives are prepared from one of two types of resin: thermosetting and thermoplastic. Thermosetting resins are synthetic organic materials that can be converted into a permanently hard, almost infusible and insoluble solid by chemical reaction. They do not support combustion and are waterproof. They resist fungi and the action of most chemicals. Heat is usually applied to aid curing; however, special resins and catalysts may make curing at room temperature possible within a reasonable time. Thermosetting resins are generally more heat-resistant than the thermoplastic types and can usually withstand occasional temperatures of 148 to 260 °C. Popular thermosetting resins include epoxies, phenolics, polyurethanes and silicones.

Unlike thermosetting materials, thermoplastic resins can be softened by heat and rehardened by cooling. They are generally weaker, less rigid than thermosetting resins and have a lower modulus of elasticity. Thermoplastics tend to distort under their own weight at temperatures of 65 to 120 °C, but certain heat-resistant grades are usable up to 260 °C. Common thermoplastic materials include polyvinyl resins, polyvinyl acetates and acrylic resins.

Filler materials

Usually a description of an adhesive states whether it is filled or unfilled resin. Fillers added to resins make a wider range of properties possible. Filler material can be organic and include silicon, asbestos, metallic powders and glass fibres. Each type of filler can be chosen to add a different property to the adhesive and several fillers can be added when a combination of the properties they contribute are desired.

Reinforcing fillers of fibrous material, such as cotton, asbestos and glass, are used to improve the flexural and impact strength of an adhesive. Clay, carbonates, silicates and bulk fillers are used to decrease cost and to help improve flow characteristics.

Special fillers of powdered metals are often incorporated to provide conductivity. However, the combination must maintain particle-to-particle contact to be effective. Mica is often employed as a filler when improved electrical resistance is desired.

Electrical insulation

Ceramics and silicones are making news because they provide good electrical insulation at high temperatures, 700 °C for instance, but other characteristics — formality, weight, moisture resistance — make epoxy resins and polyurethane foams equally important. New techniques such as transfer moulding permit substantial cost reductions in semiconductor components.

The general trend toward shrinking the volume of electrical and electronic components and circuits while maintaining their power capabilities demands insulating materials capable of withstanding severe environmental conditions, particularly high temperatures. Hence, we will begin our state-of-the-art review with the insulating materials that have the highest resistance; namely ceramics.

Ceramic insulators

Manufacturers are seeking materials with better electrical properties, particularly resistance stability, at temperatures up to 1,800 °C. Among the ceramics they have investigated are zirconia (ZrO), which almost becomes conductive above 5,037 °C and beryllium oxide which retains its high resistance values above 649 °C. Boron nitride has a maximum working temperature of 704 °C in an inert atmosphere, but in an oxidising atmosphere its maximum is 982 °C when it is in solid form and 704 °C when it is powdered.

Ordinary boron nitride hydrolyses slowly in water, but the rate increases with temperature. A newer compound is reported to be practically impervious to moisture, with high dielectric strength and a higher thermal conductivity than beryllia (above 700 °C). Hence, it is being used like beryllia as a mounting wafer for electronic devices. Both materials are readily obtainable.

Beryllium oxide is unique in that it combines the pure-oxide properties of high resistivity and low loss dissipation with a thermal conductivity equal to aluminium at room temperature. Its conductivity is eight times greater than that of high-purity alumina and it is very expensive. Nevertheless, it is used for power-transistor heat sinks to fit case outlines and as a substrate for microcircuits. Since beryllia is especially transparent to microwave radiation at high temperatures, it is being used for power antenna windows, klystron-tube envelopes and space-vehicle radomes.

Recently, a series of modified BeO ceramics have been introduced designed for high-frequency applications. Permittivities range from 6·5 to 50 for frequencies up to 10 MHz or above, and thermal conductivities range from

60 to 90 W/m.degC at 25 °C. Grades for semiconductor bases have a thermal conductivity of 145 to 150 W/m.degC at room temperature, ranging down to 29.4 kJ at 815 °C. The dielectric strength of these ceramics is 700 V.

A low-dielectric-loss material is 99·5 per cent BeO; it has a stated thermal conductivity of 0.60 W/m.degC at 20 degC.

High-alumina ceramics (originally developed for spark-plug insulators) are much cheaper than beryllia but have 1/10 to 1/20 the heat-transfer rate. The dense vacuum-tight bodies possess very high strength, a thermal conductivity 20 times that of electrical porcelain, a high softening temperature, over 1,649 °C, and excellent electrical properties at room temperatures. Alumina ceramics are being used for high-power electron-tube envelopes, RF windows, high-voltage stand-off insulators and bushings, and substrates for micro-electronic circuits.

The accepted military standard of classification for ceramic high-frequency insulation — 'Insulating materials, electrical ceramic' — grades ceramics by their loss factor at 1 MHz (see Table I.1). Note that the dielectric loss factor here is the product of the permittivity of a material and the tangent of its dielectric loss angle or dissipation factor. The aluminas are in the L-6 to L-8 grades. Their loss factors are as low as 0·0001, with dielectric constants ranging to 9·6 at 1 MHz and their dielectric strength is from 200 to over 1,000 V/mm for 98 per cent Al_2O_3 grades.

Table I.1

Grade	Maximum loss factor
L-2	0·070
L-3	0·035
L-4	0·016
L-5	0·008
L-6	0·004
L-7	0·002
L-8	0·001

One of the newest developments in electronic ceramics is a 99·9 per cent polycrystalline alumina material which is fired at high temperatures. The crystals are bonded directly to one another with neither pores nor a glass matrix between them, so that the material has a metal-like structure and is 90 per cent translucent to visible light. It has a dielectric strength of 1,700 V (average), a dissipation factor of 0·000025 and a permittivity of 9·9 and its thermal conductivity is better than that of some metals.

Ceramic substrate and flat-packs

All high-alumina ceramics, including magnesium silicate and beryllia, are widely used as substrates for thin-film microcircuits and sealed flat-pack integrated circuits. Beryllia has 240 times the thermal conductivity of electronic sealing glass (that of alumina is 10 to 20 times). Usually, ceramic substrates are used for deposited photocell, resistor, capacitor, and semi-conductor films and crystals, but there is also a wide use for hybrid thick-film microcircuits on ceramic substrates.

A typical microcircuit flat-pack consists of three ceramic parts with metallised pads and leads. Either high-purity alumina or beryllia is used for the monolithic base, frame and cover.

Injection-moulded alumina

The usual method of forming small electronic parts consists of pressing the dry powder and sintering the pack. Since the pressure is undirectional, the design of parts is limited with regard to the direction of holes, recesses and bosses. Injection-moulding techniques are used to form high-alumina parts of intricate shapes, including threads and end disks of coil forms. Temporary binders added to the normal powder mix act as particle lubricants to achieve an even flow into the moulding dies. Precision alumina parts produced by using transfer or compression-moulding techniques are said to have tolerances of 0·0254 mm. A dissipation factor of 0·0008 at 1 MHz, a permittivity of 6·5 and a loss factor of 0·004 are typical electrical properties of the material, which can be used continuously at 1,204 °C.

Glass substrates

Glass is still the most commonly used substrate material because of its smoothness and its availability in convenient form. The thermal conductivity of glass at 25 °C ranges from 0·002 to 0·004 W/m.degC.

Its dielectric strength is 2 to 10 kV/mm at 60 Hz, its permittivity varies from 4·6 to 6·6, and dissipation is from 0·0011 to 0·0062 at 1 MHz and 25 °C. The softening-points are obviously much lower than those of the ceramics.

Microcircuits and hybrid thick-film circuits are being deposited on glass substrates. Resistors, capacitors and inductors are evaporated through masks or chemically deposited using aluminium, silver, or gold leads to which discrete diode and transistor chips are soldered.

Glass-bonded mica

Glass-bonded mica bridges the gap between organic plastics and ceramic insulation. It is mouldable like plastics; it has highly stable dimensional properties like the ceramics, but it is easily machinable in those grades where the mica content exceeds that of the glass. When it has a higher glass content, machining is difficult and insulator shapes are formed by transfer moulding at very high temperatures. Because the thermal coefficients of expansion are of the same order of magnitude as those of many metals and alloys, metal inserts can be incorporated directly into the insulator during the moulding process. It is even possible to employ high-expansion brass and silver inserts.

The machinable materials are generally useful up to about 400 °C, whilst ceramo-plastics have operating temperatures up to 815 °C. Although

c

they are very satisfactory at low frequencies, the machinable glass-bonded micas are especially useful in the range of 10^5 to 10^9 Hz, where the dissipation factor reaches minimum values of less than 0·002 with permittivities around 6·5.

The permittivities of the mouldable, high-glass-content compositions are generally higher than those of the machinable glass-bonded micas. Their dielectric strengths and arc resistances are high, and their thermal endurance ranges from 300 to 649 °C.

The higher-temperature endurance of the ceramo-plastics is obtained by precipitating out a portion of the synthetic mica component to give an interlocking crystal structure. This grade is used to make high-temperature hermetic seals, bobbins, coil forms, tube sockets and commutators. Other types with lower heat-distortion temperature values are being used in gyroscope gimbals, transducer components, printed-circuit substrates and potentiometer housings. Still other grades are being used for relay parts, including contact supports, are barriers, armature carriers, coil forms and bases.

The machinable grade of glass-bonded mica with the lowest coefficient of expansion is used for brush holders, arc shields, terminal boards and HF transmitter panels. Bases for circuit breakers with stud inserts have been made with the precision-moulded, high-glass material.

High-temperature plastics

Moulding compounds that retain good electrical properties at elevated temperatures permit more compact design of electrical and electronic components. Many space applications require moulding compounds for continuous use at 500 °C; at present the polyimides are the only materials which can approach this value.

A new polyimide resin has a heat-distortion temperature slightly over 260 °C at 1·8 MN/m² and is said to retain its structural and electrical properties up to that temperature. For example at 200 °C the permittivity at 1 MHz is 3·30 (3·21 at 22 °C) and the dissipation factor is 0·0024 (0·0102 at 22 °C). The new polyimide, originally developed for aerospace use, is now being applied to industrial electrical applications where extended exposure to temperatures up to 300 °C is encountered.

In general, the silicones were the first of the semi-organic polymers to be developed for high-temperature service. They have a quartz-like (SiO_2) structure which is combined with various organic groups. Recently, glass resins have been introduced on an experimental basis; although generally classed as silicones, these resins are based on an alternating silicone-oxygen system. Moulded-glass resins have a thermal conductivity comparable to that of mineral-filled silicone, 0·08 or 0·09 W/m.degC, but their dielectric properties are generally superior to those of silicone moulding compounds and most plastics.

Regular fluorocarbon resins have a heat-distortion temperature of only 49 °C at 1·8 MN/m², but reinforced grades have a much higher heat resistance. For instance TFE has a heat-distortion temperature of 76 to 104 °C and is capable of continuous service at 260 °C. Resistance to voltage break-down is time-dependant and decreases rapidly with the duration of stress apparently caused by corona discharge in air, since the drop does not take place in a vacuum. The chief drawback of the TFE resins is the inability of most other resins to wet them.

Reinforced and flame-resistant plastics

Asbestos fibres have been used to raise the temperature capabilities of both thermoplastic and thermosetting moulding compounds. One silicone-chrysolite asbestos compound, for example, retains half its tensile strength after a 5 hr exposure to 649 °C. A new 'B' stage asbestos-silicone moulding compound retains its strength after extended exposure to temperatures up to 426 °C. Originally developed for aerospace structural use, it is now recommended for electrical or radome applications.

A flame-retardant, glass-reinforced polyester moulding compound with a heat-distortion temperature of over 287 °C at 1·8 MN/m². It has the following electrical properties: arc resistance 180 sec, dielectric strength 320 V, permittivity 4·95 at 1 MHz, and volume resistivity 1·6 × 10 Ω.cm.

Glass-fortified 30 to 40 per cent glass-fibre polysulphone is made in two grades and has a heat-distortion temperature of 182 °C at 1·8 MN/m². Glass fortified 6/6 nylon, 40 to 50 per cent glass fibre by weight, has a heat-distortion temperature of over 260 °C. Glass-fortified polycarbonate self-extinguishing resin has a heat-distortion temperature of 148 °C at 1·8 MN/m² with a dielectric strength of 490 V.

Among the high-temperature moulding compounds are the DAP (diallyl phthalate) resins, and when used with either short or long glass fibres have a deflection temperature of 232 to 280 °C. A diallyl isophthalate pre-polymer is structurally different from the DAP prepolymer and has a somewhat lower deflection temperature, 176 to 260 °C. Both these materials retain high insulation resistance (10^{13} to 10^{16} Ω.cm) and dielectric strength 375 to 400 V under high temperature and high humidity.

Electrical grades of phenolics with mineral fillers added for increased heat resistance withstand temperatures of 232 °C for short periods of time, although the stated deflection temperature is 182 °C at 1·8 N/m². The moulding compound is given a prior heat-treatment of 200 °C for 72 hrs.

Many phenolic compounds withstand 260 °C for one day, but only a few specially formulated types that are properly cured withstand this temperature for 300 to 400 hr without serious loss of structural and electrical properties.

The highest-heat-resistant grade mineral-filled phenolic has a high-temperature rating of 260 °C, with a dielectric strength of 300 V (short time). The highest-temperature electrical grade is a mineral-filled, two-stage compound with a maximum exposure temperature of 232 °C, a dielectric strength of 350 V and an arc resistance of 150 sec.

More self-extinguishing materials with better short-time chemical and heat resistance and more uniform engineering and manufacturing properties are needed by computer manufacturers. There have been some severe losses by fire of stored data in military computers where flame-resistant laminates for printed circuits have not been used. In addition, self-extinguishing materials are now available for applications that do not involve the 'sole current support' requirements. Some of the polycarbonate and nylon applications have been replaced by acetates and polyolefins. In the newest integrated circuit models, ceramic and glass substrates and flat-packs are replacing plastic laminates, but the plastic moulding compounds are still used in input-output equipment. Flexible flat cable must also be self-extinguishing.

Considerable improvements have been made recently in the properties of flame-resistant or retardant grades of plastics and those that are self-extinguishing. The properties of a wide variety of plastics compounds have been upgraded to minimise flammability. Most of these materials have been evaluated.

Casting, encapsulation and potting resins

The casting resins include epoxies (the most widely used), silicones, polyesters and polyurethanes. The silicones have the best combination of electrical properties but they are also the most expensive.

Epoxy resins

The newest of the epoxy resins is the cycloaliphatic type. These materials are prepared by the epoxidation of cycloaliphatic olefins, rather than by the reaction of materials with epichlorohydrine. The process provides a wide variety of resins. Advantages over conventional epoxy resins include outstanding arc and tracking resistance, a low and flat dissipation-factor curve as high as 250 °C, high deflection temperatures and the ability to perform well under high voltages in outdoor environments.

The arc resistance of the cycloaliphatic resins is in the same order of magnitude as the polymethylmethacrylates (acrylics), which are extremely arc resistant.

Volume resistivity at 25 °C is above 10 Ω.cm and at 240 °C is above 7.5×10^9; the permittivity is 3·10 to 3·15; dissipation factor at 10^5 Hz is 0·001 to 0·020.

Polyurethanes

The ricinoleate-polyol-based urethanes are composed of a liquid prepolymer and a liquid polyol. Recently introduced are several new types with low viscosities in the range of 650 to 700 cP to simplify application. Their properties include a volume resistivity of 30 Ω.cm, a permittivity at 1 kHz of 3·20 to 3·90 and a dissipation factor of 0·039 to 0·095.

Flexible polyurethane elastomer systems are particularly adaptable to the encapsulation of electrical connectors, components and printed-circuit boards used at very low temperatures, although they retain good electrical properties at relatively high temperatures.

Control of encapsulation stresses

Shrinkage stresses of encapsulation at low temperatures must be considered. Some ferromagnetic devices, including ferrite memory cores, are extremely sensitive to compressive stresses.

Shrinkage stress can also cause breakage in lead wire in electronic modules.

The conventional approach to stress reduction in encapsulation compounds is to use inert fillers, such as silica, so that the coefficient of expansion or contraction more nearly matches that of the component. Some incorporate fillers to provide a uniform thermal coefficient of expansion of $2 \cdot 5 \times 10^5$ mm/mm.degC throughout the operating temperature range. Internal stresses are said to be held extremely low by controlled polymer shrinkage. The material is intended particularly for encapsulating welded-lead modules.

Increasing thermal conductivity

Synthetic resins are poor conductors of heat, but their conductivity may be increased many times by the addition of such high-conductivity fillers as ceramics. In some applications, instead of embedding directly with the filled resin, the mould for the component or module is loaded with inorganic filler, and then, under vacuum, the dry system is impregnated with an unfilled resin or one with a very fine filler. The result is a body with high thermal conductivity and a low coefficient of expansion. This method has been used to increase to 1 W/m.degC the thermal conductivity of an encapsulant for densely packaged modules. The new encapsulant consists of aluminium-oxide granules held together with a polyurethane potting compound.

A much lower temperature system (148 °C maximum continuous temperature) makes use of beryllia (BeO) microspheres to provide high thermal conductivity in potting and embedment resins. A recent development is a dielectric composite containing 50 per cent BeO by volume dispersed in a high-quality silicone rubber. Thermal conductivity is 10 to 16 W/m.degC, thermal expansion is 31 to 65 from 10 to 93 °C, maximum operating temperature is 170 °C, volume resistivity is 10 Ω.cm, dielectric strength is 650 V, permittivity is 5·9 and dissipation factor is 0·002 at 1 MHz.

Castable epoxy resin systems filled with up to 60 per cent of BeO microspheres are also available for the potting and encapsulation of electronic components. Thermal conductivity ranges from 5 to 18, depending on processing. The maximum recommended temperature is 93 °C continuous.

Encapsulation by compression and transfer

Plastic encapsulation of electronic components and modules by compression- and transfer-moulding techniques is rapidly displacing other packaging methods especially the use of metal cans with glass-to-metal seals. However, moulding equipment using metal dies is expensive and, therefore, practical only when large production quantities are involved, as for resistors, capacitors and semiconductor devices.

With the latest type of transfer-moulding equipment specially designed for electronic components, it is possible to encapsulate fragile devices like glass-encased diodes, transistors with 0·025 mm diameter spring-loaded leads, glass-enclosed reed switches, and integrated circuits deposited on glass and ceramic substrates. With the increased popularity of transfer moulding, many well established encapsulating materials are being modified to allow them to be moulded at lower pressure and with faster cures. Low-pressure phenolic resins with a moulding-pressure range of 0·345 to 4·137 MN/m^2 are being used instead of higher-cost epoxies to encapsulate glass reed switches. Used as a coil encapsulant, phenolic resins provide a strong chemical bond with the nylon coil bobbins surrounding reed switches. High-flow silicone moulding compounds are also available for moulding at lower pressures: their present transfer-moulding range is 1·034 to 10·347 MN/m^2.

Epoxy resins are available in powder form for transfer moulding. A one-part system can be obtained with lower cure temperatures and shorter cure times than required by the conventional two-component liquid systems used for casting and potting. Also, the filler content can be higher in powdered-epoxy systems than in liquid systems.

Electromagnetic coils can now be transfer-moulded with epoxy resins. Uncured, the resins are solid, but they become relatively fluid at the transfer-moulding temperature. The properties of a new silicone moulding compound, which is a self-extinguishing material, have been designed specifically for encapsulation of electronic components such as semiconductor devices include moulding pressures from 1·034 to 10·432 MN/m^2, moulding temperatures of 93 to 115 °C, an arc resistance of 250 sec, a volume resistivity of 1×10^{14} Ω.cm, a dielectric strength of 270 V and a permittivity at 1 MHz of 3·8.

Diallyl phthalate (DAP) moulding compounds have also been developed for transfer-moulding encapsulation of delicate electronic components, glass- and mineral-filled grades having a moulding pressure of 3·47 MN/m^2 at 93 to 110 °C and a curing time comparable with that of phenolics. Their heat-distortion temperature is 136 °C. The material has a high arc resistance of 185 sec, a permittivity of 4·3, and a dissipation factor of 0·016 at 1 MHz. Its dielectric strength is 375 V and its normal resistance of 10^7 $M\Omega$ drops to 3,000 after 30 days at 100 per cent relative humidity. This grade is self-extinguishing.

Encapsulated semiconductors

Semiconductors now being transfer-moulded with epoxy powders include silicon transistors, silicon diodes, SCR's, varistors, flat-packs and integrated circuits. Modules containing such components are also transfer-moulded. The soft flow and low shrinkage of the new formulation do not affect glass encased diodes. Excellent adhesion to leads prevents the entrance of moisture.

Plastic foams

Very good electrical properties, excellent resistance to shock and vibration, a short cure cycle, and very light weight make the future of rigid foams quite bright for encapsulating and potting electrical and electronic parts. The polyurethane foams are gas-expanded thermosetting cellular compounds. They are non-porous and contact with water up to 426 °C is harmless to the structure, even over long periods.

Urethane foams are unaffected by mildew, rot, or fungi. They can be used in temperatures as low as −50 °C and as high as 65 to 93 °C, depending on the formulation. The CO_2-blown foams are preferred because of their good mouldability and dimensional stability.

A two-component modified-polyester CO_2-blown foam in place of styrene polyester casting compounds is used to pot terminal blocks. Special equipment is used for pouring the two liquid components which react as a mix in less than a minute. Curing is complete in 10 min without external heat. The casting material weight has been reduced by 75 per cent.

Polyurethane potting compounds are also used to pot pullout cable connectors (umbilical interconnection) on military aircraft, and medium density foams are used to encapsulate transformers. A system recently developed is a typical polyester CO_2-blown foam whose slow reaction rate provides a long working time, with a density of $13 \cdot 79 \text{ kN/m}^2$, a permittivity of $1 \cdot 02$ and a loss factor of $0 \cdot 0005$. In general, most urethane foams have permittivities ranging from about $1 \cdot 0$ to $4 \cdot 5$ at 10^7 Hz. The permittivities of other plastic foams at 10^5 Hz are from $1 \cdot 05$ for a polystyrene foam, $1 \cdot 12$ for cellulose acetate and $1 \cdot 49$ for polyethylene, to $2 \cdot 0$ for the epoxies.

Powder coatings

The insulating materials used for powder-coating applications generally consist of polymer resins, inorganic fillers and catalysts. When brought into contact with preheated objects, they flow and fuse to form smooth, continuous films. Coatings $0 \cdot 127$ mm to $1 \cdot 016$ mm thick may be applied by the fluidised-bed process, spraying or electrostatic coating. A $0 \cdot 127$ mm film is the thinnest continuous film possible with the fluidised-bed process.

A variety of resins can be applied by powder-coating methods. For thin insulating coatings, thermosetting resins are better than thermoplastics because of their ability to withstand deformation and flow at elevated temperatures. The most widely used thermosetting powder resins are the epoxies, which can be formulated with a variety of properties and form very rigid, tough films. They can be used for operation at temperatures of 130 °C. A recent addition is semiflexible epoxy material cured with an impact resistance of $1 \cdot 103 \text{ MN/m}^2$. Cure time ranges from 3 min at 140 °C to 30 sec at 160 °C.

A series of epoxy powders for impregnating coils and windings by the fluidised-bed or spray technique has also been developed; the coils themselves are heated to melt and cure the one-part resin. A no-post-bake epoxy powder has been designed especially for spray-coating of the integral insulation on stator and rotor slots of fractional-horsepower motors. Preheat temperatures are in the 200 °C range and the time-temperature relationships are such that a full cure is obtained without post-heating.

Builds of 0·203 to 0·228 mm are obtained on stator lamination slot corners; corresponding wall thicknesses are 0·0254 to 0·259 mm. Another epoxy powder is specifically formulated to retain its electrical resistance properties under extreme temperature and moisture conditions. When 4·5 gauge copper rods coated with 0·0254 to 0·259 mm of the material were immersed in water at 87 °C, their dielectric strength after 24 hr was 1,030 V.

Short-time impregnation of rotors and stators

In the trickle process of winding impregnation, a preheated wound armature or stator is slowly rotated (5 to 40 rad/sec) at an angle while a quick-curing resin mixture is dropped or trickled onto the upper end of the assembly. Under the influence of gravity and capillary action, the resin penetrates the windings and is evenly distributed without voids. When the winding is completely filled with resin, rotation is continued in the horizontal axis position until the impregnant has gelled. Successful application of the process depends on the number of variables: the reaction time and viscosity of the resin, the temperature, the rate and angle of rotation, the quantity of resin and rate of trickling, and the physical characteristics of the rotor or stator.

The Araldite trickle process, for example, makes use of a high-epoxy, low-viscosity liquid mixed with one of four hardeners to vary the gel time. A solventless polyester varnish and automatic roll-flow process equipment for applying the resin to random-wound stators or wound rotors of 1 to 50 kW a.c. motors has been developed. The entire cycle including post-curing, takes 15 min. The solventless varnish has an arc resistance of 194 sec and a short-time electric strength of 1,900 V, and can be used for vacuum-impregnation. At 150 °C it has a permittivity of 2·58 and a dissipation factor of 0·9. Insulation resistance is $2·2 \times 10^5$ at 500 V and its weight loss after 1,000 hr at 200 °C is 4·2 per cent.

High-temperature varnishes

Thermosetting baking varnishes are being modified to meet today's need for higher motor-operation temperatures. For example, polyester varnish combines high bond strength with a thermal life of 20,000 hr at 200 °C in contact with polyester magnet wire. Its dielectric strength is typically 4,200 V when dry and 3,400 V when wet. A more flexible type is thinned with naptha and prevents lead embrittlement, but its operating temperature is only 190 °C; its dielectric strength is only slightly lower.

Another has a higher operating temperature (225 °C) and higher bond strength to meet high-speed armature applications.

A polyester-type varnish formerly rated for 155 °C, has recently been upgraded to make it compatible with polyester magnet wire insulation at operating temperatures of 180 °C, It is usually applied by hot- or cold-dipping and by vacuum- or pressure-impregnation. Its dielectric strength is 2,800 V when dry and 1,300 V when wet; it takes 2 to 7 hr to cure depending on the temperature. When properly cured, some 100 per cent epoxy impregnating resin systems can be operated for extended periods at 170 °C and have a heat distortion temperature of up to 210 °C, a thermal coefficient of expansion of $5 \cdot 2 \times 10^{-5}$ degC^{-1} and a thermal conductivity of 5×10^{-4} W/m.degC.

A semi-solid one-package, unfilled epoxy resin has recently been developed for impregnating servomotors, transformers and coils operating at a temperature of 180 °C. It can be applied either by a vacuum pressure cycle or atmospheric dip. Its thermal conductivity is $24 \cdot 9 \times 10^{-4}$ W/m.degC, its dielectric strength for 3·175 mm thickness is 390 V, its permittivity at 10^3 Hz is 3·83 and its power factor is 0·0055.

Assembly materials

Attempts have been made on a production and prototype basis to examine the full advantages offered by a comprehensive microelectronic technological approach to the manufacturing of equipment modules. In being able to design and produce modules utilising the most appropriate technologies, equipment modules can be generated to optimise the performance, reliability, size and cost parameters to suit the particular equipment priorities.

Assembled function modules

The following basically covers the whole range of equipment sub-assemblies excluding the one-chip integrated circuits in a package and conventional printed circuit boards with conventional discrete components. The majority of the applications use the 'hybrid circuit' approach. This type of circuit is applied in general to a configuration containing components produced by two or more microelectronic technologies, namely thin-film, silicon integrated circuits or discrete active devices.

The hybrid circuit is not a new innovation as its dominant position in the electronics industry demonstrates. It has been claimed that in 1968, in this market, between 40 and 70 per cent of the electronics industry's gross income could be attributed to hybrid circuits. This is probably caused by differing definitions, but if only the 40 per cent figure is considered, the hybrid circuit market is vast. The impact of the hybrid circuit is beginning to be felt in equipments at the prototype development stage.

Field of application

The assembled function module is expected to penetrate nearly all fields of electronic equipment applications. The timing of the penetration, however, depends very much on the particular field. Hybrid circuits so far have been used in military applications where size and reliability are the main requirements; in airborne equipment where size, reliability and cost are the governing factors; in the industrial equipment field, power supplies, regulators, convertors and encoders where cost and reliability are the priority parameters and even in the consumer field in automobile regulators.

The assembled functions module approach can be thought of as either a basic constructional technique, or as a technique for interconnection and extending the utilisation of silicon integrated circuits, or as a cheaper prototype or small quantity method of manufacturing than a non-standard SIC product.

Although these views are all partially true, the real virtue of the assembled functions approach is its flexibility and cost-reduction potential coupled with a generally rapid turn-round time from design to prototype, and low tooling costs.

Technology capability

The technologies currently available for the fabrication of hybrid circuit modules fall basically into the following three broad categories:

1 Thick-film circuits.
2 Thin-film circuits.
3 Multiple SIC chips, interconnected and packaged.

Either thick- or thin-film circuits can overcome many of the existing deficiencies in silicon monolithic circuits. The hybrid circuit can substantially extend the limitations of the monolithic silicon circuit in respect of passive component tolerances, current and power handling capacity and voltage breakdown limitations.

Thick-film circuits

The greater virtues of thick-film circuits are in the wide range of sheet resistivity and large thermal dissipation obtainable. The substrate material, high-purity alumina, possesses a thermal conductivity characteristic closely approaching that of some metals and is capable of dealing with large quantities of heat without excessive temperature rise. The passive elements and interconnections are applied by a screen-printing process using conducting inks which are fired at a high temperature in a continuous belt furnace.

Resistive inks currently in use can be blended to produce sheet resistivities in the range 50 to 500 kΩ.sq. This allows component values to range from ten to several megohms whilst still maintaining a manageable aspect ratio on the resistor artwork. These values do not represent the limits of attainable values which are necessarily defined by the line-widths capable of being printed, the area and printing accuracy possible. A resistor width of 0·127 mm with a line spacing of 0·203 mm are the current minimum widths used in production.

The interconnection material will normally have a sheet resistivity of about 0·1 Ω.sq by solder coating during manufacture.

Thin-film circuits

The substrate is borosilicate glass and the passive elements and interconnections are vacuum deposited on to its surface. The circuit may be deposited by a number of methods. Two methods chiefly used are:

1 Evaporation through a stencil (out-of-contact mask).
2 Multiple deposition followed by photo-engraving.

The resistor material is a nickel–chromium alloy with a fixed sheet resistivity of 200 Ω.sq. This material has characteristics of a very stable nature and which allows low TCR's. The minimum resistor and interconnection track widths currently possible are 0·0508 mm with a similar amount of spacing.

Interconnection patterns are formed from a multilayer construction of nickel chromium followed by nickel and finally by gold. The total thickness of the interconnection material is approximately 3,000 Å while the resistor element material is approximately 250 Å thick. The average sheet resistivity of the gold interconnection pattern is 0·2 Ω.sq.

Because the substrate is glass it is not a good conductor of heat and in consequence the thermal performance of thin-film resistors is limited.

Stability and reliability

The thin-film component is considerably more stable than the thick-film component. It can be expected, however, that as better resistive inks, material and processing become available, this gap may be narrowed. The stability of both thick- and thin-film passive elements, particularly differential parameter changes, is greatly increased by processing all the components on a substrate at one time. Typical differential parameter changes measured over five million component hours on thin-film elements have been shown to be less than 0·05 per cent on average.

The improvement in reliability of a hybrid module on a thick- or thin-film substrate over that of a discrete component assembly is achieved mainly by the elimination of many interconnection interfaces. These can be replaced either by molecular bonds formed during the passive component manufacturing process, or by the use of unencapsulated silicon integrated circuits.

When unencapsulated integrated circuits are back-bonded to the substrate this reduces the number of interfaces by 33 per cent over the conventional encapsulated device attachment. When the back-bonded chip is compared to the flip-chip method of assembly there is a further reduction of interfaces by 50 per cent. Since interconnections are a major factor in the potential unreliability of any electronic assembly this reduction in the number of interconnections by up to 75 per cent can improve the equipment reliability very significantly.

Substrate shape and processing

It is not economic to consider shapes other than rectangular for thin-film substrates because of the difficulty in handling glass. But the alumina substrate used for thick-film circuits can be obtained in almost any shape or configuration, with holes if necessary, at little extra cost. This is a major advantage where circuits are being built inside equipment that is already densely packed. Holes enable connection to be made easily between both sides of a substrate and can provide anchors for terminations.

Capacitors

Each of the film-circuit processes can be adapted to provide capacitors. In the thin-film process the most common method is to evaporate two layers of aluminium with a dielectric layer of silicon monoxide between. In the case of both thin- and thick-film capacitors the capacitance attained per square centimetre is only of the order of 1,000 pF and is very wasteful of substrate area. Generally these film capacitors are only economic if large numbers of small-value capacitors are required.

As a more practical alternative, various types of chip capacitor are available which can be reflow-soldered, face downwards, directly on to the interconnection patterns of both thick- and thin-film circuits. These types include ceramic and glass dielectric capacitors and also, to cater for the large capacitance requirement, surface-mounted solid tantalum capacitors.

To summarise, thick-film circuits offer good-quality medium-precision components in a rugged form with high thermal dissipation capabilities. Thin-film circuits provide high-quality, high-precision, very stable components with medium thermal dissipation capabilities.

Transistors and monolithic silicon circuits

The thick- and thin-film substrates are ideal mediums to which semiconductor devices can be attached by several methods. The actual method chosen will depend upon factors such as component-density requirements, power-handling capacity and cost.

Hybrids using pre-encapsulated devices

Assemblies based on thick- or thin-film substrates fall into this category. They may incorporate flat-pack integrated circuits, usually reflow-soldered, but sometimes welded, to the interconnection pattern. Discrete devices such as leadless inverted devices or beam-lead devices, can be soldered to any of the chosen conductors.

This use of encapsulated active elements gives rise to three disadvantages: cost, size and reduced reliability. Because the value of the package is a significant proportion of the total cost of an active device, appreciable savings can be made by using chip devices. Chips can generally offer reduced size and increased reliability because the number of necessary interconnections is reduced.

Hybrids using unencapsulated devices

Well known semiconductor assembly processes are used to build multi-monolithic integrated circuit hybrids based on a thick- or thin-film passive component and conductive network. Silicon chips are eutectically back-bonded to thick-film circuits by conventional methods. To reduce the number of processes the bonding pad for the chip is produced at the same time as the conductor network: pure gold ink is preferred.

Ultrasonic wire-bonding techniques are then used to make the connections from the face of the chip to the relevant interconnection pads on the thick-film circuit. Face-up, wire-bonded silicon chips can also be incorporated in thin-film circuits. In this case the chip is secured to the film circuit by means of a conducting-resin adhesive. Another method of attaching silicon chip devices, which can be used with both thick- and thin-film substrates, is face-down bonding, i.e. flip-chips. This technique is currently at the advanced development stage.

The conductor pattern of this system is formed by photo-engraving vacuum-deposited aluminium on glass or alumina substrates. This film is 12 μm thick and is selectively etched to form pillars standing some 6 μm proud of the conductor at the edges of the conductors. The silicon integrated circuit chip is aligned with its 10·1016 mm bonding pads over the pillars and ultrasonically welded. Chips having up to sixteen bonding pads have been successfully welded to aluminium interconnection patterns. The welds are expected to provide permanent and reliable connections to the chip.

This method reduces the number of bonds that have to be made, leading to a greater reliability of the assembly. It also reduces the complexity of the production procedure and therefore the final cost. Other advantages include a reduction in the lead length and associated capacitance loading, an avoidance of the difficulties inherent in protecting highly complex active circuits and an increase in packaging density.

Crossovers are provided by special silicon crossover chips, the active chips themselves or, where large numbers are required, by evaporated or screen-printed crossovers. Because the face-down chip has only the pillars of the conductor pattern as a thermal path, a back contact may be required. This can be achieved where necessary by using a glazed ceramic lid to encapsulate the circuit. On sealing, the glaze adheres to the chip and provides a much better second thermal path.

Module encapsulation

The choice of a protective system depends on many factors:

1. The environmental and mechanical test requirements of the module.
2. The desired shape and mounting requirements.
3. The type of terminations and method of connections.
4. The assembly technique and component to be used.
5. The complexity of the assembly.
6. The quantity to be made.

Several methods of encapsulation can now be recommended because their effectiveness has been demonstrated. These cover the whole range of environmental and mechanical requirements likely to be encountered in electronic equipment. Only two general classes, hermetic and non-hermetic, are dealt with here.

Hermetic

At present three types of hermetic encapsulation are in use. The large flat-pack is particularly applicable to circuits of either screened or vacuum-deposited film with face-up-bonded silicon chips. The ceramic cap on a screened circuit is soldered on after the assembly of components. The glass-to-metal sealed header and metal can is an exceedingly robust, low-cost assembly and is the most suitable for fairly bulky assemblies.

Non-hermetic

These methods of protection have one feature in common — the use of some organic material as a barrier to moisture.

Economic considerations

The modular approach offers cost-reduction advantages over conventional current equipment manufacturing methods in four main areas, in general:

1 The use of SIC chips to a function specification rather than a general application specification.
2 The use of SIC chips rather than individually encapsulated devices.
3 A reduction of the labour content of the works cost of an equipment.
4 A reduction in the cost of multiple precision resistors.

Cost reductions are also possible from the reduction of peripheral interconnection requirements such as plugs and sockets. These cost savings must be weighed against the regions of increased cost. Ceramic or glass as a substrate material is more expensive than printed circuit boards and non-precision resistors either in thick or thin film are generally more expensive than standard resistors. In actual case studies it has been frequently found that even in the least appropriate cases, i.e. where the number of components is low and no precision resistors are required, the comparative costs of the assembled function module and the assembled printed circuit boards are not significantly different provided that the circuit function can be multiplied to occupy the full substrate area available and the quantity requirement is substantial. On balance it is found that in most cases the assembled function module offers valuable cost saving advantages in production quantities over the conventional methods as well as reduced size, increased reliability and improved electrical performance advantages.

The electronic equipment industry is beginning to show considerable interest in the assembled functions approach to equipment module manufacture where cost reduction in association with reduced size, increased reliability and the real possibility of improved technical performance is offered. It is clear that in equipment module manufacture no single technology is pre-eminent and the best results will be attained by intelligent use of all the compatible technologies.

Section One

Chapter 1.1 Materials problems in integrated circuits

The revolutionary phase in electronic development is probably nearing its end and we are entering an evolutionary phase. Even so, major changes are still to come and the results of the evolution will probably have greater economic impact than changes that have occurred so far. The revolution has concerned itself largely with the physics of the active devices and the developers have concentrated on the interior workings of such devices with relatively little effort being applied to the problems of interconnection, insulation and environmental protection. The temporary solutions we have to these problems are still unsatisfactory. Encouraging progress is being made but only through a five-fold interdisciplinary attack involving the simultaneous best efforts in the fields of electronics, chemistry, physics, optics and mechanics.

The subject of materials for integrated circuits is so vast that I can do little more in a chapter of this length than define the problem as clearly as possible. It has recently become fashionable to point out that the most significant aspect of the new electronics which is developing, is the integrated nature of the design and fabrication process. As far as materials for integrated circuits are concerned, they are not a separable problem from the problems of devices, circuits and systems.

The most general view of the integrated circuit problem is the users view. He needs improved reliability, improved complexity, and improved performance without significant increase in cost. These requirements can only be met by fabrication processes which give extreme precision and at the same time are capable of exact repetitive duplication of the product in mass production.

Without discussing integrated circuits from one end to the other, it is a little difficult to show how specialised materials problems make a contribution to the solution of the user's requirements. In any case, in addition to discussing the whys and wherefores of the actual materials which wind up as part of the product sold to the user I will try to point out some not so obvious materials problems which are involved in the fabrication of the product:

1 Photosensitive materials and their bearing on the integrated circuit problem.
2 Characteristics of available photographic lenses and photographic materials place limits on progress in integrated circuits.
3 The circuit itself and the various materials most commonly used with silicon monolithic integrated circuits.

D

4 Reasons why these materials are being replaced by a much improved second-generation materials.

5 Versions of improved metallurgical and dielectric systems in this second-generation technology.

6 Materials problems of gallium arsenide planar monolithic integrated circuits. Since it is becoming clear that flip-chip or face-bonding methods will supplant other methods of electronic assembly, the substrate on which flip-chip semiconductor devices are mounted becomes an integral part of the integrated circuit assembly.

Photographic materials

Photographic methods taken generally are the only approach to the required precision of integrated circuits and exact repetitive mass production. They lend themselves well to low-cost batch-fabrication methods. Both stencil printing, sometimes called silk-screening, and photolithographic etching, as used in silicon monolithic circuits, are basically photographic techniques. Photographic techniques can only allow fabrication work on flat surfaces and consequently all the electronic materials used must be available in one way or another in the form of films which can be processed on flat surfaces. Photographic materials themselves become critical as materials for integrated circuits. These photosensitive materials do not customarily have useful electronic properties, so the manufacturer is forced to use materials and compatible chemical processes which enable him to transfer the information or the image stored in the photosensitive material to the useful electronic material.

The photosensitive material can be helpful to the integrated circuit designer in increasing the yield of useful product. Two different types of photoresist material may be used:

1 One which is rendered insoluble by the effect of light.
2 The other rendered soluble by the effect of light.

The use of the second type of resist eliminates one source of pin-holes in the surface dielectric layer of silicon monolithic integrated circuits. Pin-holes in the oxide layer are a well known source of short-circuits between the metallising and the integrated circuit itself and are one of the basic limiting factors in the achievement of economical large-scale integration.

With the first type of resist which is polymerised and rendered insoluble by the effect of light, a dust particle, either on the surface of the integrated circuit itself or on the photographic mask, will leave a small area of the photoresist material unexposed. This will result in a hole in the photoresist when the photoresist is 'developed', as the process of removing the unpolymerised material is called. Subsequent to development, the exposure of the photoresist-covered silicon dioxide layer to the etching acid will result in holes in the silicon dioxide, not only where they are required, but wherever a dust particle has intervened between the photographic mask and the integrated circuit.

On the other hand, with the use of the opposite type of photoresist, a dust particle that occurs under the dark parts of the negative photographic mask has no effect at all. In any particular processing step most of the silicon dioxide will be left in place and only a small portion removed. That means that most areas of the photographic mask will be black, and thus a dust particle has a much higher probability of finding itself in a dark area of this type of mask as opposed to a clear area of the other type of mask. Even if a dust particle should find its way to a clear area in the second type of process, it will merely result in some silicon dioxide being left where it is not supposed to be and this produces a minor defect in most cases.

There are several problems associated with lenses and photographic plates. In spite of all the progress which has been made in lenses since the invention of the microscope more than three hundred years ago, today's lenses leave a great deal to be desired from the point of view of the integrated circuit fabricator. They do achieve the theoretical optical resolving power, but only over a limited area — too small to satisfy the designer of large-scale integrated circuits. Another way of stating the problem is that lenses are available which will cover any area the circuit designer wishes to use, but they give much poorer resolution than theoretically possible.

The small size of useful individual circuit elements is limited only by the various factors involved in photographic resolution for circuits where information processing and not power handling capability is the dominant consideration. Thus the designer is required to make his circuit elements larger than necessary and this means that any one large-scale integrated circuit must cover a larger area of silicon than it theoretically should. This means that fewer circuits can be fabricated from a single slice of a silicon crystal, with resultant increases in costs and a decrease in the practical upper limit of the degree of complication of monolithic large-scale integrated circuits.

For example, consider the use of a photographic reduction lens to prepare a circuit mask for a silicon chip having a maximum dimension of 5 mm. It certainly will eventually become possible to photographically produce over the entire area of such a silicon chip circuit details having dimensions of 50 μm.

Such detail has already been achieved, but only over the extremely small areas of the highest frequency transistors available and such photographic resolution is well within the theoretical capability of lenses and photographic materials for circuit details having maximum dimensions of 5 μm.

Since three or four photographic plates are often involved as masks in the fabrication of any one element of an integrated circuit, the details of these various plates must naturally be correctly aligned with respect to each other with an error which is consistently less than the dimensions of the smallest important portion of the circuit. Thus successive photographic plates must be correct with respect to each other to perhaps 10 μm. Any lack of flatness, or bowing, of one photographic plate with respect to the others will result in lateral displacement of the image created on the plate by an amount that may be intolerable. In particular, if we are to achieve registry from one plate to the next of 10 μm the photographic plates and their emulsion coating must be flat to better than 10 μm.

Metals and insulators

There are still some problems with the conventional aluminium/silicon dioxide/silicon system. Silicon was originally chosen as the semiconductor for integrated circuits partly because of the compatibility of its oxide with the available photographic processing methods, and partly because the tremendous purity of the silicon itself offered the hope that the silicon dioxide produced by oxidising the surface of the silicon could be equally pure and thus offer protective material for the circuit.

Unfortunately, silicon dioxide has turned out to have a serious defect which results from its tendency to become positively charged from a variety of sources. The most common source of this positive charge is sodium, which diffuses readily through silicon dioxide in the form of a positive ion. This diffusion takes place at temperatures well within the operating range of silicon integrated circuits. Other positive charges are induced in the oxide by X-rays, γ-rays, high-energy electrons and the like. Also, the exact details of the processing by which the oxide is produced give rise to greater or lesser amounts of positive charge.

The behaviour and influence of sodium in this case is reasonably well understood, whereas the effects of radiation and processing variables remain to be fully explained. For various reasons, sodium contamination tends initially to be on the outside of the oxide and adjacent to the aluminium metallisation. While it is in that location, the electric field of the ions merely terminates on the metal and has no detectable effect on the integrated circuit. However, a positive bias applied to the metal with respect to the semiconductor will lead to a drift of the sodium ions away from the metal and cause an accumulation of positive charge just outside the surface of the silicon. As one might expect, the presence of this charge leads to an accumulation of negative electrons in the silicon at the surface, but for some reason neither the positive sodium ions or the electrons can cross the interface in order to neutralise each other.

The result is a semi-permanent condition in which the surface of the silicon is made either more n-type if it is already n-type, or is converted from p-type to n-type by the sodium-ion accumulation. This results in spreading out of the effective area of a p-n junction with a resultant increase in junction capacity, an increase in junction leakage current at reverse biases, and a generally lossy behaviour at high frequencies. For the opposite type of junction in which one has a small diffused p-region in an n-type semiconductor body, the presence of the sodium positive charge, or for that matter the positive charge resulting from radiation or imperfect processing, results in severely reduced junction breakdown voltage.

The origin of the well-known 'purple plague' arises from the use of gold-bonded connections to the aluminium metallising on integrated circuits; gold and aluminium, even at temperatures within the range to which assembled integrated circuits are sometimes exposed, undergo mutual solid-state interdiffusion to form a number of brittle intermetallic compounds which can either lead to complete rupture of the gold to aluminium bond or to open circuits in the aluminium metallising which surrounds the bond.

This particular combination of materials — silicon, silicon dioxide, aluminium and gold — are the gift from nature which made possible the discovery of the feasibility of integrated circuits. They do not by any means result from the team effort of a group of knowledgeable specialists, each contributing his best to an over-all desirable result. Rather, silicon dioxide, aluminium and gold are some of the first things that would be thought of in a semiconductor laboratory by a research worker who knows all there is to know about the semiconductor and how to make junctions.

Silicon dioxide is nothing but quartz and this is probably the best known high-grade inorganic insulating material. Aluminium is the purest metal available at reasonable cost and was the first or very nearly the first and the easiest material to be laid down in thin-film form by vacuum evaporation. Gold was the high conductivity metal which was readily available in the form of the required thin wires that also had the ability to be bonded to aluminium or other metals without great difficulty.

Thus it is fair to say that the integrated circuits which have been available on the market until recently are examples of extremely good laboratory technique rather than well developed mass-production items. Now that the economics of integrated circuits is indisputable, it has proved worthwhile to find alternatives for each material and for each step in the manufacturing process in order to achieve a more nearly optimum product from an over-all point of view.

A completely different material system for providing the required insulation, protection and interconnection has been introduced. The semiconductor device is inverted compared to the previous configuration since the method of assembly which is evolving for integrated circuits is one in which the metallised silicon chips are placed upside down onto geometrically matched metallised patterns on glass, on ceramic, or on plastic substrates.

Starting first with the insulators, the layer in contact with the silicon is still silicon dioxide produced by direct oxidation of the silicon itself. This layer can remain on the surface throughout the entire processing of the semiconductor. One of its functions is to control the distribution of impurities during the diffusion processes by which the p-n junctions are formed, another of its functions is to terminate the chemical bonds of the silicon surface atoms in such a way that the electronic properties of the semi-conductor are disturbed to the minimum possible degree, the silicon dioxide must be protected from sodium contamination. However, for this purpose it is overlaid with silicon nitride deposited from a gas-phase reaction between silicon compounds and nitrogen compounds. The nitride deposition is carried out at elevated temperatures, which are, significantly lower than the processing temperatures for the silicon dioxide and for impurity diffusion.

Contact to the p-n junction is made by depositing a layer of platinum on the exposed silicon. In subsequent heating, the platinum and silicon react to form the compound platinum silicide, which makes an excellent ohmic contact to p-type silicon and low resistivity n-type silicon.

All metals deposited on the insulating layer must adhere well to it; for this reason a bonding layer of one of the reactive metals is always used.

One of the fortuitous advantages of the aluminium/silicon dioxide system is that aluminium itself is an active metal and adheres well to the silicon dioxide. Other metals with good adherence are titanium, chromium, molybdenum, tantalum and nickel, titanium being far superior to the others. The beam-lead system is used for titanium for this purpose.

The only metal with sufficient ductility, stability and corrosion resistance to be used for the external contact of the device is gold. Unfortunately, just as gold produced harmful compounds when it was in direct contact with aluminium, it does the same thing with titanium. For this reason, a second layer of platinum is used as an intermetallic barrier layer between the titanium and the gold. The platinum also makes it possible to electroplate or electroform the gold in the required shapes.

It can be seen why this metallurgical system is referred to as the beam-lead process. The gold leads extending beyond the edge of the silicon chip resemble cantilevered beams and eventually, when connected to the substrate, do become the beam-like structures which support the semiconductor device or integrated circuit itself.

Beam-lead devices and integrated circuits

An advantage of the beam-lead metallurgical system is that common-anode diode arrays can be fabricated using the beams to provide structural integrity between air-isolated sub-units of the array. The previously mentioned positive charge tendency in silicon dioxide makes common-anode diodes hard to fabricate monolithically. On the other hand, beam-leads have been applied effectively to monolithic circuits.

Molybdenum–gold system

A different metallurgical system which accomplishes many of the same purposes has been developed. The essential metals used in these systems are molybdenum and gold: molybdenum as the bonding layer on the dielectric and gold for the current-carrying conductors and for the external beams.

The metal-bonding layer is molybdenum, and the intermetallic barrier layer is omitted since molybdenum and gold do not readily interdiffuse at temperatures likely to be encountered after integrated circuit fabrication.

In some cases, the platinum silicide ohmic contacting layer can also be omitted, but where the contact resistance must be reduced to a minimum, the platinum silicide is retained between the molybdenum and the silicon. The molybdenum–gold metallurgical system is not only applicable to beam-lead structures but has also been used with gold-wire-bonded assemblies.

Photomicrographic cross-sections of gold-wire bonds to gold–molybdenum metallised layers show no visible changes after extended exposure to elevated temperatures. On extended heating, similar cross-sections of the gold-aluminium system always show extensive visible damage from the formation of 'purple plague'.

Another use of the molybdenum–gold system has been in multilayer interconnection wiring on integrated circuits. Here molybdenum is used to promote adherence between the metals and the two layers of silicon dioxide, since the adherence of gold to the insulating layers is very poor. The metallurgical stack inside the sandwich then becomes molybdenum–gold–molybdenum.

The outside layer of silicon dioxide is deposited from a pyrolytic gas phase reaction between silicon compounds and oxygen or oxygen bearing compounds, much as was the silicon nitride referred to earlier.

Other possible metallurgical systems

In all of these material systems, mutual compatibility among all of the various materials which have to be used in making up the completed electronic assembly is necessary, not only must the materials be chemically and electronically compatible with each other, but they determine the types of circuits which are feasible and in the end dictate system design and have a major effect on software choices. For example, the question of whether future computers will consist largely of extremely-high-speed time-shared central installations or medium-speed, special purpose, broadly dispersed, smaller-capacity units will, in the final analysis, be answered when we have a clear picture of the compromises forced on us by the best possible systems we can find.

Table 1.1.1 is an example of only a few of the materials characteristics which have to be considered. Here we see, for example, that gold and platinum have low adherence to silicon dioxide and that titanium, chromium

Table 1.1.1   **Selected properties of pure metals**

Metal	Resistivity at 20 °C, $\mu\Omega$	Melting-point, °C	Adherence to SiO_2
Silver	1·50	961	Low
Copper	1·67	1,083	Low
Gold	2·35	1,063	Very low
Aluminium	2·65	659	Very high
Magnesium	4·45	650	High
Rhodium	4·51	1,960	Low
Iridium	5·3	2,443	Low
Tungsten	5·6	3,380	High
Molybdenum	5·7	2,610	High
Zinc	5·8	4·9	Low
Cadium	7·6	321	Low
Nickel	7·8	1,455	High
Cobalt	9·8	1,490	High
Iron	10	1,535	High
Platinum	10	1,770	Very low
Palladium	11	1,555	Low
Tin	11·5	232	Low
Chromium	13·0	1,903	Very high
Tantalum	15·5	2,977	High
Lead	22	327	Low
Vanadium	25	1,730	High
Zirconium	40	1,852	High
Titanium	42	1,812	Very high

and aluminium are extremely adherent. One of the next things we will need to know is the adherence of these various materials to silicon nitride. The ability to process the materials by photoresist and etching methods is not something which can be determined in absolute terms, but depends in a major way on the cleverness of the engineer, and thus a material which is a poor choice today may become quite suitable as a result of improvements in photoresist technique.

Ductility, corrosion resistance, bondability and compound formation with other materials, are additional characteristics which the development engineer must consider.

Gallium arsenide and integrated circuits

Gallium arsenide inherently has several attractive attributes for use in integrate circuits. In the first place, because, electron mobility in gallium arsenide is extremely high, it should be possible to make higher speed circuits with it than with silicon. Also, gallium arsenide's energy band gap is wider than for silicon and this should lead to the possibility of higher operating temperatures and hence higher power densities in gallium arsenide integrated circuits than with silicon. Since gallium arsenide can be fabricated in a form known as semi-insulating, in which it is more of a dielectric than a conductor, there is a possibility that integrated circuits could be prepared in which the semiconductor substrate electrically isolates the active and passive circuit components from each other. It is too early to state whether or not the system is completely suitable for mass production of integrated circuit fabrication. At the moment, the potential high speed of the gallium arsenide circuits has not been realised as a result of what appears to be electron trapping associated with impurities or imperfection in the gallium arsenide itself.

Substrates and passive components

We can pass on briefly to some of the materials problems associated with the other parts of the integrated electronic assembly. Since two dimensions are not enough in which to perform complex electronic network functions, we achieve a degree of three-dimensionality by stacking photoprocessed layers, each with its own function. This first results in planar semiconductor circuits, thin-film circuits, thick-film or screen-printed circuits, multilayer wiring boards, and so on. The most recent advance in integrated circuitry has been the development of surface protection layers such as silicon nitride, which make separate packaging of each semiconductor device or integrated silicon chip unnecessary. Thus, it becomes possible to mate one film laminate face to face with another until as many layers of circuitry as current technology permits on one substrate have been deposited.

The third dimension is added to by bonding that assembly face down to a second substrate, which itself carries one or more layers of circuitry. In some cases a third substrate is sandwiched between the other two. This technique of face-to-face bonding in various forms promises to proliferate and fill most of the needs of the electronic industry.

Not all of the desirable characteristics for passive circuit substrates are within the capabilities of today's materials technology. Briefly, the substrate should be inert, refractory, smooth (or polishable), of highest possible heat conductivity, compatible with silicon in thermal expansivity, inexpensive and — in common with the surface dielectrics used with silicon — immune to sodium migration. Various glasses have all these properties except thermal conductivity; glasses are poor thermal conductors, typical glasses being 40 times worse than high-alumina ceramic. On the other hand, ceramics give less than optimum surface smoothness, either in the as-fired condition or when ground and polished.

As fired ceramics are suitable for thick-film work and for medium-tolerance thin-film resistors. They are not yet good enough for the precise thin-film work required in the fabrication of capacitors. The state of the art in as-fired ceramics is advancing rapidly.

For inexpensive substrates which must have optimum smoothness, glazed ceramics are used. The glaze layer, being glass, introduces some thermal problems and may have sodium drift problems as well. Just as sodium causes shifts in transistor parameters, it can be electrolysed out of the glaze to cause serious drifts in thin-film resistors. The cause of the resistor drift is presumably some sort of chemical reaction unrelated to the effects of sodium on transistors.

Two interesting substrate materials now being used in limited quantities are beryllia and single-crystal sapphire. Beryllia has approximately eight times the heat conductivity of pure aluminium metal. Sapphire is chemically identical to alumina, but being single crystal it has somewhat better thermal characteristics. Its main advantage over the alumina ceramics is that it can be polished perfectly flat and smooth, while alumina when polished always has pits which result from the chipping away of small crystallites. The price of sapphire can be expected to decrease as time passes. It is already within reach for critical applications, such as space-borne electronics.

Will silicon continue to serve as the basic material for integrated circuitry, and if so, for how long? Since we have learned that silicon nitride is at least partially suitable as the surface dielectric layer on semiconductor integrated devices, and since the silicon nitride is not grown from the underlying silicon itself but deposited from the outside as a result of a gas-phase chemical reaction, we ought in the long run to be able to find materials and methods which are as economically suitable for germanium and gallium arsenide as the current processes are for silicon. The primary limiting factor has been that silicon is the only known semiconductor which has an autogenous oxide which is also impervious to most contaminants, resistant to most reagents, and at the same time an excellent dielectric.

However, the introduction of a new semiconductor as the basic material for integrated circuits will either take a long time or require a vast effort in which the skills of chemist, physicist, electronics engineer and optical scientist are fully co-ordinated and integrated. It is probable that no single company has the resources to carry germanium, for example, as far as silicon has been carried in the last ten years by the concerted efforts of many research workers and development engineers, supported by all of the resources of the laboratories of a great number of companies.

On the other hand, except for the semiconductor, we can be certain that many alternatives exist for all the materials, none of the materials in present use seems to be optimum from all viewpoints. In particular, the over-all materials systems used are very complex, and it is necessary to find new materials which can fill several functions simultaneously and thus reduce the number of elements in the system and simplify the processes of fabrication.

The extreme difficulty of replacing silicon with another semiconductor may guarantee it the leading role for many years. All the other materials currently in use, however, are likely to change repeatedly as new knowledge is brought to bear on solid-state electronics.

 Chemicals in semiconductor manufacture

Many chemicals and materials are used in the fabrication of semiconductor devices. Of utmost importance is the purity of these chemicals, the reason being that semiconductor performance is easily affected by impurities. One intolerable result is low yield.

To fabricate high-purity silicon, germanium, or gallium compounds into semiconductor devices, makers commonly use high-purity acids to etch away contaminants. Most often used are nitric, hydrofluoric and acetic acids; others are hydrochloric, phosphoric and sulphuric acids. Solvents such as highly purified trichlorlethylene, acetone, methanol and isopropyl alcohol remove moisture and impurities.

Plating chemicals include gold chloride and nickel chloride. Gold is plated on device-encapsulating components for corrosion protection, while nickel forms the ohmic contacts for attaching leads to large-area silicon devices. Among other chemicals for the plating process are ammonia, citric acid, sodium hypophosphite and acetic acid.

Emulsion compounds, or photo-resists, are required in the photo-engraving technique for making individual microcircuit elements. Information on these compositions is proprietary in nearly all cases. Strippers and solvents used to remove the photo-resists are often chemicals such as toluene or xylene. For tough stripping jobs, proprietary solvents are used.

Dopants, which are chemicals added to the silicon or germanium crystals to produce the desired electrical properties, include phosphorus pentoxide, phosphorus oxychloride and boron tribromide. They are used only in small amounts, but their performance is vitally important.

Traditional packaging materials for semiconductor devices are ceramics, glass and metal. However, they are being replaced by epoxies and silicones.

Other chemicals and materials include passivators, grinding agents and inert gases. The semiconductors are covered with junction coatings, usually silicone resins, to protect them from the environment. Grinding agents such as silicon carbide, diamond, aluminium oxide and silicon dioxide are employed for surface polishing. Argon and helium provide an inert atmosphere for various phases of device making.

Acids, etchants and solvents

These categories of basic chemicals are of vital importance to the semiconductor industry since they are used to etch, clean, or otherwise remove impurities from semiconductor devices during fabrication. Consequently, chemical purity and uniformity are absolute requirements because impurities of one part per million can change semiconductor electrical properties by many orders of magnitude.

Sampling procedures by the manufacturers maintain the average quality requirements to ensure that results in parameters are meaningful to their production engineers.

Electronic-grade standards have been established that are often more stringent than maximum permissible impurity concentrations. As an example, reagent-grade nitric acid is limited to 0·00005 per cent chloride by the normal specification, whereas specifications for 'Electronic grade' call for a maximum of 0·00001 per cent chloride. Purity levels of two acids, two solvents and

Table 1.2.1   Purity levels for two 'Electronic-grade' acids

Acetic acid	*Chemical/other tests*	
	Assay (CH_3COOH)	99·9% min
	Residue after evaporation	0·0008% max
	Substances reducing dichromate	passes ACS test
	Substances reducing $KMnO_4$ (as SO_2)	0·015% max
	Physical tests	
	Colour (APHA)	10 max
	Dilution test	passes ACS test
	Specific gravity at 60/60 °F	1·049–1·052
	Impurities	*Maximum limits, %*
	Chloride (Cl)	0·0001
	Phosphorus (as PO_4)	0·0001
	Sulphate (SO_4)	0·00005
	Arsenic (As)	0·0000005
	Copper (Cu)	0·00001
	Heavy metals (as Pb)	0·00003
	Iron (Fe)	0·00001
	Nickel (Ni)	0·00001
Hydrochloric acid	*Chemical/other tests*	
	Assay (HCl)	37·0–38·0%
	Residue after ignition	0·0003% max
	Physical tests	
	Colour (APHA)	10 max
	Specific gravity at 60/60 °F	1·185–1·192
	Impurities	*Maximum limits, %*
	Bromide (Br)	0·005
	Sulphate (SO_4)	0·00005
	Sulphite (SO_3)	0·00008
	Free chlorine (Cl_2)	0·00005
	Ammonium (NH_4)	0·0003
	Arsenic (As)	0·0000005
	Copper (Cu)	0·000005
	Heavy metals (as Pb)	0·00001
	Iron (Fe)	0·000005
	Nickel (Ni)	0·000005

etchants that are widely used in the semiconductor industry are shown in Tables 1.2.1, 1.2.2 and 1.2.3. For optimum specific etching objectives, high-purity mixtures of nitric, hydrofluoric and acetic acids are made by chemical suppliers. Some typical formulations and their associated purity levels are shown in Table 1.2.4.

Etching improvements

For etching thick aluminium layers, a system with minimised bridging and undercutting has been developed for semiconductor device contacts. The system sought was required to yield maximum contact surface uniformity as

Table 1.2.2   Compositions and purity levels for two 'Electronic-grade' solvents

Trichloro-ethylene	*Chemical/other tests*	
	Residue after evaporation	0·0005% max
	Acidity (as HCl)	0·0005% max
	Alkalinity (as NaOH)	0·001% max
	Water (H_2O)	0·010% max
	Physical tests	
	Colour (APHA)	10 max
	Specific gravity at 25/25 °C	1·458–1·463
	Boiling range* 1—95 ml	0·5 degC
	95 ml—dryness	0·5 degC
	Resistivity (MΩ/cm)	5 (min)
	* Recorded boiling-point, 87·1 °C	
	Impurities	*Maximum limits, %*
	Phosphate (PO_4)	0·0001
	Arsenic (As)	0·0001
	Copper (Cu)	0·00001
	Heavy metals (as Pb)	0·00005
	Iron (Fe)	0·00001
	Nickel (Ni)	0·00001
Xylene	*Chemical/other tests*	
	Residue after evaporation	0·001% max
	Acidity (as HCl)	0·001% max
	Substances darkened by H_2SO_4	passes ACS test
	Water (H_2O)	0·020% max
	Physical tests	
	Colour (APHA)	10 max
	Specific gravity at 25/25 °C	0·860–0·870
	Boiling range	137–140 °C
	Impurities	*Maximum limits, %*
	Sulphur compounds (as S)	0·003
	Copper (Cu)	0·00001
	Heavy metals (as Pb)	0·0001
	Iron (Fe)	0·00001
	Nickel (Ni)	0·00001

Table 1.2.3   Compositions and purity levels for two 'Electronic-grade' etchants

Ammonium fluoride solution	*Chemical/other tests*	
	Assay (NH_4F)	39·6–40·4%
	Ammonium bifluoride (NH_4HF_2)	0·6% max
	Insoluble matter	0·002% max
	Residue after ignition	0·004% max
	Neutrality	passes test
	Impurities	*Maximum limits, %*
	Chloride (Cl)	0·0004
	Nitrate (NO_2)	0·001
	Phosphorus (as PO_4)	0·00005
	Sulphate (SO_4)	0·001
	Arsenic (As)	0·00005
	Copper (Cu)	0·00005
	Heavy metals (as Pb)	0·0005
	Iron (Fe)	0·0004
	Nickel (Ni)	0·00005
Ammonium fluoride	*Chemical/other tests*	
	Assay (NH_4F)	95·0% min
	Ammonium bifluoride (NH_4HF_2)	1·5% max
	Silicofluoride (H_2SiF_6)	0·3% max
	Insoluble matter	0·003% max
	Residue after ignition	0·005% max
	Impurities	*Maximum limits, %*
	Chloride (Cl)	0·0005
	Nitrate (NO_3)	0·002
	Sulphate (SO_4)	0·005
	Calcium (Ca)	0·002
	Heavy metals (as Pb)	0·0005
	Iron (Fe)	0·0005

well as minimised or eliminated bridging and undercutting of contacts due to the thickness gradient across a wafer of the deposited aluminium.

Definition problems are the limiting factor in determining how much aluminium may be used for contacts or semiconductor devices. If a single-shot etch is used on thick aluminium layers, gross undercutting occurs; this undercutting becomes more critical as the contact area patterns become narrower. When two separate aluminium evaporations are performed, each followed by a photo-resist and etching operation, two problems are evident:

1   Alignment of the second photo-resist step.
2   Interface problems between the two aluminium operations.

A convenient four-step process was developed that produces a uniform contact surface and, at the same time, eliminates bridging and minimises undercutting. The steps: after aluminium deposition, the wafer is photo-resisted with an inverse mask using a thin-film resist system; the wafer is then placed in an aluminium etch solution for a time sufficient to remove approximately

30 per cent of the aluminium; the photo-resist is then removed and the wafer is again photo-resisted by the same method; again it goes in the etch solution where the remaining aluminium is removed. As the aluminium deposit increases, additional photo-resist and etch steps may be required for optimum results.

Table 1.2.4   Compositions and purity levels for some 'Electronic-grade' etch mixtures

| Code* | Composition, weight % of anhydrous acid | | | Total acidity, meq/g |
	Nitric	Hydrofluoric	Acetic	
0121	19·8–21·0	23·5–24·7	21·0–22·2	18·5–19·1
0211	38·5–39·8	11·2–11·8	20·2–21·2	15·3–15·8
0311	45·2–46·4	8·6– 9·4	15·9–16·5	14·1–14·7
0410	57·1–58·8	8·0– 8·8	—	13·2–13·7
0413	38·9–40·3	5·5– 6·1	31·0–31·8	14·1–14·7
0531	41·2–43·1	14·7–15·3	8·4– 9·4	15·3–15·9
0533	35·4–36·4	12·3–13·1	22·6–23·2	15·5–16·1
0920	58·5–59·7	7·2– 8·0	—	12·9–13·5
1235	45·7–47·3	6·1– 7·3	19·7–20·9	13·7–14·3
1254	43·5–44·7	10·2–11·2	14·8–15·4	14·6–15·2

* *Code:* First two digits represent parts by volume of nitric acid (70 per cent HNO_3), third digit represents parts by volume of hydrofluoric acid (49 per cent HF), and fourth digit represents parts by volume acetic acid (99·8 per cent CH_2COOH).

Specifications:

Colour (APHA)	10 max
Residue after ignition	0·001% max
Iron (Fe)	0·0001% max

Impurities:

Each lot of etch mixture is tested for the following impurities:

	Actual analysis, %
Arsenic (As)	0·000005
Boron (B)	0·000001
Copper (Cu)	<0·00001
Heavy metals (as Pb)	0·00004
Nickel (Ni)	<0·00001
Chloride (Cl)	0·00005
Phosphorus (as PO_4)	0·00005
Sulphate (SO$_4$)	0·00005

Application of this process can be made for contact materials other than aluminium. Any thickness of contact material can be used provided the number of photo-resist and etch steps is adjusted accordingly.

Alkaline etching process for preventing over-etching

It is said to represent an improvement over standard etching techniques by producing more beam-leaded integrated circuits than previously possible in a silicon slice. Slots etched by this method are narrow, precisely defined and wedge-shaped, tapering to the bottom instead of being undercut on the sides as with standard methods.

The process depends on a unique orientation of the etching mask on a specific lattice plane of the silicon crystal and on the use of a special 'preferential' etchant. Precise control of the rate and direction of etching is possible because the special etchant attacks the three main crystal lattice planes of silicon at different relative rates.

Before this development-improved etchant, a mixture of hydrofluoric and nitric acids was commonly used to etch silicon. Preferential etchants consist of strongly basic, or alkaline, solutions. One such etchant is a mixture of potassium hydroxide, propanol and water.

For fabrication of beam-leaded integrated circuits, slots must be etched in a semiconductor slice to separate the circuits, or to isolate components within each integrated circuit. Fewer circuits are possible in a slice when standard etchants are used. With non-preferential etchants, variations in the thickness of silicon slices cannot be economically reduced to less than 0·00025 mm.

For complete penetration of the thickest parts of the slice, considerable over-etching must be tolerated at the thinner portions. Since the isotropic (non-preferential) solutions etch as fast sideways as downward, the active devices on the opposite side of the slice must be spaced three to four times farther apart than the average slice thickness to guard against over-etching. With the improved technique the possibility of over-etching is almost eliminated, despite thickness variations in the slice. This is because the slot cut by the new etchant is shaped like a flat-bottomed wedge. The slot sides maintain a constant angle of about 55° from the top surface of the slice to the bottom of the cut. The slot, therefore, narrows as the etching progresses.

Dopants and epitaxial chemicals

Yield in the manufacture of high-quality semiconductor devices is a function of the controlled presence or absence of certain impurity elements in the semiconductor material used. Thus it is absolutely essential that the basic material be of the highest purity so that the products made have a high degree of predictable quality. Since chemical processing and facets thereof constitute a substantial portion of the solid-state manufacturing cycle, it is extremely important that processing chemicals be pure.

Dopants and epitaxial chemicals can be impure for reasons other than that they contain foreign solid material. Moisture, for instance, if present (always a possibility with hygroscopic chemical compounds such as phosphorous pentoxide) will adversely affect the doping activity of a compound by diluting its chemical concentration. A severe chemical imbalance results in chemical vapour during high-temperature diffusion, so great care must be exercised in handling chemicals to ensure there is no moisture contamination introduced either during manufacturing or in the final packaging operation.

Purity of chemicals can often be preserved by the use of special packaging methods. One method is to package the chemicals as 'unit charges'. Whether the chemical be a liquid or a solid, it is packaged in unit charges determined by specific customer requirements. This packaging is completed in an inert nitrogen atmosphere and eliminates tedious and unnecessary weighing and the possibility of contamination during processing.

In addition to the above-mentioned high-purity chemicals, other high-purity materials for research and development in semiconductor work are prepared by zone-refining techniques (for example, one wide line of zone-refined single crystal and polycrystalline compounds). (See Table 1.2.5.)

Table 1.2.5   **Purity levels for some important dopants and epitaxials**

Solid	Purity, %	Liquid	Purity, %
p-*dopants*			
Boric acid, H_3BO_4	99·99	Boron tribromide, BBr_3	99·99
Boron hydride, B_2O_3	99·99	Methyl borate, $B(OCH_2)_3$	99·99
Boron nitride, BN	99·9		
Aluminium, Al	99·999		
Gallium, Ga	99·999		
Indium, In	99·999		
Tellurium, Te	99·999		
Boron, B			
n-*dopants*			
Phosphorus pentoxide, P_2O_5	99·9	Phosphorus oxychloride, $POCl_3$	99·99
Antimony, Sb	99·999	Phosphorus tribromide, PBr_3	99·99
Arsenic, As	99·999	Phosphorus trichloride, PCl_3	99·99
Bismuth, Bi	99·999		
Epitaxial chemicals			
Iodine (double resublimed), I	99·999	Silicon tetrachloride, $SiCl_4$	99·999
		Trichlorosilane, $SiHCl_3$	99·99
		Germanium tetrachloride, $GeCl_4$	99·999

Gold-plating of semiconductors

Gold is perhaps one of the most widely used plating materials for semiconductor devices. The electroplater of gold is concerned with two aspects of semiconductor manufacture. Gold is deposited on areas of the device itself, as well as on the base upon which the device is mounted. For high-reliability components, gold is the one material which has all of the necessary functional characteristics required.

The gold electrodeposit employed for the device itself must be pure so that 'poisoning' of the chip will not occur. Gold will not, in itself, poison the semiconductor. But any impurities co-deposited with the gold will adversely affect the electronic functions required of the finished device. Gold is plated on a semiconductor so that electrical connection can be made to the surface of the semiconductor. For this application, the gold electrodeposit must be 99·99 per cent pure and must be soft so that the gold wire can be readily bonded to its surface. In some cases the semiconductor device is gold-plated on its 'support' side for mounting purposes; this gold electrodeposit must be

E

soft so that the device may be mounted by compression bonding to a supporting substrate.

The electroplating of the semiconductor device itself is relatively straightforward:

1 Alkaline soak.
2 Immerse in 10 per cent (by volume) hydrofluoric acid to remove oxides of silicon and/or germanium.
3 Gold-strike.
4 Gold-plate.

Adequate rinsing between each step is essential to avoid intercontamination of systems.

Ninety per cent of all high-reliability semiconductors produced today use Kovar as the base upon which the semiconductor device is mounted. Kovar is an iron–nickel–cobalt alloy which has good matching expansion characteristics with the classes of ceramic used to produce hermetic seals. There are four areas on the base which require a gold electrodeposit:

1 Chip-bonding area.
2 Whisker wire.
3 Encapsulation area.
4 Component leads.

Gold is deposited on the Kovar to impart surface characteristics to the Kovar. The gold plating must be pure in the area where the semiconductor chip is mounted, so that chip bonding may readily occur. This chip bonding is, in reality, the formation of a gold–silicon alloy which has a melting-point of about 400 °C. Organic impurities present in the gold itself will cause frothing or foaming at the silicon–gold interface and reduce the strength of the bond between the gold and the silicon.

Metallic impurities in the gold electrodeposit will 'dope' the gold and thereby change the electronic function of the device. Impurities will also change the melting-point, usually upwards, of the gold–silicon eutectic formed, thereby weakening the bond between the gold and the silicon. The thickness of gold required for direct chip bonding is normally 2–3 μm. Where a preform is used (a preform is a gold–silicon alloy wafer, 25–50 μm thick, used to aid the bonding of the silicon to the gold-plated Kovar) the gold thickness can be reduced to 0·5–0·75 μm. It has been determined that devices with coatings thinner than 2 μm do not have true reliability, because during manufacture and use of the device, the operating temperature causes diffusion of the substrate material through the gold. Subsequent oxidation of the diffused material interferes with the semiconductor function.

Gold is required on the whisker-wire bonding posts for the same reasons that were discussed regarding gold on the semiconductor device itself. The gold must be pure and soft to accept whisker-wire bonding and should also have a minimum thickness of 2 μm. Ball-bonding and chisel-bonding have both been used successfully to make whisker-wire attachments to the base leads.

The area used to seal the device also requires a pure, soft gold so that compression bonding may be effected between the base and the cover or lid of the device. The gold thickness must be 2–3 μm minimum.

The component leads require gold to improve the soldering of these leads into the completed electronic package. Gold thickness requirements are of the order of 0·5–0·75 μm and the gold itself should be relatively pure so that it is readily solderable.

Not only must gold have all of the above properties, but it must also withstand high-temperature ageing (300 °C for 45 hr minimum), to stabilise the electronic characteristics of the semiconductor device. Without this ageing stabilisation, the device would be erratic in its electronic output. In order for the electrodeposited gold to be functional its density must be sufficiently high to retard the diffusion of substrate materials through it and so prevent their interference with the function of the semiconductor.

The gold film itself must also be continuous so that it can withstand the various etchants used on the semiconductor device.

In summary, the properties required of the gold coatings are:

1 High purity.
2 Continuity.
3 Corrosion resistance.
4 Weldability.
5 Solderability.
6 High density.

Electrodeposits possessing these properties may be obtained through the use of any of the three chemical systems available:

1 Alkaline cyanide system.
2 Neutral system.
3 Acid system.

Three chemical systems

The conventional hot alkaline cyanide system operates at a pH of between 9·0 and 13·0 and a temperature of 60 to 70 °C, with gold present as the potassium cyanide complex and with an excess of free potassium cyanide. This bath, when freshly made, will produce sound electrodeposits for semiconductor manufacture.

However, industry has found that continuous use of this system is costly because organic decomposition products form readily and co-deposit with the gold, thereby seriously interfering with the gold-plating requirements for semiconductor manufacture. Activated carbon purification is frequently required to maintain the system. One disadvantage of this is that the carbon will absort 10–15 per cent of its weight in gold. Not only is this gold lost for electrodeposition, but there is also a considerable loss in production time. The hot cyanide gold system has thus become obsolete in semiconductor manufacture.

The second plating system includes the neutral gold baths, which operate at a pH level between 6·0 and 8·0; temperatures are between 60 and 70 °C.

The neutral system utilises gold as the cyanide complex; conducting salts and chelating agents provide bath stability and conductivity. The deposits obtained from these systems meet all of the requirements set forth by the semiconductor industry. These electrolytes are 95 to 100 per cent efficient and will consistently produce sound, dense gold electrodeposits with a minimum of care and/or purification of the bath.

The third system is the mildly acidic gold electrolyte. These baths operate in a pH range of 3·0 to 6·0 and a temperature range of 60 to 70 °C. Gold is present as a cyanide complex, and conducting salts and chelating agents again provide bath stability and conductivity.

In some cases additives are made to the bath to assist in the formation of fine grain deposits. Acid gold baths will meet all of the requirements of the semiconductor industry and require infrequent bath purification. These baths are 90 to 100 per cent efficient and have been found to produce the soundest electrodeposits for semiconductor manufacture.

Resists and strippers

Resists

Photosensitive resists are thin coatings produced from organic solutions which when exposed to light of the proper wavelength are chemically changed in their solubility to certain solvents (developers). Two types are available, negative-acting and positive-acting. The former is initially a mixture which is soluble in its developer, but after exposure to light becomes polymerised and insoluble to the developer. Exposure is done through a film pattern. The unexposed resist is selectively dissolved, softened, or washed away, leaving the desired resist pattern on the clad laminate. Positive-acting resists work in the opposite fashion — exposure to light making the polymer mixture soluble in the developer. The resist image is frequently dyed to make it visible for inspection and retouching. The pattern that remains after development (and post-baking in some cases) is insoluble and chemically resistant to the cleaning, plating and etching solutions used in the production of devices.

Most photo-resist materials are available as proprietary materials. These have for the most part been formulated with well characterised chemical-reaction products, remain non-crystalline during exposure to light, and have good chemical resistance and dimensional stability to the chemical agents involved in device processing. Typical resists used in the semiconductor industry are all negative-acting resists.

In general, both positive- and negative-acting resists offer the greatest protection in acidic rather than alkaline solutions. The negative-acting types are more tolerant of alkaline solutions.

The negative-acting resists, once exposed and developed, are no longer light-sensitive and hence can be processed and stored in normal white light. The positive resists remain light-sensitive, even after developing, and must be protected from white light.

Increasing the temperature of processing solutions also increases the chemical and swelling action on photo-resists. Photo-resists that can be baked at high temperatures are the least affected by higher solution temperatures. Negative-acting photo-resists that cannot be baked at high temperatures lose their adhesion as solution temperatures increase. Thicker coatings of these materials offer greater resistance, but care must be taken not to sacrifice fine-line definition or acuteness. The positive-acting resists are generally affected by high temperatures, particularly in alkaline solutions.

Thin-film resist is a leading resist specifically formulated to meet the very-fine-line, high-resolution requirements of microelectronics. Its viscosity at 25 °C is 465–535 cP. It is the same type of resist as KMER (in general, it can be used the same way) except it is capable of giving thicknesses of 0·0076 mm down to fractions of a micrometre.

Processing is similar to that for KPR except that pre-bake is performed at 82 °C for 20 min or less. Higher temperatures adversely affect adhesion. Lower temperatures and longer times are preferred for copper. Less scumming and staining are evidenced with KTRF on copper than with KMER. Modification of the exposure time and developing procedure may be required since coatings are generally thinner than KMER coatings. Common practice is to expose the KTFR, bake at 82 °C for 5 min, cool and develop. When this is done less lifting occurs, particularly on smooth surfaces.

That KTFR films can be made thinner than KMER is illustrated by the fact that a coating of 16 parts (by volume) thinner to 1 part (by volume) breaks during whirl-spinning and does not completely cover, while dilution of 60 parts thinner to 1 part KTFR remains continuous and completely covers the foil being coated.

Photo-resist stripping

The negative-acting resists, KPR, KPR-2, KLP and mixtures of KPR and KPL, are removed using trichloroethylene, Stoddard solvent, methylene chloride and commercial strippers. With these solvents the resist does not dissolve but rather softens and swells, breaking the adhesive bond to the substrate. Once this has taken place over the entire coated area a water spray can be used to flush away the film. With resists that are more difficult to remove (baked resists for example), about a 3-min soak may be required. Thick coatings are generally removed more readily than thin coatings.

The other negative-acting resists, KMER, KOR and KTFR, can also be removed relatively easily by commercial strippers. Again, ease of removal depends on degree of baking. The stripping of KTFR is the most difficult of these, since these resists are used most commonly on metal masks or other solid metal parts. The consideration of substrate chemical resistance is not a problem. A good practice is to allow the chemicals to do the removing, rather than trying to scrub off half-softened resist. Soaking, swabbing and water-spraying will remove the swelled film.

Efforts are constantly underway on developing improved strippers. One recently developed stripper has stability up to 160 °C and is said to be capable of removing nearly all resists. Further, it is formulated to high-purity specifications to reduce residual contamination. This phenomenon has always been a problem in semiconductor processing.

Processing cleanliness

In addition to rigid requirements for ultra-clean, high-purity materials, the processing operation itself, especially the work with resists, demands cleanliness.

The installation of air conditioning with an effective dust-filtering system provides the best control of three factors that affect photographic quality: temperature variations, relative-humidity variations and atmospheric dust. Such a system, apart from increased comfort for personnel, can maintain a constant relative humidity and temperature, thereby practically eliminating dimensional changes in the photographic materials and emulsion-speed variations.

Dust particles, the bane of all semiconductor processing, can be controlled effectively by several methods, as mentioned above. One of these is air-conditioning. Freestanding or wall-mounted filter machines, which provide rapid filtration of the air in a room, can also be used. A third method involves a 'clean bench' and the laminar-flow principle in which filtered air is directed as a 'block' several inches thick, and the full width of the filter, toward the person standing in front of the device. Each of these methods has an application in microphotography, but they cannot overcome the effects of poor house-keeping or extreme conditions of dust and dirt.

In addition to strict cleanliness in work areas, the other dirt-causing factors should be controlled. For example, personnel in the work area should wear surgical masks, hair covers, shoe covers and lint-free uniforms that cover as much as possible. Street clothes should be excluded from the area and, of course, should not be worn under uniforms.

As a general rule, humans generate more objectionable particles than any other element within the area. Persons with flaking skin conditions (dandruff, peeling sunburn, etc.), should not be allowed in critically clean areas. Also, the use of cosmetics should be controlled if the area is to remain clean.

Frequent washing with a mild soap such as pHisoderm is recommended as an effective method of controlling skin oil and skin flaking, both of which physiological conditions can seriously affect semiconductor device quality. Gloves should be worn with care because they get dirty rapidly, and there is always the possibility that the operator will touch the emulsion.

Although not directly associated with work areas, the condition of the water used for processing is important. With extreme microphotographic reductions, particulate matter, such as suspended solids, rust and scale from iron pipes, and even micro-organisms (diatoms, bacteria, etc.), can cause problems. Most often, particulate matter can be eliminated (or at least reduced) by using water filters in combination with ion-exchange media. The number of methods used generally depends on the condition of the water in the surrounding geographic area. An accurate analysis of the water usually will reveal the number of corrective steps needed.

Coatings and encapsulants

Coatings

Protective coatings of microcircuits, especially semiconductor circuits, poses special problems. In addition to the chemical incompatibility of semiconductors with many resins, breakage of fine leads and joints can be a problem, especially in thermal cycling. Consequently, flexible coatings are commonly used either alone, or as a barrier coating. Special lines of high-purity, non-ionic coatings have been developed for protection of semiconductors. These are high-purity semiconductor coating resins designed specifically for coating transistor, diode, and rectifier junctions. Used on such junctions, they offer the following benefits:

1 Electrical breakdown across the junction surfaces is prevented.
2 Junction surface leakage and reverse current is minimised and peak inverse voltage stabilised.
3 Junction surfaces are protected from environmental effects that occur during high-temperature operation and thermal cycling.

Each lot of coating resin is made in glass equipment under controlled laboratory conditions. The extremely high purity of the resin is controlled by spectrographic analysis specifications which guard against metal contamination. One method is to measure the effect of coatings on semiconductors and plastic coatings applied to bare micro-diodes.

Because of the high insulation resistance values for most plastics, ohmic leakage through the plastic coating is assumed to be negligible compared with the reverse conduction current exhibited by the device. With some plastics, however, such as cast epoxy, ohmic leakage could be a factor.

Fluorocarbon-coated and silicon-coated diodes generally exhibit smaller changes in reverse current than the epoxy-coated diodes. However, large changes in reverse current — from nanoamperes to microamperes — indicate that the commercial epoxies tested contained sufficient amounts of ionic or other contaminants to affect the semiconductor. Epoxies with non-stoicheiometric resin-to-hardener ratios display high values of reverse current. Diodes coated with phenolic material exhibit high insulation resistance, in the presence of large amounts of such impurities as ammonium ions. The reverse-current characteristic decreases after heating for a time without electrical bias, presumably because ordered ions on the surface are moving to a more random state (see Table 1.2.6).

Table 1.2.6 **Effect of temperature on reverse current of plastic-coated diodes**

Coating systems	Initial reading, I_R, nA, at 50V	I_R, nA, at 50V after 20 hr at 150 °C and 75V reverse bias	I_R, nA, at 50V after additional 20 hr at 150 °C, no reverse bias
Epoxy, stoicheiometric purified	10	548	14
Epoxy, non-stoicheiometric purified	9	10×10^3	10×10^3
Epoxy, stoicheiometric technical	12	237×10^3	121×10^3
Epoxy, non-stoicheiometric technical	12	31×10^3	14×10^3
Fluorocarbon, resin	17	18	16

The excellent results with fluorocarbons may be attributed to the inherently inert, stable and symmetrical structure of these plastics. In fact, TFE and FEP plastics are unique among plastics in their purity, thermal and oxidative stability (up to 260–315 °C), chemical inertness, low rate of moisture permeability and electrical stability; these plastics have a zero temperature coefficient of resistance to temperatures above 200 °C, in contrast to the high negative values of most plastics.

Encapsulants

Cost reduction is the principle advantage to plastic encapsulation. Wafer yields are increasing and individual dice are being produced at lower factory costs. Package and assembly costs are becoming the most important elements of cost in integrated circuits. The desirability of low-cost packaging is obvious; combined with lower assembly cost, it would be so much better.

One problem with plastic encapsulation is that the coefficient of thermal expansion of plastics differs markedly from those of silicon and the bonding wires. While the plastic is hardening, it can actually tear off lead wires, thus destroying the integrated circuit. In addition, the silicon is pressure-sensitive and changes in electrical characteristics can result if sufficient pressure occurs due to differences in temperature coefficient between the plastic and the die.

Another, and potentially more serious, problem involves the semiconductor surface. The desirability of hermetic seals for semiconductors is based on the need to protect the surface from ionic contamination and/or moisture. Moisture may damage the aluminium interconnecting pattern and ionic contamination may change the apparent silicon doping level immediately below the surface of the silicon dioxide. This has the effect of reducing breakdown voltage in *npn* transistors and increasing leakage in *pnp* structures. Severe contamination may also alter the resistance value of the diffused resistors. MOS-type devices because of their relatively light doping levels, are most sensitive to this type of contamination. They are designed to be sensitive to the charge induced through the gate electrode. For the same reason, they are highly sensitive to ionic contamination.

Because of these problems, much study has been devoted to development of more compatible encapsulants. Such studies are currently at a high level, both by encapsulant producers and end users. This is essential, since often the encapsulant system must be matched to specific devices, device manufacturing processes and device end uses.

Plastic encapsulation as a feasible means of mass-producing semiconductors made its initial market impact in 1962. Since that time, plastic encapsulated semiconductors have captured a significant portion of the market. As a result, a growing emphasis is being given to reliability, means of evaluating reliability and process variables influencing performance and reliability of these devices.

In the signal transistor market, almost every type of device having a significant volume of demand has been offered in a plastic encapsulated version. As with the mass-produced automobile, the lower cost has resulted in a substantial increase in transistor market unit volume, potential uses having burgeoned with low-cost availability.

Following the acceptance of the plastic encapsulated signal transistor has come the plastic power transistor with a substantial market developing for it. This product has required considerable engineering work on the basic encapsulation technique to permit substantial heat dissipation. Parallel with this development has come the low-power SCR (silicon-controlled rectifier) with similar engineering requirements. As technology develops, use of encapsulation will be increasingly developed and used for the higher power semiconductor devices. Whenever high package costs have been a factor, plastics have found application. No example is more striking than the integrated circuit, where the multiplicity of through-connections and glass seals has been a cost barrier since their inception. Plastic encapsulation has reduced prices sharply.

Chapter 1.3   **Materials for conductor elements and connections to thin films**

This chapter discusses the development and evaluation of bonding techniques for fine wire microbonds made by 'split-tip' resistance welding and ultrasonic bonding. Gold and aluminium wire microbonds to three thin-film terminations commonly used in hybrid integrated circuits were evaluated on various substrate materials. The terminations were molymanganese–nickel–gold, gold–chromium and aluminium. The substrates studied were alumina, glazed alumina, glass, oxidised silicon, sapphire and beryllia. The effects of thermal ageing on the microbonds studied was analysed. The interrelationship of wire, film, substrate and bonding technique on the quality and reproduceability of the microbonds will be discussed and a statistical analysis of the reliability of the various microbonds studied presented.

There is a little argument that the monolithic integrated circuit, as we presently know it or in the more advanced multiple-function array, represents the ultimate in microelectronic devices. However, just as the transistor never completely replaced the vacuum tube, it appears that the monolithic integrated circuit, for the forseeable future, will not completely replace discrete, active part hybrid thin-film integrated circuits.

A primary source of failure in silicon semiconductor devices has been the thermocompression bond of gold wire to the aluminium metallising on the silicon chip. The fabrication of reliable silicon devices requires stringent control of this bonding process. The variables associated with state-of-the-art thermocompression bonds to silicon devices are relatively small in number when compared with those encountered in the microbonds associated with the fabrication of hybrid thin-film circuits. In the case of hybrid integrated circuits, the basic substrates may be alumina (with varying degrees of surface roughness), glazed alumina, glass, sapphire, oxidised silicon, or beryllium oxide. The thin-film terminations most commonly encountered are aluminium, chromium–copper–gold, chromium–gold and, in the case of thick films, a fired-on metallising such as molymanganese–electroplated nickel–gold, platinum–gold, or silver.

A high degree of reliability has been achieved for the thermocompression bond as a result of the mass of experience that has been accumulated in the semiconductor industry. No similar bulk of data exists for the microbonds utilised in hybrid integrated circuits.

This chapter discusses the results of an evaluation undertaken to develop quantitative data for a series of microbonds currently in use in hybrid integrated circuits. Attention has been paid to the following:

1 Effects of materials.

2 Methods of fabrication.
3 Control of bonding-machine variables.
4 Effects of thermal ageing.

Bonding techniques

Two bondings techniques were employed: (*a*) 'Split-tip' resistance welding was performed and slightly modified to allow the use of a smaller bonding force; (*b*) the ultrasonic microbonder.

Conductors

The conductor materials employed were gold and aluminium wires of the following diameters: 0·01778, 0·0254, 0·0508 and 0·1270 mm. Soft, fully annealed wires are most commonly used for microbonding applications. However, the ease of deformation and attendant low strength of annealed materials suggested the use of conductors in harder conditions.

Substrates

The two essential characteristics of a substrate for 'split-tip' resistance welding are minimal surface roughness and the ability of the substrate to withstand thermal shock. This was found to be especially true when welding 0·1270 mm diameter wires because of the higher energies required. The substrates which were most susceptible to thermal shock were the glass and glazed alumina substrates and, to a lesser degree, the oxidised silicon. With respect to ultrasonic bonding, the surface roughness was the predominant substrate characteristic.

The choice of substrate materials should be based, in part, on the following criteria:

1 AS-FIRED ALUMINA Optimum mechanical strength and thermal conductivity leading to its wide usage in thick film, thin film and chip and wire applications.

2 GLAZED ALUMINA Optimised mechanical strength, thermal conductivity and superior surface smoothness characteristics leading to its usage in thin-film hybrid integrated circuits; also its compatibility with flip-chip bonding.

3 GLASS Superior surface finish, wide spread and extensive use in thin-film integrated circuit applications.

4 BERYLLIUM OXIDE High mechanical strength, superior thermal conductivity and compatibility with thin- and thick-film conductors for chip and wire applications.

5 SAPPHIRE Mechanical strength, thermal conductivity and superior surface finish possible because of its single-crystal structure and its potential in thin-film integrated circuits containing both active and passive devices.

6 OXIDISED SILICON To determine the practicability of replacing thermo-compression bonds on integrated circuits where the termination pads are distributed around the periphery on the oxidised silicon surface.

Film terminations

Three basic terminations were applied to the various substrates discussed above. A limited effort was applied to the evaluation of fired-on molybdenum manganese with an intermediate film of electroplated nickel and an electroplated gold surface film. This was restricted to alumina substrates, gold wire and 'split-tip' bonding. The electroplated film thicknesses were 0·00288 mm sulphamate nickel and 0·00508 mm gold electroplated from a cyanide bath. The other terminations were thin-film terminations of chromium–gold and aluminium. The chromium film thicknesses were of the order of 1,000 Å and the gold thicknesses greater than 15,000 Å. The aluminium films were 7,000 Å thick. Although bonding to metal films as thin as 300 Å has been reported, evaluations show that films of either aluminium or gold must exceed 5,000 Å if the reproduceability required in practical applications is to be achieved.

In the case of gold wire to gold terminations, welding machine parameter control is found to be extremely critical on bonds made to terminations of 5,000 to 15,000 Å. Up to 50,000 Å variations in weld parameters are not critical. It was concluded that 20,000 Å thick terminations were the optimum for gold-to-gold microbonds. Aluminium films of 7,000 Å proved to be suitable for ultrasonic bonded aluminium wires.

Tables 1.3.1 and 1.3.2 summarise the material combinations and also present the average pull strength and joint efficiencies of the microbonds.

'Split-tip' welded microbonds

In general, a 'split-tip' series-resistance welder is defined as a welding machine designed to join metal wires to a thin-film metal surface under controlled pressure by passing the current through the electrically insulated electrode that makes contact with the fine wire only. The term 'split-tip' is applied when the electrodes are of one-piece construction separated by an organic insulator as opposed to parallel-gap series welding where an adjustable air gap serves as the insulator in the immediate area of bonding. 'Split-tip' welding is a special case of parallel-gap welding. The prime requirements for 'split-tip' welding are fine adjustment of the applied energy and pressure as well as minute dimensional electrodes.

Electrodes

The 'split-tip' electrodes consist of two molybdenum electrodes bonded together and separated by organic insulation. The condition of the electrode tip is of prime importance in achieving consistent weld results. Dirty, pitted, or oxidised electrodes can be cleaned with a fine abrasive or burnishing tools. Metal remaining embedded in the insulation gap after dressing can usually be burned out by firing the welder at maximum energy settings. Electrode life during the development of weld schedules was proven to be relatively short, necessitating frequent reshaping of the tip. Hand-shaping by filing has proved impractical. Good results can be obtained with a surface grinder in conjunction with a dividing head providing the tip can be properly positioned;

le 1.3.1   Summary of results for microbonds formed with a 'split-tip' resistance welder

strate	Film	Film thickness, Å	Wire	Wire diameter, mm	Wire tensile strength, g	Pull test sample size	Average pull strength, $\bar{x}$, g	Pull strength, % of tensile strength
fired umina	Au-Ni on fired-on Mo-Mn	190,000	Au	0·0178	2·8	779	2·7	95·0
				0·0508	26·7	106	22·8	85·5
				0·1270	182·5	89	126·7	69·4
phire	Au/Cr	10,000		0·0254	6·7	779	6·5	97·0
		45,000		0·0508	26·5	783	24·6	92·8
		19,000		0·1270	200·0	782	159·0	79·5
llia		10,000		0·0254	6·7	785	6·5	97·0
		45,000		0·0508	26·5	788	25·0	94·4
		24,000		0·1270	200·0	782	141·0	70·5
or		40,000		0·0254	7·0	783	6·6	94·3
		20,000		0·0508	26·7	869	21·4	80·0
		20,000		0·1270	201·0	129	135·0	67·2
fired umina		23,500		0·0254	6·7	790	6·5	97·0
		25,000		0·0508	27·5	772	22·7	82·6
		25,000		0·1270	182·5	99	139·8	76·5
zed umina		32,000		0·0254	7·0	789	6·3	90·0
		25,000		0·0508	27·5	786	20·6	75·0
		25,000		0·1270	182·5	136	52·9	29·0
dised icon		10,000		0·0254	6·7	776	6·6	98·5
		30,000		0·0508	26·7	810	20·2	77·0
ss		30,000		0·1270	182·5	79	102·8	56·3
		25,000	Al	0·0508	56·4	111	14·0	24·8
fired umina		32,000		0·0508	56·4	101	13·4	23·7
zed umina		30,000		0·0508	56·4	94	13·8	24·6
dised icon		30,000		0·0508	56·4	106	16·9	30·0
ss	Al	20,000	Au	0·0508	26·7	97	15·1	56·7
		20,000		0·1270	182·5	104	90·9	49·7
dised icon		25,000		0·0508	26·7	105	16·8	62·8
		25,000		0·1270	182·5	101	84·8	46·5

le 1.3.2   Summary of results for microbonds formed with ultrasonic bonder

strate	Film	Film thickness, Å	Wire	Wire diameter, mm	Wire tensile strength, g	Pull test sample size	Average pull strength, x, g	Pull strength, % of tensile strength
phire	Au/Cr	24,000	Al	0·0254	14·5	100	9·2	63
		24,000		0·0508	54·2	100	31·6	58
		36,000		0·1270	310·0	100	172·0	55
llia		43,000		0·0254	14·5	100	5·8	40
		36,000		0·0508	54·2	100	17·4	32
nina		23,000		0·0254	14·5	100	6·8	47
s	Al (sintered)	7,000		0·0508	54·2	100	27·1	50
dised icon		7,000		0·0508	54·2	100	21·5	40
phire		7,000		0·0508	54·2	100	21·5	40

however, since the tips are manufactured in two halves and then bonded together, unsymmetrical tip shapes often occur.

This problem becomes increasingly acute as tip sizes diminish. Unequal tip areas result in welding only under the smaller leg or with increased energy, burning of the conductor and/or the smaller tip. Special features are necessary to effect proper electrode reshaping.

Weld schedule development

Welding parameters for many combinations of materials requires extensive evaluation of a large number of variable combinations. The principle variables which must be considered are:

1 Pulse width.
2 Pulse height.
3 Electrode size.
4 Electrode configuration.
5 Electrode force.
6 Film thickness.
7 Film type.
8 Conductor composition.
9 Conductor size.
10 Conductor conditions.
11 Substrate material.
12 Substrate surface condition.

In order to fully bond in terms of just pulse height and width by means of the weld-profile method (where individual parameters are varied and plotted versus bond strength), several thousand bonds would have to be made and pull-tested.

If another variable such as film thickness is introduced, the study for one bond type becomes monumental. Therefore, once the effects of the various parameters are understood, it becomes practical to employ this experience to narrow down the number and range of significant variables studied in the selection of the basic bonding equipment parameters.

Microbonds produced by 'split-tip' resistance welding differ from other joining techniques in that they consist of two distinct welds, one under each electrode. Joint failures resulting from tensile testing may occur in one or more of the following modes:

1 Base wire.
2 Heat-affected zone.
3 Faying surface.
4 Film adhesion.
5 Weld zone.

The stress-strain curves shown in Fig. 1.3.1 demonstrates the typical failure of bonds subjected to the 90° pull test occurring in the heat-affected zones and at the faying surfaces. These curves clearly demonstrate that the strength can vary significantly at the two portions of the bond. This is the result of slight variations of the electrode contact-surface area. The most common failure observed was in the heat-affected zone and was sometimes the result of excessive electrode pressure, particularly with soft-annealed wire. However, this failure type is to be expected as a result of probable grain growth and reduction in section. Figure 1.3.2 shows the distribution of failure modes of 0·0508 mm gold wire to gold–chromium terminations on a glazed alumina substrate.

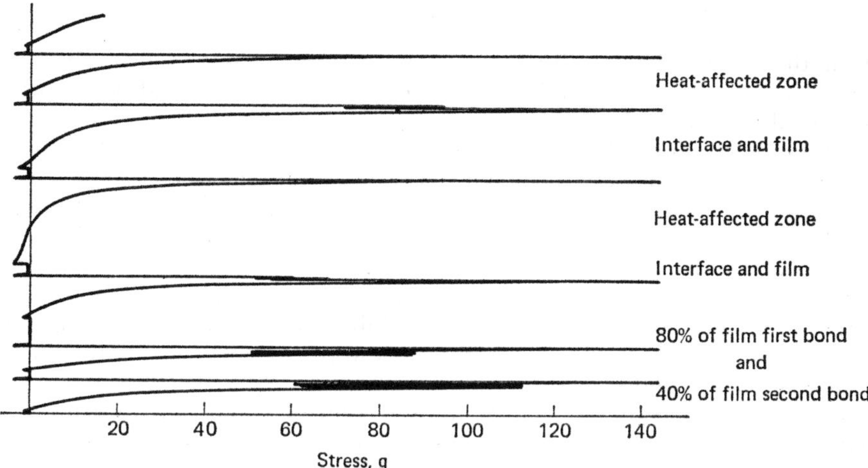

Fig. 1.3.1   Typical stress-strain curves for 0·13 mm diameter hard gold wire bonded to 25,000 Å Cr-Au film on as-fired alumina: strain rate, 112·7 m/min; chart speed, 2·1 m/min; average wire strength, 360 g

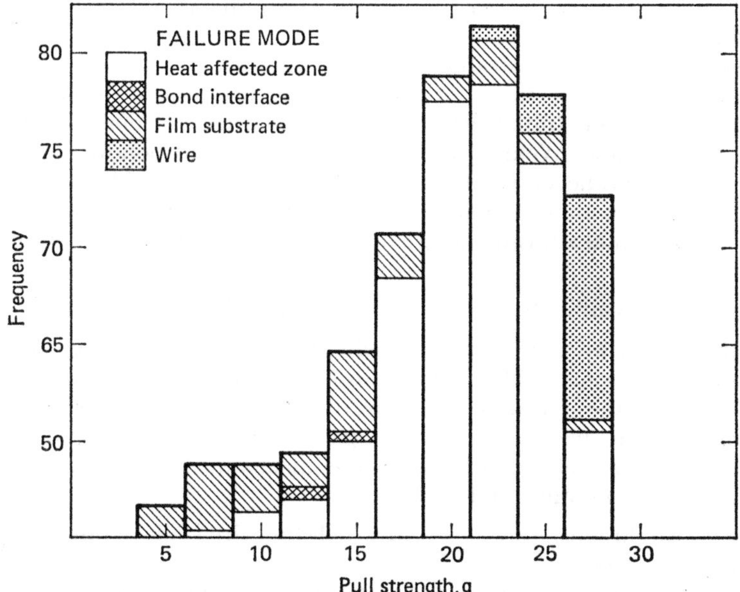

Fig. 1.3.2   Pull strength distribution (gold wire/gold film/glazed alumina)

It was observed that many parameter combinations yield strong comparable bonds. Consequently, as previously mentioned, other factors such as the mode of failure and physical appearance must be considered during final selection.

For example, the greatest average strength may occur at settings where failures in the base wire predominate while an occasional failure at reduced strength occurs through the interface. Other settings with increased energy and/or electrode force might exhibit a lower average but smaller range of strength and consistent failure at the heat-affected zone. Obviously, the parameters resulting in a high degree of reproduceability should be adopted for the fabrication of the bonds.

This evaluation has indicated that the microbonds obtained by 'split-tip' welding are of the thermocompression or diffusion (molecular) type. Consequently longer bonding cycles might seem in order; however, optium results are generally achieved with low ratios of pulse-width-to-height in conjunction with moderate electrode force. Normally, the latter is set to yield a 50 per cent upset with soft materials such as gold or aluminium. Short pulses limit the total energy input thus minimising heat transfer or sinking effect of the conductor, film, and substrate and produce less deformation and annealing of the conductor (heat-affected zone).

In general, the criteria used in the selection of bonding parameters — pulse height, pulse width and electrode force — are reproduceability of optimised mechanical properties coupled with minimal heat input.

Substrates subject to thermal shock generally require shaping of the energy pulses or addition of low-energy pulses before and after bonding to avoid damage to the substrate. These side pulses do not contribute directly to the bond strength, but merely permit the use of optium energy without substrate degradation.

Weld schedules for substrates sensitive to thermal shock can be developed by establishing the electrode force at a setting to produce the desired conductor deformation. Initial deformation varies with the conductor material hardness and shape. Welding energy is increased until bonding is obtained by using the middle or weld pulse controls. The pre-weld and post-weld energy cycles are then increased generally to approximately one-half the weld cycle and final adjustments are made to shape the pulse as required.

Weld schedule verification

The selection of an optimum weld schedule inherently requires arbituary judgements. Before a weld schedule can be introduced into production, it is necessary that the schedule be proved using operation and equipment variations which are normal to the production environment. The most common test for weld evaluation is the mechanical pull-test. Two basic techniques are generally employed:

1 The straight tensile-shear test in which the wire is placed in tension and the weld in shear.
2 The 90° (or some lesser angle) tensile test where the wire is pulled perpendicular, or at some lesser angle, to the substrate surface.

Tests were conducted comparing the two test techniques. The frequency of failure modes for bonds produced by a typical set of development parameters, and tested by straight pull and 90° pull-test is shown in Fig. 1.3.2. This clearly points out the insensitivity of the straight pull-test, since in this test all joints failed either in the base wire or the heat-affected zone.

Although the majority of the 90° pull-test bonds also ultimately failed in the heat-affected zone, initial failures started at the faying surface or film-to-substrate bond. It is noted that the majority of failures by either test occurred in the heat-affected zone — a reflection of the hardness and high strength of the base wire. With annealed wire a more even distribution of failures between the heat-affected zone and base wire would have resulted.

The 90° pull-test was utilised for the verification of weld schedules for the 'split-tip' welded microbonds. The test imposes more severe conditions than the bond would encounter in service, however, inherent to this test is a high degree of sensitivity to bonding parameters. This is basically a 'peel test', and (because of the nature of the bond) the initial loading is concentrated at a fine point which progresses to a line across that area of the bond. In addition to revealing the strength of the weakest area of the bond, the test is very sensitive to the adhesion properties of the film and will disclose any significant deterioration of the substrate itself. All pull strength measurements for 'split-tip' microbonds were carried out on a tensile testing machine employing a constant strain rate of 12·7 mm/min. This machine graphically records the pull-strength to an accuracy of 0·5 per cent of the full-scale reading (50 g for 0·0508 mm conductors). A mechanical test fixture was fabricated for the mechanical pull-test. This fixture employs a vacuum substrate holder. The conductors are clamped in pneumatic jaws. The substrate holder pivots to allow straight tensile or peel testing at angles up to 90°.

'Split-tip' welded microbonds with gold conductors

General discussion

Molybdenum electrodes are uniquely suited to bond gold wires. Electrode pick-up, or smearing the adhesion of the conductor to the electrode surface, is minimised; conduction of heat away from the joint is significantly less than with copper-based alloys; and the inherent hardness of this material provides long tip life without deformation or degradation. The recommended electrode gap spacing is one and one-half times the conductor thickness or diameter, permitting an even distribution of pressure at the bond. Larger gaps allow excessive material to be forced into the opening while insufficient weld areas are obtained with narrower gaps.

Microbond resistance

The only practical test for the quantitative evaluation of fine wire microbonds is the destructive pull-test. The desirability of establishing a non-destructive test technique led to the study of the resistance and thermal ageing characteristics of the microbonds studied in this evaluation. Joint resistance measurements for this programme were accomplished by the four-point probe

F

method. Since thin films were used, it was decided to limit the test current to approximately 100 mA and, thus, prevent overheating of the 0·0178 mm wire. Plots of resistance versus force were made for each type of microbond to determine if there was any correlation between bond pull strength and electrical resistance. No correlation was observed.

Gold wire to MoMn–Ni–Au terminations

The fired-on, thick-film terminations on alumina substrates demonstrated a much greater tolerance to pulse amplitude variations than the thin-film gold system; this is undoubtedly due to the relatively thick films employed. Joint efficiencies, those of breaking strength (mean of samples) and tensile strength of the wire were in excess of 69 per cent.

The prime consideration in obtaining reproducible strong bonds to thick films is proper positioning of the conductor and electrode. The relatively thick metallisation presents a rather sharp step at the termination edge and with slight mislocation of the conductor, the larger electrode used with 0·1270 mm wire may extend to the edge and cause excessive deformation. Bond failures were principally through the heat-affected zone. Film failures were non-existent.

Gold wire to gold–chromium terminations

As-fired alumina substrates: microbonds of 0·0254, 0·0508 and 0·1270 mm, wire-to-gold-to-gold chromium films on alumina substrates have been evaluated, bond efficiencies of 97 per cent, 82·6 per cent and 76·5 per cent, respectively, were observed. The results obtained were very similar to the thick-film microbonds in that the primary failure mode was in the heat-affected zone; however, film-to-substrate failures were observed and these were, in general, of low-pull strength. The surface finish of the substrate is a factor in this type of microbond, and it is concluded that maximum reliability would require substrate finishes of 30 μm or less.

Substrates

Glass and glazed alumina substrates

Glass and glazed alumina substrates offer essentially the same characteristics with respect to 'split-tip' bonding. Thermal shock to the substrates is the primary problem. The weld requirements for glass substrates require a relatively high bonding pulse amplitude so that lower amplitude pulses can be utilised for preheating and postheating. It should be noted that if the preheating and postheating pulses are of similar amplitude to the bond pulse this would essentially have the same effect as increasing the pulse duration of the bonding pulse and, therefore, have been experienced in both glass and glazed alumina systems. The first and more obvious type occurs with a relatively large chip of glass being pulled from the substrate. In this type of failure the portion of the film over the chip and the glass chip are usually pulled out at very low bond strengths. Frequently, cracks are visible after bonding and the conductor with the chip attached often sticks to the electrode and is raised with it. In other instances, no cracking is revealed until the bonds are subjected to 90° pull testing. Unless the bonding parameters are unusually severe, a failure of this nature may not be revealed by straight (bond in shear) pull testing.

The second type of glass substrate failure is less obvious. In fact, these fractures can be easily mistaken for film-to-substrate failures when examined at the normal magnification of ×30 to ×60. Increased magnification and/or transmitted light is usually necessary to reveal these failures which appear as a series of very shallow, small individual chips and are limited to the area immediately under the bond. Generally, no evidence of cracks exists after bonding, and no failures of this nature are revealed by straight pull-testing.

In the case of gold conductors to chromium–gold terminations on glazed alumina substrates, a high failure frequency of the glazed surface was encountered. These failures initially appeared to be film-to-substrate failures, but were, in fact, failure of the glaze, which was in evidence as fine grains of glass on the back of the weldment. The addition of preheating and postheating pulses did not significantly reduce the frequency of the occurrence of glaze failure modes. Anticipated failure rates for glazed alumina substrates indicate that this type substrate is the least compatible with 'split-tip' welding.

Oxidised silicon substrates

Microbonding of both 0·0508 and 0·1270 mm gold conductors and gold–chromium films on oxidised silicon requires considerably more energy than the other substrate materials, a phenomenon associated with the relatively high thermal conductivity of silicon. As a result of the high thermal conductivity, the bonding energy is rapidly dissipated throughout the substrate and the energy of the single bond is sufficient to raise the temperature of the entire substrate. This distribution of thermal energy throughout the substrate might appear to beneficial when using preheating pulses; however, the use of side pulses did not eliminate failures through the film and oxide layer. Preheating and postheating pulses are used only in the fabrication of 0·1270 mm diameter gold wire bonds. The incidence of film and oxide failure types decreased as the bonding energy was lowered, but could not be eliminated entirely. The increased energy required for bonding 0·1270 mm gold conductors often resulted in all pull-test failures occurring through the film or oxide layers.

The examination of each bond after pull-testing at magnifications in excess of ×200 was necessary to distinguish between film failures and oxide layer failures. In general, these failures exhibited small chips in the oxide coating in conjunction with large areas of film being pulled off, although occasional failures were noted between the film and oxide. In a very few instances in which near-maximum bonding energies were employed, actual chipping of the silicon occurs.

Polished sapphire substrates

Polished sapphire substrates used in this study had a surface finish of from 1 to 3 μm. Weld schedules used for the sapphire substrates were similar to those used on the as-fired alumina. The primary failure mode encountered was 'film-to-substrate' on the low-bond-strength samples.

The sapphire substrate was found to be one of the most compatible with 'split-tip' welding. Improvements in the consistency of film adhesion on the substrates would improve significantly the bond efficiencies observed.

Beryllium oxide substrates

Microbonds to beryllium oxide substrates were readily accomplished by 'split-tip' welding. Again the primary failure mode encountered was 'film-to-substrate' on low-bond-strength samples. The significance of the film adhesion problem is shown out by the fact that all the low-bond-strength samples observed in the gold wire evaluation occurred on only one wafer of the sixteen tested.

Bonding

Aluminium wire to gold–chromium films

The bonding of aluminium wire to gold–chromium films, regardless of substrate, proved to be very difficult. These bonds have shown themselves to be very sensitive not only to the bonding energy parameters but also to the variations of the aluminium surface condition. These microbonds require the maximum allowable energy with essentially no range of settings. Although the bonding energy was less than the optimum parameters established for the systems of gold wire to gold–chromium films, bonds could only be obtained at pulse amplitudes just below those which resulted in severe melting of the wire. Even at the optimum parameters, limited melting of the wire occurred very frequently. Melting of the wire during bonding resulted in extreme deformation of the wire and adversely affected the bond strength.

Electrode pick-up, i.e. adherence of the wire material to the electrode surface, commonly occurred with aluminium wire, and it was found necessary to clean the electrodes after every three to five bonds and frequently after only one bond. When melting of the wire occurred, molten aluminium would often be forced up between the electrodes which shorted them. Cleaning and reshaping were then necessary. This condition did not occur with new electrodes where the insulation material was flush with the ends; however, repeated cleaning operations and the heat of bonding caused rapid deterioration of the glass-fibre-type insulation, leaving an air gap.

All of these aluminium wire bonds were marginal since low strength interface failures could not be eliminated. Some of these bond interface failures were very likely the result of brittle intermetallic compound formation at the bond interface; however, microscopic examination after failure revealed that many of the lower strength failures showed no evidence of having been bonded. Furthermore, the established settings were so sensitive to parameter changes and electrode gap size that they were not repeatable from one period to the next.

No practical bonds were obtained with 0·1270 mm conductors. The increased energy necessary to obtain bonds invariably resulted in complete fusion of the wire, severing it. Although occasional electrical contacts were obtained, the resultant mechanical strength of these bonds was nil. It was concluded that 'split-tip' welding of aluminium wires to gold–chromium films was not practical.

Gold conductors and aluminium films

The bonding of gold conductors to aluminium films was found to be relatively difficult. The difficulties are associated with the inherent oxide layer present on aluminium, and the age of the films was a significant factor in the ability to obtain bonds.

It is commonly known that a thin oxide film forms on aluminium immediately upon exposure to the atmosphere and retards further oxidation. Thus it might be assumed that ageing beyond this initial period would have no appreciable effect. However, the aluminium films employed during this study were deposited on substrates at ambient temperatures and would, therefore, be expected to contain residual porosity. It is believed that the marked change in bondability of wafers with time results from oxide diffusion into the porous films.

The energy requirements for effecting bonds were relatively high and substantially longer pulse durations were necessary than with gold films. As mentioned previously, the effective range of energy was narrow and near the maximum or 'burn-out' point of the wire. Visual examination of these bonds occasionally revealed small areas of fusion at the top surface of the gold wire. However, no intermetallic compound formation or other abnormalities were observed.

Ultrasonic bonding

Microbonds between aluminium wire and films of gold–chromium and aluminium on sapphire, beryllia, alumina, glass and oxidised silicon substrates were evaluated using ultrasonic bonding techniques. The ultrasonic wire bonder used for this study consists of a 10-W, 40-kHz, vertical-action, ultrasonic transducer (associated power supply, 360° rotational manipulator) and a vacuum-operated clamp chuck. The machine variables are:

1 POWER The range setting is continuously variable from 0 to 10 in two ranges (the high range is ten times the low range).

2 PULSE TIME Continuously variable from 0·10 to 0·35 sec.

3 BONDING FORCE Approximate range from 25 to 600 g for 0·0254 mm to 0·1270 mm diameter wire. The 45° pull test was selected to avoid preliminary weakening of the bond at the weld-affected zone. This weakening would occur due to the hardening of the aluminium wire as it is cold-worked.

Alumina substrates

The alumina substrates were in the as-fired condition and had a surface roughness of 9 to 13 μm. The failures encountered in the ultrasonic microbonds to alumina substrates were primarily in the weld-affected zone although the minimum pull strengths were predominately interface failures. The aluminium wire to gold–chromium film microbonds on alumina exhibited a joint efficiency of 47 per cent. However, the distribution of joint strength did not exhibit any skewness indicating that it is a highly reproducible microbond.

Sapphire substrates

Microbonds were made with aluminium wires to gold–chromium terminations on sapphire substrates. The sapphire substrate exhibited a surface roughness of 1 to 3 μm.

In the case of sapphire microbonds, the failures were again in the weld-affected zone. However, as in the case of the alumina microbonds, the minimum pull strength failures were in the interface. The percentage of failures was relatively equal. However, the low joint strengths were predominately interface film failures.

Beryllia substrates

Microbonds were made to gold–chromium terminations on beryllia substrates. The surface roughness of the beryllia substrates was from 16 to 24 μm. As expected, the thicker films, which would tend to be smoother, had a higher bond strength. Also, the bonds on burnished films showed an even larger increase in bond strength. A comparison of the optimised bond pull strength was made for similar bonds of alumina, sapphire and beryllia as a function of surface roughness. The increase of bond strength with surface roughness was obvious. It was concluded that the ultrasonic bonding technique is much more sensitive to substrate surface roughness than 'split-tip' welding and that stronger bonds are achievable either by obtaining smoother surfaces or increasing film thickness.

Ultrasonic microbonds between aluminium wire and aluminium terminations

Aluminium wire microbonds were made to aluminium terminations on Vycor, oxidised silicon and sapphire substrates. This film thickness was of the same order as that commonly used for aluminium interconnections on integrated circuits. In order to improve film-to-substrate adhesion, the films were sintered. The films on Vycor and sapphire substrates were sintered in dry nitrogen, in order to simulate integrated fabrication techniques. (See Table 1.3.3.)

Glass substrates

Three failure modes were encountered on the glass substrates. They were weld-affected zone, interface and film. The bond strength failures were

Table 1.3.3 **Comparison of pull strength of bonds made by ultrasonically bonding 0·1270 mm aluminium wire to various substrates**

Substrate	Film	Surface roughness, μm	Average pull strength, g
Glass	Au/Cr	1–2	133
Oxidised silicon	Au/Cr	1·5–2	129
Oxidised silicon	Al	1·5–2	57
Sapphire	Al	2–3	98
Beryllia	Au/Cr	16–24	15

predominately film and interface failures. The joint efficiency observed for the microbonds on the glass substrate was 50 per cent.

Oxidised silicon substrates

Two failure modes were encountered on the oxidised silicon substrates. They were: weld-affected zone and interface (93 out of 100 of failures were in the weld-affected zone and interface failures were in the low-joint strength). The joint efficiency for the wire microbonds on the oxidised silicon was 40 per cent.

Gold wires can be reliably 'split-tip' welded to gold–chromium films on substrate such as alumina, sapphire, beryllia and Vycor. 'Split-tip' welded microbonds to glazed alumina and oxidised silicon appear limited.

Improvements in film adhesion might permit an extension of the technique. However, the inability of these substrates to cope with extreme thermal shocks precludes the 'split-tip' welding of diameter wires. Aluminium wires of 0·0254 and 0·0508 mm diameters can be reliably ultrasonically bonded to gold–chromium films on alumina sapphire, and beryllia substrates; however, none of the aluminium wire to aluminium film microbonds evaluated in this programme meet reliability requirements. Aluminium wire to aluminium film microbonds on oxidised silicon and glass demonstrated a probability that no more than one bond in a thousand would be less than the established minimum bond strength.

Microbonds containing either aluminium wire or aluminium film or both are not compatible with 'split-tip' resistance welding does not include a method for removal of the oxide coating on the aluminium.

Film thicknesses of 7,000 Å or greater proved to be sufficient for consistent production of ultrasonically formed microbonds on substrates with surface roughness of less than 3 μm. Substrate surface roughness has a significant effect on microbond strength. Microbond pull strength decreases as surface roughness increases regardless of the bonding technique; however, the decrease is more pronounced in the case of ultrasonic bonding. It was concluded that surface roughness greater than 30 μm was highly undesirable and ideally, a 3 μm or less surface roughness was preferred.

Thermal ageing is a valuable tool for determining compatibility of materials being joined but has no application for the prediction of microbond failures which result from the process variation; thermal ageing of the ultrasonic bonds, formed with no supplemental heat, between gold films and aluminium wires showed that intermetallics could form as low as 200 °C in less than 1,000 hr.

Improved uniformity of film adhesion is the prime requisite for increased reliability of the microbond types.

Chapter 1.4   **Adhesive bonding of microelectronic packages**

The use of adhesives in connection with microelectronic concepts such as multilayer circuits has focused on material and processing characteristics not considered critical in the past.

A specialised area in which considerable work has been done is the development of adhesive systems for bonding integrated circuits (IC) to multilayer circuit boards. ICs as used in these assemblies are small $15\cdot494 \times 10\cdot414 \times 0\cdot254$ mm silicon chips that have been processed to perform several circuit functions. The chips are mounted inside metal cans; multilayer assemblies may contain over one hundred integrated circuits in a few centimetres. Adhesive for this application must adequately position the device prior to soldering and provide a path for the dissipation of heat. An ideal adhesive must also relieve high stresses developed by dissimilar expansion rates during temperature changes.

Thermal conductivity

The presence of a large number of integrated circuits in a small area has necessitated the use of heat sinks for the dissipation of thermal energy. As a consequence, the requirement has been imposed on the adhesive that it be sufficiently heat conductive. Frequently, the adhesive's primary function is transfer of heat and not necessarily mechanical attachment of components to the substrate, i.e. the existence of an air gap or of inconsistent contact between an electronic device and the substrate greatly reduces the rate of heat flow from the device. Air with a thermal conductivity coefficient of $0\cdot6 \times 10^{-4}$ W/m.degC is approximately one eighth as conductive as the most heat-insulative organic adhesives having heat conductivities around 5×10^{-4} W/m.degC. The adhesive provides a stable and more reliable path for the flow of heat than the air gap which in addition to being a poor heat conductor, might vary in thickness with vibration or shock if the component or integrated circuit is not securely attached.

Because filled adhesives exhibit a higher thermal conductivity coefficient, initial evaluation and experimentation were performed with these substances. Heat-conductive fillers such as silica, aluminium and beryllia were used. A comparison of the resin and filler thermal conductivities was made. In addition to the chemical nature of the filler, other factors which contribute to the heat transfer ability of the adhesive are filler characteristics such as density and particle size. Very fine particle fillers generally increase the viscosity of the filled adhesive.

The method used consisted of a boiling-water heat source in contact with the plastic test slab which in turn was in contact with a copper plug heat sink; thermocouples were placed on either side of the test slab and the current produced by resultant voltages plotted against time. The slope of the curve was used to calculate the material's conductivity. As was seen from the data, degassing improved the thermal conductivity of the polyamide epoxy adhesive.

Evaluating unfilled adhesives

Early in the evaluation it was reasoned that a thin bond line was essential since the heat transfer rate through a substance is inversely proportional to the thickness. It was therefore decided at that time to concentrate on the evaluation of unfilled adhesives, since their bond line thicknesses are likely to be thin enough to overshadow the drawback of poorer innate thermal conductivity because of the absence of filler. As a result of this approach, an alternate procedure had to be developed to determine thermal conductivity, not as a thick slab but as an adhesive bond line. In this technique, the temperatures on either side of the bond line were monitored while a constant rate of power was applied to the heater coil.

The difference is between the two temperatures as they approach a steady-state value. This steady-state value was used as a measure of the adhesive's ability to transfer heat, i.e. the lower the temperature difference, the better the heat conductivity of the adhesive.

The importance of adhesive bond line thickness on heat transfer rate was demonstrated experimentally; several unfilled and one filled adhesive were tested for thermal conductivity by the bond line method.

As was seen from the test results, cyanoacrylate adhesive exhibited the lowest temperature drop. Good heat transfer was displayed consistently by this adhesive and even if the temperature-drop values varied substantially from test run to test run, they were always below those of other adhesives.

The calculated thermal conductivity coefficients for the cyanoacrylate compound were unbelievably high (10 W/m.degC). This may have been due to erroneous adhesive film thickness measurements which resulted from unevenness in the bond line. Cross-sections of integrated circuits depicting different bond line thickness exhibited by two different adhesives were examined.

Thermally induced mechanical stresses

Adhesives may induce and transmit stresses. Integrated circuits are delicate devices which can break or at least behave abnormally when excessively stressed. Thermally induced stresses are therefore of great concern and must be considered in the selection of adhesives. They can induce stresses by shrinking during cure. Once cured, they can transmit stresses produced during thermal cycling. Thermal stresses are a function of the elastic moduli and the differences in coefficients of thermal expansion of the materials comprising the assembly. The stresses can go beyond the ultimate strength and produce rupture or in some cases, e.g. semiconductor devices, the component element may be altered in electrical performance.

(The resistance of silicon changes directly with the stress and some strain gauges are actually made from silicon.) Thermal stress studies were conducted to investigate this problem on integrated circuits. Strain-guage measurements on the silicon chips in several integrated circuit designs brought out the importance of the adhesive as a factor in determining the magnitude of the stresses produced. Initial work revealed that stresses induced during cure in the silicon chip after bonding the integrated circuit to the copper heat sink went as high as 20·68 kN/m² in compression. The adhesive in this particular case was a rather stiff acrylic material. The compressive stresses were as anticipated since the copper heat sink and the adhesive should contract more than the silicon chip on cooling from the 72 °C curing temperature because of their coefficients of thermal expansion.

Stress measurements during thermal cycling were next conducted. Strain-guage measurements were made on silicon chips mounted in 'Kovar' cans which were in turn adhesive-bonded to copper heat sinks. (This is typical material combination for integrated circuits.) Several types of adhesives were used and the stresses measured after the bonded integrated circuits were cooled to −65 °C. It was obvious from the data that use of the more resilient adhesives such as the nitrile and the neoprene-based compounds resulted in lower stresses.

These results substantiated the belief that low-elastic-modulus adhesives will deform easily at low stress levels and will, therefore, minimise the stresses in the bonded materials. The results on the polyamide epoxy, however, were surprisingly poor in view of its relative softness at room temperature. The high stresses appear to be the result of the adhesive becoming brittle at the −65 °C temperature.

Adhesive processing characteristics

Another important aspect of adhesives to be considered, particularly for its effect on automation, is the mode of application. Since adhesive bonding in electronic assemblies requires the attaching of hundreds of ICs or components in a few square millimetres, the processing requirements are particularly important. The adhesive must develop a fast tack strength (1 sec) so that the devices may be rapidly and securely positioned, thus permitting further assembly operations.

Strength requirements

Since electronic assemblies are subjected to shock and vibration tests after exposure to high humidity, the desired adhesive must have adequate strength after. Fortunately, the strength needed for attaching integrated circuits and most electronic components to substrates is low since these items weigh so little. Even with accelerations many times that of gravity such as are encountered in vibration tests, the unit loads on the adhesive bond are not more than a few kilogrammes per square millimetre. It is, therefore, not difficult to meet these requirements with most adhesives. The low strength requirements are particularly true in the case of integrated circuits since they have a relatively large bonding area and low centre of gravity. A component's centre of gravity removed an appreciable distance from the adhesive bond line can result in increased peel and shear loads on the adhesive since lateral forces on the component produce moments of force.

Strength data on adhesives for integrated circuit bonding have been obtained by measuring the total shear force on the adhesive necessary to push the integrated circuit off the substrate. The substrate is gold-plated copper and is primed with a very thin application of an epoxy flash primer.

Since the integrated circuits all have the same nominal bonding area, $3 \cdot 175 \times 6 \cdot 35$ mm, the total force in newtons is used to grade the adhesives.

It subsequently became evident that the adhesives containing solvents — neoprene, nitrile and single-component polyurethane — actually gained strength during exposure to humidity. The increase in strength may possibly be attributed predominately to the loss of solvents with ageing. The cyano-acrylate and the photogelling acrylate lost additional strength on the 24-hr ageing after humidity while the photogelling compound exhibited a slight gain in strength.

The strength tests show that the solvent-based adhesives performed better than the 100 per cent reactive cyanoacrylate and acrylic compounds.

Summarising the properties of adhesives tested, in reviewing the data, it was observed that the neoprene cement was particularly good in its ability to relieve thermal stresses. This property is especially desirable since thermal stresses are capable of inducing failures in integrated circuits. The other neoprene properties are also adequate. In view of the favourable test perform-ance and the fact that it is particularly suitable for use in an automatic production machine, this adhesive has emerged as the best of those tested in the programme.

Chapter 1.5 **The application of film materials and components**

The electronics industry is making increasing use of solid materials in the form of thick and thin films. Design engineers are exposed to publicity for all sorts of film components and techniques, each promised to be better than the last.

A working knowledge of the possibilities and limitations of available film technology will assist the designer in the objective evaluation of this fast-developing field. As a starting point it is convenient to classify film materials and devices. This can be done in a number of ways, each of which highlights one aspect or another of particular products and processes. Table 1.5.1 shows

Table 1.5.1 **Classification of film materials for use in electronics**

Attribute of class	Sub-divisions	Examples
Thickness dimension	1 Thick films about 0·0254 mm	1 Screened and fired 'cermet' films for resistors and conductors
	2 Thin films about 10 μm thick	2 Evaporated nickel-chromium alloy films for resistors
Application	1 Functional	1 Resistive dielectric semiconducting and magnetic films
	2 Process aids	2 Etchable films, solderable contact finishes, etch 'resists'
	3 Protective	3 Surface coatings for corrosion protection
Composition	Elements, alloys, inorganic compounds, organic polymers	Silicon films, gold films, nickel-chromium alloys, silicon monoxide, poly-p-xylylene
Preparative method	1 Deposition	1 Vacuum evaporated dielectric films, 'electroless' plated nickel resistor films, electroplated gold coatings
	2 Conversion	2 Anodic oxide films for electrolytic capacitors, thermal oxide films on silicon
Structure	1 Continuous	1 Evaporated thin film conductor
	2 Discontinuous	2 Evaporated nickel-chromium resistor films
	3 Homogeneous	3 Epitaxially deposited single crystal silicon films on sapphire substrates
	4 Heterogeneous	4 Thick- and thin-film 'cermets'

classifications based on a number of attributes with examples of topical interest.

The simplest classification is by linear dimensions; in particular thickness. Layers from a few micrometres average thickness to a few ten-thousandths of a metre are commonly employed. Where the thickness is of the order of 0·0254 mm, the film is conventionally described as thick. Anything much below this is called a thin film.

Of more direct interest to users are classifications based on purpose and, if possible, cost/effectiveness. Unfortunately the latter feature is not expressible in general terms. Applications are, however, shown in Table 1.5.1, subdivided into functional applications, process aids and protective coatings.

The functional or primary classification covers applications where the operation of the device takes place wholly or partially within a film material of suitable composition or shape.

As the name implies, process aids are obtained from films which facilitate achievement of some aim during electronic device manufacture. In so doing such films may themselves be destroyed or may remain embedded in the structure of the finished device.

Table 1.5.2   **Some applications of film materials in electronic technology**

Application	Example	Class
Conductors and contacts	Gold, copper, aluminium, chromium, nickel, tantalum, silver, platinum/gold, palladium/gold	Functional
Resistors	Nickel-chromium alloys, tantalum nitride, tin oxide, palladium/silver/glass mixtures ('thick film cermets')	Functional
Semiconductors for active devices	Silicon, cadmium sulphide, indium antimonide, tin oxide	Functional
Magnetic memory devices	Nickel-iron, ferrites	Functional
Dielectrics for capacitors and insulation	Silicon dioxide, silicon monoxide, tantalum pentoxide, titanium oxide, aluminium oxide, siloxane polymers, poly-p-xylylene	Functional
Superconducting devices	Tin, lead, niobium	Functional
Semiconductor junction passivation	Silicon dioxide, silicon monoxide, silicon nitride, aluminium oxide	Protective
Semiconductor diffusion barrier	Silicon dioxide, silicon monoxide	Process aid
Intermetallic diffusion barrier	Nickel	Process aid and protective
Etch resist for film shaping	Proprietary light-sensitive polymer films (photoresists)	Process aid
Etch assistant for film shaping	Lead sulphide, copper, magnesium	Process aid for difficult-to-etch films

Another secondary classification covers surface coatings which protect the underlying material by isolating it from its environment, stopping undesired electronic processes or chemical reactions. Some materials are used for all three applications.

Table 1.5.2 shows some specific applications of film materials in electronic technology. Note that consideration is restricted to films formed from raw materials on the final support or substrate. This substrate gives the very fragile film mechanical strength and has a great influence on the other physical properties and behaviour of film materials formed and used upon it. Materials preformed, e.g. foils, and subsequently bonded on to substrates are not usually considered to be part of film technology, although it is, of course, sometimes appropriate to employ them to solve a given design problem.

Composition is the responsibility of the film technologist seeking to fulfil a given design requirement. It is important for the user to recognise that the composition and properties of a given film may differ significantly from those of the bulk material. Minor impurity constituents can be introduced deliberately or accidently during film preparation and may exert a major influence on the film properties and stability. A competent film component supplier will be aware of this and maintain rigorous quality control procedures.

The method of film preparation also effects film composition and structure, which exert major influences on physical and chemical properties of films in device manufacture and service. The economics of most types of electronic device incorporating film layers are also largely influenced by the film preparation used.

Many chemical and physical techniques have been evaluated for film formation and several are in large-scale commercial use. Nearly all are based on one or other of two main approaches. The first is condensation of the required constituent atoms from vapours, gases or liquids surrounding the substrate to be coated.

The second approach is conversion of pre-existing coatings or surface layers on the substrate by exposure to chemical reagents and/or energy. The selection of a preferred film forming method is complex, involving, *inter alia*, substrate size, shape and throughput, film properties and tolerance and, if required, the film-patterning methods available. Two aspects of film structure influence film properties. On the atomic scale of dimensions the arrangement of the constituent atoms of the film relative to one another is predominant. If the atoms are orientated in a stable regular array throughout the sample the material is a single crystal. Single-crystal films display optimum properties in semiconductor devices so that they are preferred for making integrated semiconductor circuits by the silicon-film-on-sapphire technique.

In contrast, many of the more useful dielectric films show highly disordered structures at the atomic level and are described as amorphous or 'glassy'. Between these two forms is that of the polycrystalline grains of various orientation and sizes. Most metal film resistors have been found to have a polycrystalline atomic structure thought to be partially responsible for the desired low-temperature coefficient of electrical resistivity.

On the scale of dimensions corresponding to the film thickness, i.e. from about 30 atomic diameters upwards, structural features manifest themselves as a departure from the ideal model of a film as a flat slab of material with parallel sides. In thin films, electron microscope and other studies reveal that some continuous polycrystalline films contain local variations in thickness corresponding to the individual crystal grains mentioned above. Other films are discontinuous, particularly on rough substrates. Here the crystal grains appear as islands separated from one another rather like raindrops on a dry pavement. If such islands change in shape, number or position during processing or use, drastic changes in film properties occur. This is particularly the case for conducting films when the islands are just touching or just separated. Another type of heterogeneity arises when particles of conductor are embedded in an insulating matrix. This is the case with thin-film 'cermet' (ceramic-metal) resistor films. On a much coarser scale of dimensions 'thick-film' metal glaze resistive films are of the cermet type, based on more or less aggregated conductor particles dispersed in a matrix of insulating glass (glaze).

Another important aspect of structure arises from film surfaces and interfaces. To some extent the properties of films are determined by their surface atoms rather than those in the thickness of the film. The thinner the film, the higher the proportion of surface atoms and the more behaviour will be dominated by surface effects. As well as the free surface, the interface between the film and the substrate determines behaviour. Also, in heterogeneous films such as the above cermets, the interface between the metal particles and the glass matrix modifies many of the observed properties. For all these reasons the determination and understanding of film microstructure is a vital aid to the film technologist aiming to produce stable products under good control.

A special structural feature is the local absence of film material, the so-called pinhole. Rigorous removal of dust particles and other contaminants in film processing environments is essential in order to minimise the occurrance of pinholes. The presence of pinholes in completed film structures will frequently nullify the designer's objectives, leading at best to a degradation of some device parameter in service and at worst to a catastrophic failure. Strict quality control during manufacture will eliminate potentially serious pinhole defects before they can cause trouble.

All features of thick and thin films so far discussed are exemplified in their applications to the fabrication of microelectronic and integrated circuits. Thick- and thin-film layers are used in microelectronic technology for their own functional characteristics and as process aids and protective layers in monolithic silicon integrated circuits (SICs).

Film (FIC) and film hybrid (FHIC) integrated circuits are built around networks of deposited film passive components such as fixed resistors and capacitors. These (lumped) components are isolated from one another by an insulating substrate such as glass sheet or ceramic tile. They are connected by metal film conductors. A flat substrate facilitates the use of printing methods such as silk-screening and photochemical etching to form the more or less intricate film patterns required to realise electronic circuits. FHICs increase the versatility of FICs by the bonding of separately made electronic components into the film network. Examples of such additions include transistors, SICs and other semiconductor devices as well as inductors and adjustable components.

The range of film component ratings and characteristics which can be formed by a given film forming and patterning method is also limited so that special-purpose resistors and many types of capacitors must be assembled on FHICs when required.

In general the film circuit designer tries to minimise hybridisation because of the expense and bulk of the added discrete components; also because of the extra soldered or welded joints introduced to bond additional components into the film network. These joints are a potential source of failure in service and hence detract from the original objective of increasing reliability by forming the circuit as an integrated unit.

Minimum hybridisation for the realisation of a given analogue or digital function will tend to make for complex FIC processing. From the standpoint of economical production, it is desirable to minimise the number of process steps and film materials. Moreover the only permissible steps and materials are those which can be applied sequentially to the substrate without damaging previously formed film structures. Such steps and materials are said to be compatible ones.

Favoured FIC technologies are those with a high degree of compatibility with few process steps and materials. The three well known FIC technologies exhibit different kinds of compatibility. Thus the thick-film technique makes repeated use of two dry processes, silk-screen film deposition and high-temperature firing in air ovens. The screened and fired films must be selected for their compatibility.

The all-vacuum process is based on a single complex film deposition and

Table 1.5.3 **Representative properties of film circuit resistors**

Technology	Thick-film circuits	Vacuum deposited thin-film circuits	Sputtered tantalum film circuits	
			(a)	(b)
Material	Noble metal dispersion in glaze	Nickel-chromium	Tantalum nitride	High resistivity tantalum
Sheet resistance range	30 Ω.sq to 30 kΩ.sq	200–300 Ω.sq	40–100 Ω.sq	Up to 1,000 Ω.sq
Resistance range for individual resistors	10 Ω to 1 MΩ (using several sheet resistivities)	10 Ω to 100 kΩ	10 Ω to 100 kΩ (single sheet resistivity)	1 kΩ to 1 MΩ
Manufacturing tolerance (with adjustment)	$\pm$1%	$\pm$1%	$\pm$0·5%	$\pm$1%
Temperature coefficient, ppm/degC	< $\pm$300	< +150	< −100	< −250
Stability at 70 °C and full rated load	<1% drift in 1,000 hr	<0·1% per 1,000 hr	<0·1% per 1,000 hr	No data
Maximum power rating, watts per cm of ventilated substrate	1·53 (on alumina)	0·23 (on glass)	0·15 (on glass)	No data

G

shaping step carried inside a high-vacuum process chamber. All the film materials used are compatible with deposition by vacuum evaporation and condensation.

The tantalum film circuit process is essentially a single material technology, tantalum steel and tantalum compounds are prepared by a combination of wet and dry processes to form a wide variety of passive components.

The exact schedules used in production will depend to some extent on the application. Thus for the highest precision analogue applications hermetically sealed cans will be required with individually adjusted R's and C's. For some digital applications neither of these processes is necessary and encapsulation can be limited to a conformal resin coating.

Bearing in mind these reservations, the resistor and capacitor properties obtainable from the three FIC manufacturing technologies are specified in Tables 1.5.3 and 1.5.4. The values given are for readily manufactured components, not extreme limits. An exact comparison is difficult because of the rapid changes now occurring in all these technologies. For the sake of completeness some information is given on thick-film glaze capacitors although most thick-film units produced commercially at the present time attached 'chip' capacitors bonded to the substrate in the same way as transistors and other discrete components.

Any of the technologies can be used for many FHIC designs incorporating fixed resistors and capacitors. In such circumstances the user's decision can be reached on the purely commercial grounds of price and delivery. For non-standard circuits there will be a design charge to cover the costs of film masks and other jigs. This will be much the same for all three technologies.

The 'best buy' differs for different applications. If miniaturisation is not important and the circuit function is a digital one working at comparatively

Table 1.5.4 **Capacitors**

Technology	Thick-film circuits	Vacuum deposited thin-film circuits	Sputtered tantalum film circuits
Material	Dielectric is a proprietary mixture of ferroelectric materials dispersed in a glaze	Aluminium/silicon monoxide/aluminium	Tantalum/tantalum pentoxide/gold
Maximum capacitance density, pF/cm^2	1,000	925	10,000
Range of capacitance, pF	50–500	10–5,000	100–50,000
Manufacturing tolerance	Probably ±20%	±20% to ±1%	±20% to ±1%
Temperature coefficient, ppm/degC	Non linear	±100	< +250
Stability (sealed) at 70 °C	No data	<0·1%/1,000 hr	<0·1%/1,000 hr
Power factor at 1KHz	<2%	<1%	<1%
Insulation resistance	>1,000 MΩ	>10,000 MΩ	>10,000 MΩ (Ta positive)

high power levels or in very hot environments, the thick-film technology may be optimum (Tables 1.5.3 and 1.5.4). If high-stability precision resistors are required as in attenuators and a.c. converters, the vacuum-deposited film type may be preferred. If high-stability precision resistors and high-value capacitors are needed as in active filters, the tantalum-based technology is favoured. This method can be used for both miniature and larger sized applications.

Another point to be borne in mind by the electronics designer is that straight forward translations of existing discrete component designs rarely give the cheapest possible equipment cost.

A better approach is to consider partitioning the system or equipment functions into a sub-system or systems best realised in SIC form, a sub-system or systems best realised in FHIC form, and a residual portion for discrete components. This over-all view of systems or equipment cost effectiveness can be based on criteria such as minimisation of sub-system interfaces (to minimise interconnections) rather than the minimisation of components.

Within each sub-system advantage can be taken of the particular component technology selected to obtain the optimum relationship between performance and ease of manufacture. Thus, in the tantalum film technology, precision resistors can be used *ad lib*. In thick-film technology high current and voltage resistors can be used without restriction. This design philosophy can be termed the 'what can be done, will be done' approach. The rapid development of such an approach does require continuous liaison between the IC technologist and the circuit and systems designers. In this respect the captive or 'in house' microcircuit facility of large equipment manufacturers can be most valuable. Indeed such a facility stands in the same relation to the electronics engineer as does the machine shop to the mechanical engineer.

Turning to future prospects for applications of film materials, a major limitation which requires to be overcome is the essentially two-dimensional nature of existing film circuit techniques. There is a need for cheap reliable multilayer methods so as to permit the realisation of complex three-dimensional networks. For the lowest possible parasitic capacitance such as will be required at very high operating speeds air dielectric crossovers using self-supporting conductors are under development.

Other improvements can be expected in ratings and characteristics of film components as improved understanding of the physics and chemistry of materials is applied to film technology. At the same time process and product optimisation will reduce the cost of film components. Thus the electronics designer will be presented with still further opportunities to devise more sophisticated and versatile electronic circuits and systems for new and expanding markets.

Section Two

Chapter 2.1   **Encapsulation materials: plastics**

Electronic components depend on the predictable and reliable maintenance of electrical potentials across dielectric media for their successful operation: and, whilst current conductors are normally of metal and relatively stable, exposure to a hostile environment frequently affects dielectric performance. Application of a protective coating is therefore necessary and though this may introduce useful secondary characteristics such as strength and shock resistance for the improvement of mounting and handling, exclusion of moisture is of paramount importance.

For extreme reliability, recourse to hermetic sealing is necessary. This means the exclusion of air by the enclosure of components in glass containers fused round emerging conducting leads. Variations include metal cans with leads retained in glass to metal seals, the melt-application of glass at temperatures necessarily about 400 °C is, however, impracticable with heat-sensitive materials and composite structures of glass and metal are expensive. The resistance of glass to heat and mechanical shock may be poor.

The traditional alternative materials — waxes, bitumen and drying oils — are now largely replaced by synthetic plastics. Thermosetting mixtures may be used cold in liquid form and then set by heating. Alternatively, they may be compounded into moulding powders and applied under heat and pressure. Thermoplastics on the other hand, like glass are used in the melt, but in either case the temperature to which components are exposed is rarely over 200 °C, costs are low and impact strength high. Why then are they inferior to glass as sealants?

Moisture will penetrate all plastic materials and the typical seal to emerging leads is poor or non-existent. Moisture penetration is not always obvious and it may be only under critical conditions of test and operation of equipment over extended periods that this is revealed. With microcircuitry, where unit costs are already high, protection must be perfect.

It is, therefore, of interest to examine the permeability of plastic encapsulating materials more carefully and to relate the bond-strength of plastic to metal with the resistance to moisture penetration along protruding leads. Despite the amount of testing for the effects of moisture that is currently undertaken, little reliable information is available and results are influenced by sample variability and test methods. Absorption is determined, for example, by a test which specifies measurements of water absorbed by 3·175 mm thick specimens immersed for one or more days.

Transmission of moisture through plastic films is important to the packaging trade and is measured by the gravity cup method. Increase of weight of the assembly in a controlled environment allows permeability to be calculated.

Data from published permeability values are given in Table 2.1.1. The results vary significantly within materials due to differences in thermal history, crystallinity, molecular weight, methods and conditions of measurement.

Table 2.1.1   **Comparison of published data on permeabilities, water absorption and diffusion coefficients for various plastics**

Material	Diffusion coefficient, $cm^2/sec \times 10^{-8}$	Water absorption, % w/w	Permeability, g. cm/cm^2. sec (cmHg) $\times 10^{-12}$
Nylon 6·6	0·027	8·6 (85% RH)	5–50 5–50 6–10 50
		0·8–1·8 (55% RH) 95 °C	2·7–17
Polyethylene (HD, LD)*		0·5 (LD)	0·75 1–6 0·25–0·6 2·2 3–13 1·2–8·7 5·6–22 4·3 HD 4·0 LD 6·6
Polypropylene		5·5	3·8 3 0·3–0·6 1 3–4·5
Mylar/Melinex	0·4	0·5	10 12 15–17 12 7
PVdC (Polyvinylidene chloride)			0·22–0·66
Polystyrene	16·3	1·32 (37·7 °C)	>5 30 50 80 70–110
Silicone rubber	300		800 1000

* LD — low density = 0·913–0·925 g/ml ; HD — high density = 0·921–0·965 g/ml

Table 2.1.1 *continued*

Material	Diffusion coefficient, cm²/sec × 10⁻⁸	Water absorption, % w/w	Permeability, g. cm/cm². sec (cmHg) × 10⁻¹²
Epoxides:			
Araldite 753/951	0·22	3·3	30
Araldite 753/951			30
			470
			1270
Araldite FRL	2·96		130
Araldite 750/972			22
PVC			30–150
plasticised		0·5	8
	0·044	0·9	5
			14
		0·3–1·0	15–45
			20
			111
PTFE			2
			4
Transfer-moulding powder:			
Hysol MG6 epoxy			7
Silicone			80

The variable known as 'permeability' is related to diffusion and absorption by the equation:

$$P = DC/L$$

where P the permeability, D the diffusion coefficient at a particular temperature, C the water absorption in the environment being considered and L the thickness of the specimen.

The diffusion coefficient can be considered as a measure of the rate of reaching equilibrium when a material is exposed to a new environment. Consequently, a statement on permeability of an encapsulant without knowledge of water absorption or diffusion coefficient conveys little information regarding the amount of water absorbed or the time elapsing before it is transmitted through to the vicinity of the component.

Since the cup determination of moisture penetration depends on repeated direct weight determinations, results are obtained in a reasonable time only with very thin membranes. Obviously these are difficult to produce without pinholes which would give erroneous figures. Sealing into the desiccant cup also requires skill.

The disadvantages have largely been eliminated by building a cell in which vapour diffusing through a plastic test piece from a controlled environment is swept by a dry gas stream through an electronic device that determines the moisture content with extreme accuracy.

Consideration of a fully bonded encapsulant indicates that at worst, the component will be in an environment of plastic saturated with water present to, say, the extent of 1 per cent and therefore a proportionately small dielectric loss. But where the bond between component and encapsulant is non-existent, voids may fill completely with water and dielectric losses could be disastrous. A similar situation exists with polythene-insulated underground telephone cables. With varying temperatures and levels, water may actually be 'pumped' into voids in the cable core until it becomes saturated.

It is important therefore that besides having low moisture absorption and permeability, the encapsulant must bond to the component. The value of epoxides as adhesives is well established and this feature commends them for precision encapsulation also.

Thermoplastics with good moisture repellent characteristics rarely adhere to substrates, but the high shrinkage of polyolefines on cooling from the melt may reduce, to some extent, the tendency for interfacial voids to occur.

Thermosetting materials are often applied as 'B stage' moulding powders. Application is ultimately by heat and pressure but without post-cure it is probable that the hardening reaction is incomplete. This means that the hardener or catalyst dispersed in the resin mass is partially free and moisture absorption and penetration are therefore likely to be high. The water can carry with it soluble residues from impure fillers and degraded plastic, and ionic materials will be transported to interfaces between encapsulant and component. Acids and bases used as hardeners for thermosetting resins can lead to progressive corrosive deterioration of metals and changes may be induced in microcircuitry.

The main reason for seeking encapsulants that are also good bonding agents is to prevent moisture ingress along projecting leads. These are normally of copper, plated with tin, but microcomponents can be made of more exotic metals. The success of glass as a sealant is due to its ability to establish a form of chemical bonding with oxides naturally present on thermally matching leads of nickel–iron. The most that can be expected with organic adhesives on metals is that secondary or Van der Waals forces may be involved and a strong physical bond should be established.

In order to investigate ingress of moisture along leads, 'bungs' are prepared with typical encapsulants including wires and exposed in the same apparatus as for moisture penetration of plastic diaphragms. Contaminants on leads such as oils, lubricants and mould release agents are typically shown to increase moisture penetration but certain 'primers' promote improvement. The relationship of permeability to strength of adhesion of plastic to metals is so striking that simultaneous adhesion tests may indicate a simple solution to an otherwise long term problem.

Studies such as those described contribute to the information available for the choice of encapsulants for microcomponents. And while it is essential to choose materials which will be easy to apply by standard production methods, it is possible also to ensure that stability in aggressive environmental conditions will be optimised. The efficiency of the hermetic seal is being approached with inexpensive and reliable plastics.

Moisture absorption and permeability for different encapsulants have been closely compared and bond-strength between encapsulants and leads is being related to moisture penetration.

Encapsulation of resistors

Resin encapsulation has for many years been one of the standard methods of protecting components. A wide variety of materials is used for this purpose, including polyurethanes, polyesters, PVC and silicones, but the most widely used are the epoxies despite the well known difficulties in maintaining adequate standards of process hygiene. The technique is generally more advantageous for the encapsulation of delicate components.

Various methods of overcoming the disadvantages have been tried but no complete solution has been found. Until recently the biggest step forward has perhaps been the use of transfer moulding, in which the epoxy resin is purchased as a solid B stage powder and compressed into a pellet of the required size. By the application of heat and pressure the pellet is liquified and forced into a heated mould cavity where it is converted into a solid cured resin.

New problems

Transfer moulding, while overcoming the worst problems associated with liquid resins, introduces a new set of problems. For example, even though the moulding pressures are low by normal transfer-moulding standards, they are often high enough to cause damage to delicate components. Also, the process does not lend itself to achieving the degree of impregnation required for components such as coils and resistors. Finally, the tools required for transfer moulding are relatively complex and expensive.

Two years ago a contact cooled resistor was introduced, this was basically the well-tried glass-fibre core, wire-wound resistor, resin-encapsulated in an aluminium tube. It soon became apparent that a more heat-resistant and stable resistor would be obtained if the glass-fibre core could be impregnated with resin.

Various types of casting and transfer moulding machinery were examined and a new machine for vacuum casting emerged as being more suitable than any other. In this machine, an upper chamber serving as a reservoir for resin and a lower chamber accommodating the moulds can both be evacuated to the same low pressure for the purpose of de-gassing the system. When the desired vacuum is reached, the moulds are positioned under the resin chamber and a simple cone valve is opened which allows resin to flow from the top chamber into the moulds. To assist impregnation and to hasten the flow of resin, a small predetermined amount of air is introduced into the resin chamber to give a differential pressure of 13–20 kN/m^2. The vacuum is then broken and the moulds removed to an oven or hotplate to cure the resin. The completed casting cycle takes about 3 min and as many components can be encapsulated as can be accommodated in the lower chamber. Resistors made by this new technique proved superior in every way to the earlier type, particularly in respect of breakdown voltage which increased by a factor of 5.

Increased range of tooling materials

Because of the low clamping and moulding pressures involved in this technique the range of tooling materials which can be used is considerably increased, whereas conventional moulds are almost invariably chrome-plated steels, moulds manufactured from vacuum-formed thermoplastic materials such as polypropylene and TPX have been evaluated for use in this casting technique and shown to be eminently suitable for many applications. Low-temperature fusible alloys have proved to be similarly successful. Moulds such as these possess the inherent advantage of being low in cost and may be discarded if damage occurs. Mould tools of these types are portable, cheap to manufacture and require no setting.

As a result of the success of the resistor when encapsulated by this method, it was decided to apply the method to metal film resistors. In the manufacture of resistors of this type, a thin layer of metal usually a nickel–chromium alloy is deposited on a ceramic former. The metal film, which can be as thin as 200 Å, becomes the resistor element which is helically ground to the desired resistance value. The resistor is very stable to heat, but like all metal film resistors is very sensitive to moisture since electrolysis may take place if the slightest trace of moisture reaches the metal film.

Good protection is therefore most important, and the best moisture protection is given by metal, glass or ceramic containers into which the resistor can be hermetically sealed. Because these have certain inherent disadvantages fluorocarbon and epoxy resins were examined but processing difficulties with fluorocarbons ruled in favour of epoxies. Casting epoxies are generally less permeable to water vapour than solid transfer-moulded types: also there is less moisture ingress via the junction between the termination and the moulding since transfer moulded epoxies contain built-in-mould-release agents which impair lead adhesion. This problem does not occur with liquid systems since mould release is applied externally.

Extending the field

As the vacuum-casting technique is ideally suited to the encapsulation of metal film resistors, the field was extended to include attenuators and thin film circuits. New problems encountered during the latter development were mainly associated with shrinkage of resin during the change of state from liquid to solid. This produced undesirable stresses on the glass-based resistive elements and the delicately attached discrete components. A buffer coating over the completed assembly prior to encapsulation provided the solution. The buffer coating consists of a flexible high purity silicone elastomer.

Thick-film circuits are being developed based on cermet materials and these will be encapsulated like the thin-film circuits. The encapsulation of SICs in the same way is also being evaluated.

Process hygiene

The main problem connected with the use of liquid casting resins is process hygiene. Recently epoxy resin systems have become available in a pre-mixed, pre-weighed and pre-evacuated form. This has eliminated operator contact at all stages apart from the final cleaning of the machinery. This method has proved ideal protection of a large variety of components and shares the following comparisons with transfer moulding:

1 Better moisture protection.
2 Lower pressure on the component.
3 Speeds comparable with those of transfer moulding.
4 Quickly adaptable to mould changes. In fact, all moulds are interchangeable and require no setting time — an essential feature for any equipment which is to be considered for short-run-work.
5 Cheaper tooling.
6 Moulds can easily be designed to avoid pinching of the leads at the exit from the body of the component.
7 Resin formulations may be easily modified to suit a particular component.
8 No need for refrigerated storage for casting materials.

The vacuum-casting method has one residual disadvantage — operator contact — but this has now been reduced to a ten-minute cleaning period at the end of the shift.

Encapsulation of capacitors

The encapsulation of capacitors for protection against mechanical and environmental conditions can be carried out by a number of techniques, depending on the application requirements of the equipment in which they are used. One of the earliest methods used was to dip the component in a microcrystalline wax or bitumen or, for the larger types, to enclose the elements in metal containers sealed with bitumen. This provided a seal which proved satisfactory for use in equipments such as radio sets where the component was not subjected to extreme tropical or arctic environmental conditions.

The next step was to enclose the components in metal containers and to provide terminal insulation of moulded phenolic resin for the rectangular metal containers and natural rubber for the tubular types. In certain instances the unit was enclosed in a phenolic moulding, but this again was not satisfactory in extreme environmental conditions. In more recent times the rapid development of epoxy resins, phenolic dipping compounds, synthetic rubbers and ceramic insulated terminals has improved the standard of protection a great deal.

Encapsulation must provide an adequate seal against moisture since the ingress will degrade the properties of a capacitor, depending on the type of dielectric. An unsealed capacitor, such as ceramic or mica dielectric, can be used in sealed equipments provided a temporary drop in insulation resistance can be tolerated before the equipment warms up and allows the capacitor to dry out.

Paper capacitors must be sealed against moisture since any ingress will result in a permanent lowering of the insulation resistance. Modern specifications include both environmental and mechanical tests.

Rectangular metal containers

The container must be capable of operating satisfactorily between the extreme temperatures of -55 to $+100\,°C$. This is achieved by using the rolled seam construction, the body being made by folding the strip in one piece and joining by means of a folded joint. This joint is then soldered and the lid and bottom, which have shallow depressions just fitting into the body, are attached by rolling or swaging their edges onto the body of the container. Sealing of the joints is achieved by soldering with an iron or by a dipping method in a shallow solder bath.

The terminal insulators are either of ceramic or phenolic moulding depending on the environmental conditions. Ceramic terminals provide a 56-day humidity category whilst the phenolic type are suitable for the 21-day humidity category. This construction is suitable for paper, paper/plastic, plastic mica and aluminium electrolytic dielectrics.

Cylindrical metal containers

The material of the case is normally aluminium or tin-coated brass. After the capacitor element is inserted in the tube a sealing disc of SRBP faced with a specially developed synthetic rubber is placed in each end and the edge of the aluminium can is turned over on to the rubber to effect a seal.

Encapsulation in synthetic resins

Dipped resin capacitors

For this type of capacitor the element is dipped in a slurry of loaded epoxy resin which is then oven cured. This method is widely used for the protection of plastic, ceramic and mica dielectric capacitors and is capable of passing the long-term damp heat test. (Dielectrics—mica, ceramic and plastic films.)

Epoxy moulded or resin filled plastic cased capacitors

In this class of capacitor the element is coated by being placed in a mould and encased in a thermosetting epoxy resin. Alternatively, the element is placed in a premoulded plastic container which is then filled with the thermosetting resin.

These capacitors have a generally similar performance to the resin dipped but are usually able to withstand a 56-day long-term damp heat test. In the resin filled plastic case construction the top temperature may be limited to $+85\,°C$ depending on the material used for the premoulded case. (Dielectrics—mica, ceramic and plastic films.)

Moulded plastic construction

The capacitor unit is moulded in materials such as polypropylene or alkalyd resin. These capacitors are manufactured in either cylindrical or rectangular form. The capacitors can be obtained with radial connectors making them suitable for mounting on a printed circuit board. (Dielectrics — mica, ceramic and plastic films.)

Plastic wrapped with resin end seal
The capacitor element is wrapped in a plastic film material which extends beyond each end of the element. The space at each end is then filled with a suitable end seal such as epoxy resin. It is suitable for temperatures and humidity conditions as described in the previous construction. (Dielectrics — mica, ceramic and plastic films.)

Wet sintered anode tantalum capacitor
This capacitor has posed a rather special problem since it was not only important to prevent the ingress of atmospheric moisture but also vital to prevent the escape of the acid electrolyte. It has a cup-shaped container which contains the anode and the acidic electrolyte and the sealing of the unit is effected by means of a PTFE gasket clamped between coined plates of tantalum and silver by a work-hardened nickel ring. The capacitors are suitable for extreme temperatures of -60 to $-200\,°C$ and 56-day damp heat test.

Tubular metal case with glass-to-metal seal
This is mainly used for the solid tantalum electrolytic capacitor although it can be used for other types of dielectric such as paper. The metal case is made from brass and subsequently solder or tin-coated. The glass terminal is fused on to a cylindrical tin coated plate which in turn is soldered to the open end of the case. This construction is suitable for operating at temperatures of -55 to $+125\,°C$ and 56-day damp heat test.

Wax protected capacitors
Although these capacitors have been largely superseded by the introduction of dipped resin types, they are used in telephone exchanges and computers but are normally limited in temperature to -30 to $+70\,°C$ and have a humidity category of H4. (Dielectrics — mica, and ceramic.)

Encapsulation of wound components

Demand for quality and reliability in components has increased greatly in recent years as indicated by the transition, in military electronics, from traditional trade practices. The trend in components is for more compact, efficient items, with higher degrees of proof against environmental hazards. In the commercial field there are increasing demands for both batch and quantity production of components to somewhat less exacting standards; enhanced environmental status is sought compared with that of their traditional counterparts, and the components have to be engineered carefully to low price targets. The economics of the design must determine whether or not the traditional gives way to the new, and there is wide scope for value engineering to meet these often conflicting demands.

During World War Two, electronic equipments were subjected to climatic variations including tropical, marine, desert and arctic conditions. High equipment breakdown incidences, particularly under Pacific conditions of high humidity, were experienced; this was hardly surprising considering the wide variety of components included in the equipments.

Many of the components had evolved by normal trade practices and were inadequate against such environmental hazards. Post-war investigations into the causes of component failure eventually led to increasing component reliability under climatic extremes.

Transformers and inductors

Transformers and inductors came under particularly close scrutiny and subsequent investigations, plus the introduction by industry of new, improved materials, led to the development of a standardised range of hermetically sealed oil-filled transformers, using grain-oriented 'C' core type magnetic circuits.

This was a significant advance in the demand for higher reliability standards and when the 'C' core standardised range was introduced, service equipment failure rates showed a significant decrease attributable to transformer failures. Despite this, oil-filled 'C' core transformers had certain disadvantages careful attention to sealing was vital since the sealing regions were vulnerable points. Oil-filling needed to be carefully controlled because, under faulty conditions, generated heat could cause case rupture, allowing oil to escape, increasing fire risks under certain circumstances. Mechanical parts were numerous and expensive in material cost and assembly labour.

Resin casting

The shortcomings cited stem from the basic concept of the oil-filled design and the sealing difficulties inherent therein. If the liquid impregnant and its container are replaced by a block of solid insulating material, these adverse factors are removed, reliability is enhanced and other advantages accrue. These shortcomings inspired early investigation of resin casting approaches.

Favoured materials initially investigated were the epoxide resins, which gave comparative ease of processing and excellent electrical and mechanical properties. However, the process of casting with resins introduced certain problems of control, the most significant being dimensional inaccuracy due to shrinkage during curing, harmful effects of the encapsulation on the magnetic core materials and parameter changes of the windings arising from encapsulation.

Nevertheless, despite these particular difficulties, resin-cast coil-wound components are in wide use. The rest of this chapter outlines some basic problems in designing reliable resin encapsulated electromagnetic components and discusses various other encapsulation techniques in media such as plastics applied to some components.

Dimensional accuracy

Problems of dimensional accuracy have to a large extent been overcome by the careful choice of epoxide hardener systems, combined with controlled curing cycles, which give a known degree of shrinkage for components of certain sizes and shapes. Tool design is important in influencing the final dimensions, position or fixing bushes, and terminations; also the temperature gradients due to suitable fillers in the resin systems assist in this.

Controlling temperature gradients and selecting appropriate fillers are largely a matter of practical experience and systematic recording of the results as a guide for future control. Dimensional accuracy is related to the size of the component.

Effects of encapsulation on magnetic core materials

A magnetic core may be affected in three ways. The first type of core defect is particularly relevant to nickel–iron materials, where the magnetostrictive character of the material causes alterations in dimensions with changes of magnetisation; such changes are restrained by a rigid encapsulating medium. A direct result of this restriction is to increase the magnetising current of a transformer, or reduce the inductance of a choke carrying a given d.c. current. This effect can be reduced by use of flexible epoxide resin systems. Various systems have been successfully used, e.g. the inclusion of dibutyl phthalate to plasticise standard mixes, and chemically 'building-in' flexibility to the resin macromolecule with duodecenyl succinic anhydride, polyamide hardeners and thiokol modifiers. The incorporation of aliphatic poly-epoxides gives good all-round results in reducing the effects of magnetostriction and the consequent permeability alteration in the core material. Flexible encapsulants assist when resistance to thermal shock is required, although their high-temperature performance and electrical properties are somewhat inferior to the more rigid encapsulants.

Secondly, it is possible in the encapsulation processing for the resin to become interposed between magnetic core faces, particularly if heated mixes and vacuum techniques are employed. Magnetising current changes due to the alteration of the effective gap can result. The magnetic reluctance of the gap will alter if the encapsulated device is temperature cycled and the defect applies, in general, to both iron and ferrite magnetic circuits. Careful core construction, which may include the closely specified use of adhesives at the gap region prior to potting, is often essential in controlling this effect. With this method, a sacrifice of inductance is accepted to achieve predictable changes during encapsulation. An alternative method, to overcome the difficulty of the gap alteration, is to encapsulate the coil only. This also solves the magnetostriction problems and is a desirable procedure if design parameters will permit. A third source of core effect is adhesion between the core and the encapsulating medium for, since the expansion coefficients are significantly different, the core tends to be prone to fracture, a problem which applies particularly to ferrite and iron-dust cores. Flexible epoxide systems help to alleviate the trouble but an alternative approach is to avoid adhesion by using an essentially non-polar material such as silicon, or modified phenolic gel. In this method, the potting cannot, of itself, constitute the mechanical structure and some form of container is necessary.

Gel-encapsulated ferrite devices with critical temperature coefficient performance specifications have been processed in production quantities for several years.

H

Injection moulding of ferrite-cored components with polypropylene, or similar thermoplastic materials, combines to a large extent the desirable properties of non-adhesion and mechanical viability. This method has been applied to a wide range of ferrite and iron cored components, for military and commercial application extended life tests show that polypropylene is a stable material capable of meeting the most stringent humidity testing. It is not always possible, however, to use the injection technique since the moulding pressures necessary have a marked effect on core materials. Application of injection techniques requires a careful evaluation of the design and is normally associated with large-scale production where effects are understood and under control.

Effects of encapsulation on winding parameters

In the use of resin systems, fine wires can be displaced or possibly broken, due to various strains set up as the potting resin cures. Pre-impregnation combined with an accurately formulated resin system, plus carefully controlled curing schedules to alleviate excess evolution of heat and undue shrinkage, normally solves this problem.

Injection techniques, in particular, demand exacting component and tool design allied to control of temperatures and pressures for the appropriate shots. Another main effect is on the Q of the coil. The power factor of a resin can markedly affect the parameters and due design allowance is necessary to allow for 'drift'. Dependence of the loss angle of many thermosetting resin systems on time, arises from the fact that the 'cure' continues for days, or weeks, after gelation and this can create production testing problems and quite often, end-user acceptance problems. Again, care in resin formulation and curing schedules, including a post cure can alleviate difficulties. Different cure schedules can affect the power factor when plotted against time. The dielectric constant of the resin is similarly dependent on the percentage molecular cross-linking of the encapsulants. This property can give unpredictable changes in the self-capacitance of the coil if the curing is not under strict surveillance. There is no short cut to experience in this field of control, and command of events must, of necessity, invoke strict recording of data relating dielectric constant to appropriate curing cycles for the particular encapsulating medium involved.

Examples of current production

Some of the design considerations mentioned in this chapter are now illustrated in the following sample of items. The selection, all taken from current production is representative of military, professional and commercial applications, each being interesting in specificational requirements versus target costs.

The solution to most problems involving encapsulating techniques requires a closer liaison and understanding between designer and end-user than that normally experienced with non-encapsulated wound components.

Miniature laminated transformer
This is a fully encapsulated transformer which incorporates a flexible resin system to overcome magnetostriction problems. The terminations are moulded into a glass filled nylon bobbin to minimise soldering difficulties when low temperature resistant flexible mixes are used. The bobbins are large to reduce the resin needed, reducing overall shrinkage on gelation. The laminations are afforded some protection by the bobbin against the resin shrinkage.

Semi-encapsulated miniature transformers
The laminations are in a flexible epoxy system, the flexibility being achieved by an aliphatic poly-epoxide. Particular grades of ferrite pot cores which will not stand injection moulding pressures are encapsulated in this material.

RF pot core assemblies
These printed circuit assemblies use a gel system supported by a copper screening case. The design, which was produced against particularly stringent demands in respect of temperature coefficient requirements, has now been extended to form a standard range of pot core assemblies. The gel places very little strain on the ferrite since its adhesion is poor and its volumetric change small, whilst its flexibility serves to minimise magnetostriction effects.

Resin-dipped transformer
This component uses an oil-modified phenolic impregnant and dipping resin. Less harmful effects on the core material result from the process, but special brackets and fixings are required and the environmental status of such components is wholly dependent on the over-all construction. Hence design is all important and compliance with stringent humidity specifications is less easy to achieve than with cast-resin encapsulated versions.

Toroidal transformer
This toroid developed for a nuclear application, has to withstand working temperatures up to 220 °C. The epoxide resin gives the sealing quality provided as, in this instance, other design materials are compatible. The temperature compatibility was obtained by use of a nadicmethyl anhydride hardener with a carefully formulated accelerator and a well controlled curing cycle. A high loading of closely graded and clean silica flour is used to reduce the coefficient of expansion of the encapsulant, in order to prevent cracking due to thermal shock.

RF inductors
Typical RF inductors utilising ferrite or dust cores from a standard range are encapsulated in polypropylene. These components give improved environmental protection compared with 'open', 'dipped' or 'plastic sleeved' counterparts.

Ignition control assembly

A transformer-ignition unit is designed to produce an encapsulated, homogeneous whole, i.e. a mains transformer, microswitch, pilot gas valve and terminals; Polypropylene is chosen as the medium. The assembly eliminates all supporting structural items and at the same time gives complete environmental protection. The cover and activating push-lever are produced in the same medium. Pressing the push-lever activates the chain of events that leads to the gas central heating system being ignited and operative.

Against the background of such diverse requirements, research and development proceeds on encapsulating materials, both thermosetting and thermoplastic, that are being made available by the resin and plastic producing industries. These include, for example, resin variations, such as powders and foams, diallyl phthalate, phenolics, modified nylons, polypropylene, polycarbonates, acetals, polyurethane, etc.

As an example of particular interest, assemblies containing foam encapsulated laminated transformers have successfully passed stringent specifications including additional shock and acceleration tests and the technique lends itself readily to many applications in the ferrite field. It may be of specific interest to engineers to learn that miniature pulse transformers, utilising ferrite pot cores and toroids, have been processed to the same specifications.

The diverse basic materials and their many variants will, in the future, reveal an even wider variety of possibilities appropriate to the spread of requirements. In techniques, low-pressure transfer resin moulding will have relevance for many larger scale production items. Injection moulding techniques utilising thermoplastics have a particularly rewarding approach to automatic large scale production.

The technique, initially thought by many to be ill-suited when applied to coil wound components — particularly to fine wire components using say wire (0·295 mm) by large-scale customer acceptance in diverse industries. It is not claimed however, that the technique is the universal panacea for all encapsulation problems that fall within the temperature scope of thermoplastic materials such as polypropylene. Each problem must be analysed on its merits and all factors taken into account before a decision can be made as to the most appropriate medium, and method, versus performance parameters, quantity and target cost.

Chapter 2.2 **Encapsulation compatibility**

Epoxy encapsulation and impregnants have been economically used for many years for environmental protection of electronic components such as magnet coils in motors, transformers, solenoids, reactors, etc. Recently, the use of magnet wires has increased greatly in electronic functional blocks: many epoxy-impregnated components containing magnet wire are used in modern electronic systems and the physical properties of the epoxy resin compounds used are ideally suited, in most cases, for this purpose. These properties can be varied from rigid to flexible, transparent to opaque and from high to low viscosity; they may also be formulated to specific degrees of toughness and impact resistance, low shrinkage and high moisture resistance to a host of solvents and chemicals, notably alkalies.

With a proper formulation, it is possible to apply an epoxy compound by:

1 Moulding.
2 Casting.
3 Dipping.
4 Vacuum impregnation.
5 Spraying.

The properties may be formulated to provide a larger variety of applications than the more widely used but less expensive impregnating varnishes.

An additional advantage of epoxy-impregnating materials is their solvent-free nature; this permits thicker, void-free films and deeper sections than can be obtained with varnishes. In spite of the numerous advantages of these materials, epoxies have one feature that can cause concern in their applications, namely magnet wire film insulation.

It has become known that some epoxy compounds have an adverse effect on the thermal life of certain wire enamels at elevated operating temperatures and this effect cannot always be predicted.

It is not the purpose here to detract in any way from the already proved excellence of the epoxies; their outstanding properties have been demonstrated and accepted by the manufacturing industry; rather it is my purpose to present data for both favourable and unfavourable combinations of epoxies with wire enamels.

The thermal life of any system is dependent on many factors, acting singly or (more reasonably) in combinations, to effect the ultimate thermal life of the material combination.

The following is a discussion of some of the factors affecting thermal life.

Stability of wire enamel

Many independent laboratories have established that some wire enamels have a high degree of inherent thermal stability whether tested with or without varnish. High-temperature polyester enamels and aromatic polyamides are good examples. It is desirable to have an enamel with good inherent thermal stability so that it may be used with a wide variety of compatible impregnants.

Stability of impregnant

It is recognised that epoxy compounds have different degrees of thermal stability and that this stability may be affected not only by the particular epoxy resin, but also by the hardness, dilutants and fillers used in the formulation. At this time, there is no industry-wide or general test procedure for thermal stability of such impregnating compounds to permit rating by thermal classifications.

Reaction between the wire enamel and impregnant

Several tests have established that inherently stable wire enamels and impregnants can adversely affect each other to give a comparatively unstable combination when tested at elevated temperatures.

The exact mechanism has never been fully explained, but it is thought that a synergistic chemical-thermal effect between the wire enamel and the impregnant may produce a combination with lowered thermal resistance. This is akin to the many cases where a combination of two resins exhibits poorer properties than either component examined singly.

Solvent-type attack of thinners and dilutants

Some of the components of epoxy impregnating compounds exhibit a strong solvent effect on wire enamels. Volatile thinners used to reduce the viscosity of epoxy resins may show this action, but the non-volatile reactive dilutants such as the glycidal ethers are possibly the worst offenders in this respect.

While the exact mechanism of solvent attack on wire enamels is not fully understood, it is believed that low-molecular-weight fractions of the wire enamel polymer are extracted to leave weakened areas within the enamel film. Another possible explanation is that the molecules of solvent may be absorbed into the wire enamel, and upon subsequent heating or testing at operating temperature, may depolymerise or may simply soften the enamel polymer.

Attack of the epoxy catalyst on the enamel film

The catalysts used to harden the epoxy compounds are usually either organic acids, anhydrides or amines, and each of these materials can have an adverse effect on the wire enamel. For instance, an acid catalyst may attack a wire enamel containing an excess of hydroxyl (basic) groups while a strong amine catalyst would be likely to attack and destroy a polyester enamel. It is likely that a room-temperature curing impregnant would show less reaction with wire enamels than epoxy impregnants that require high-temperature curing, possibly due to increased reaction or attack as a function of temperature.

Attack of wire enamel by its own decomposition products

At elevated temperatures, wire enamel tend to release gaseous products which may consist of free radicals of amines, acids or alcohols. These fragments can attack the epoxy at the interface between the enamel and the impregnant and decompose the epoxy so that its insulating value is decreased. In some cases, entire molecules of such materials as creosols can be released to exert a solvent type of attack on the epoxy impregnant.

Stress-cracking of the epoxy on thermal shock

It is known that some epoxy compounds shrink on curing as well as in subsequent ageing. Dimensional changes may also occur during temperature cycling in a finished electronic component. These dimensional changes based primarily on the thermal coefficient of expansion tend to crack the more brittle compounds and destroy their insulating properties. The dimensional changes combined with the strong adhesion properties of epoxies on wire enamels often lifts the enamel from the conductor to form areas which can ionise under test voltages and give premature failures. Incipient cracks which may extend to the outer surface are vulnerable points which induce moisture or dirt.

Adhesion between the wire enamel and the encapsulant

As mentioned before, the epoxies produce excellent adhesion to the surfaces of wire enamels; such adhesion is advantageous as it gives a strong continuous body of insulation without voids which may ionise or absorb moisture. A strong bond is also desirable to prevent vibration or other mechanical movement of the wire. It must be remembered that very hard impregnating compounds adhere so strongly to the enamel that differential dimensional changes will actually strip the enamel coating from the copper. Therefore, it would seem correct to use a flexibilised compound that yields slightly under thermal or mechanical stress yet still maintains its adhesion to the wire enamel.

Oxidation of enamel and impregnant

It is well known that wire enamels are subject to degration by oxygen; one method of preventing this might be to apply a heavy coating of epoxy over the enamel. However, several factors must be borne in mind when selecting such a coating. Depending on the coating thickness, oxygen may diffuse through the epoxy and lead to a porous structure which allows oxygen to reach the wire enamel; microscopic bubbles of air may be trapped within the impregnant when it is compounded with fillers or hardeners. Some of these bubbles will be in contact with the wire enamel where they can react to degrade the film; air may also be trapped within the windings of an electrical unit. A remedy for these 'bubbles' is the thorough evacuation of the compound before impregnation, and before the compound is cured.

Thickness of the epoxy impregnant

In the development of an epoxy impregnant, the thickness of the epoxy should be studied as one of the variables of the system. A thick coat may provide a barrier against moisture or oxidation, but at the same time it may act as a barrier against the release of harmful decomposition products.

Brittle materials when exposed to temperature extremes are more likely to crack in thick sections. Furthermore, the cost of using the material increases as a function of thickness; it is obvious that there is probably an optimum thickness for each compound, and this is affected by such things as resistance to cracking, permeability to gases and water and coverage of sharp corners and windings. It must be borne in mind that two or more of the factors listed may be operating simultaneously and the effect of interaction is likely to be much more severe than that of any one factor. For example, either oxidation or cracking alone might have only a minor effect on the life of a given system; cracking on the other hand allows oxygen to penetrate deeply into the epoxy and makes it more prone to cracking. Thus, the combined effect becomes a major factor.

It is quite possible that unsuspected combinations of factors may also cause degradation of these systems, and an understanding of the problems that allows us to predict the behaviour for a given system is very necessary.

Chapter 2.3 **Rubber seal materials**

Although rubbers are among the most important materials used for sealing duties, it is probable that their properties are not well known to seal users, and even to seal designers, and accordingly their possibilities and limitations are not fully appreciated.

It is the intention in this chapter to survey the properties of some of the various types of rubbers used for the manufacture of seals and to discuss those properties that are considered are of particular importance in sealing applications.

The essential characteristics of a rubber can be simply stated as follows:

1 High deformability — in extension, of the order of hundreds of per cent.
2 Forcible and rapid retraction.

If a material has these properties it may be classed as a rubber, in spite of its origins or of its chemical composition.

The properties of high deformability and forcible retraction are of importance in almost all rubber products, and the rubber seal is no exception. Seals are invariably deformed on assembly and the tendency to recover their original shape results in a loading between the rubber and the adjacent metal parts, thus preventing the passage of gas or liquids.

Many of the additional properties which are looked for in a rubber are required so that the basic rubberlike characteristics can be retained in service, and these additional properties are usually the reason for the selection of a particular type of rubber for a given duty. We may require, for instance, resistance to heat because high temperatures bring about chemical breakdown which eventually and permanently destroys the elasticity of the rubber, or we may require resistance to low temperatures because low temperatures bring about a physical and temporary change which results first in increased stiffness and finally in brittleness and again in complete loss of rubberlike properties.

It is of interest to consider why the great majority of rubber seals, and indeed rubber products for engineering duties generally, are black in colour. In Table 2.3.1 the effect on strength of the addition of carbon black on two rubbers is shown, the gum-stock in each case being the rubber with only the necessary chemicals added to vulcanise it. In the case of natural rubber, a 30 to 40 per cent increase in strength can be brought about by the addition of carbon black, but in the case of nitrile rubber the increase may be up to 600 per cent. It is true to say that most rubbers of the hardness normally used for seals are relatively weak materials if they are not loaded with carbon.

Table 2.3.1   **Effect of compounding with carbon black on strength of rubbers**

	Tensile strength, MN/m²	Breaking elongation, %
Natural rubber:		
Gum stock	17–20·5	750–850
Black-loaded	24–27·5	550–650
Nitrile rubber:		
Gum stock	3·5–5·5	400–500
Black-loaded	20·5–24	450–550

Types of rubber

The types of rubbers to be considered here are those given in Fig. 2.3.1 and as each of these types represents a group of materials rather than an individual, it will be appreciated that the comparative properties, to be given later, are necessarily generalisations. A brief description of each type is given below.

Butyl rubber

Butyl rubber or isobutylene–isoprene copolymer is possibly the least rubber-like of the types considered; nevertheless, it is used on a somewhat limited scale where low gas permeability or resistance to specific chemicals and

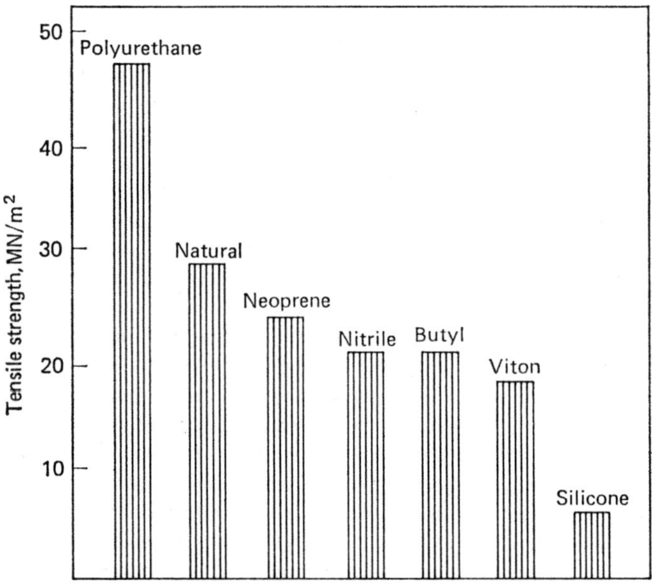

Fig. 2.3.1   Tensile strength of rubbers

fluids is required. Butyl rubber is, for instance, currently used in seals for the phosphate esterbased hydraulic fluid. It is not resistant to mineral-based oils.

Natural rubber and general-purpose synthetic rubber

The general-purpose synthetic rubber, butadiene–styrene copolymer or SBR is here grouped with natural rubber for, although differing chemically, its salient properties are similar.

These rubbers are characterised by their high strength and resistance to wear, particularly where cutting and tearing are involved. Resistance to mineral oils and petroleum hydrocarbons is poor, however, and so their use in the fluid sealing field is somewhat limited. An exception is the automobile hydraulic brake system where natural rubber or SBR seals are used in conjunction with a castor-based fluid.

Neoprene

Neoprene or polychloroprene rubber was the first of the synthetic rubbers to be produced on a large scale — in the early 1930s. It combines abrasion resistance, moderate oil resistance and weather resistance to a marked degree. Although weather resistance is not normally an important factor in fluid seals, in such allied items as protective gaiters or bellows, it is of great importance and a reason for the frequent selection of neoprene for such duties.

Nitrile rubbers

Copolymers of butadiene and acrylonitrile nitrile rubbers are a most important class of materials for the manufacture of rubbers seals and it is probable that something approaching 80 per cent of all fluid seals are made from nitrile rubbers. Among the moderately priced materials they are outstanding in oil resistance.

These rubbers are available with acrylonitrile contents ranging from 18 per cent to 50 per cent, the higher contents giving greater oil resistance, but inferior performance at low temperatures. In many applications involving low temperatures, an increased swelling in oil must be accepted therefore to achieve adequate performance, and this is well known to the designers of aircraft hydraulic systems.

Polyacrylic or acrylate rubbers

These materials have a definite if somewhat marginal superiority over nitrile rubber in respect of heat resistance and have similar resistance to oils. They are, however, inferior to nitrile rubbers in respect of most mechanical properties and, therefore, have rather limited applications. The polyacrylics are particularly resistant to the embrittlement which many extreme-pressure oils tend to produce in high-temperature service.

Polyurethane rubbers

These materials are available under several names. They are relatively new materials which have markedly higher strength and resistance to tearing and abrasion than any other type of rubber.

The polyurethanes have not, however, been widely exploited as seal materials and this may be because their behaviour in oils is somewhat unpredictable. Although swelling is not excessive, chemical breakdown may occur resulting in surface softening, particularly at high temperatures and if traces of water are present in the oil.

Silicone rubbers

These materials, unlike the others discussed, are semi-inorganic in constitution, their basic molecular chains being composed of silicon and oxygen rather than carbon atoms. The most striking property of silicone rubbers — their retention of rubberlike behaviour over a very wide temperature range —stems from this fact.

Silicone rubbers are adversely affected by the presence of carbon and so the earlier comments on the general superiority of black rubbers do not, in this case, apply. Resistance to mineral oils is only moderate and the comparatively low strength presents difficulties in dynamic sealing duties.

A recent development in this class is the fluorinated silicone types. These rubbers have oil resistance approaching that of nitrile rubbers with only minor adverse effect on high- and low-temperature stability. They are, however, the most costly of the commercially available rubbers.

Fluorinated rubbers

Several fluorinated rubbers are now available, but of these a copolymer of hexafluoropropylene and vinylidene fluoride, is probably the only type in current use on any scale. This rubber may be regarded as an attempt to confer rubber-like properties on polytetrafluorethylene (PTFE) by molecular modification. This has been achieved, but not, however, without some penalty. In the case of the chemical resistance for instance, although it is markedly superior to any other rubber, its resistance is not so complete as with PTFE. The high temperature resistance, although second only to that of silicone rubber, also falls short of that of PTFE.

Comparative properties

The important requirements of a seal material are considered to be as follows:

1 Strength and wear resistance.
2 Elastic recovery.
3 Stability at high and low temperatures.
4 Resistance to fluids.

It is not suggested however, that these are listed in order of importance as this order will vary according to the particular duty. These requirements are discussed in terms of those measurable properties which are currently used by the rubber technologist in the evaluation of rubberlike materials and comparison is made of the various types of rubbers.

Strength and wear resistance

The determination of tensile strength is among the most widely used tests performed on rubber, although it is appreciated that rubber in general, and possibly seals in particular, are seldom highly stretched in service. The importance of tensile strength lies in its relationship to abrasion or wear resistance and it may be regarded as a general index of toughness.

The levels of strength which may be expected from the various rubbers are shown in Fig. 2.3.1 and in each case something near to maximum obtainable has been indicated.

The range 6·865 to 34·474 MN/m^2 is considerable, and the superiority of polyurethane in respect of strength is apparent and certainly in the case of moving seals particularly where rough shafts and doubtful lubrication are concerned, this will be reflected in reduced wear. In many applications, however, where conditions of lubrication and shaft finish are good, strength in excess of 13·790 MN/m^2 appears to make little difference to seal life. Silicone on the other hand falls markedly short of this level of strength and where movement is concerned, wear may be expected to occur at a comparatively rapid rate.

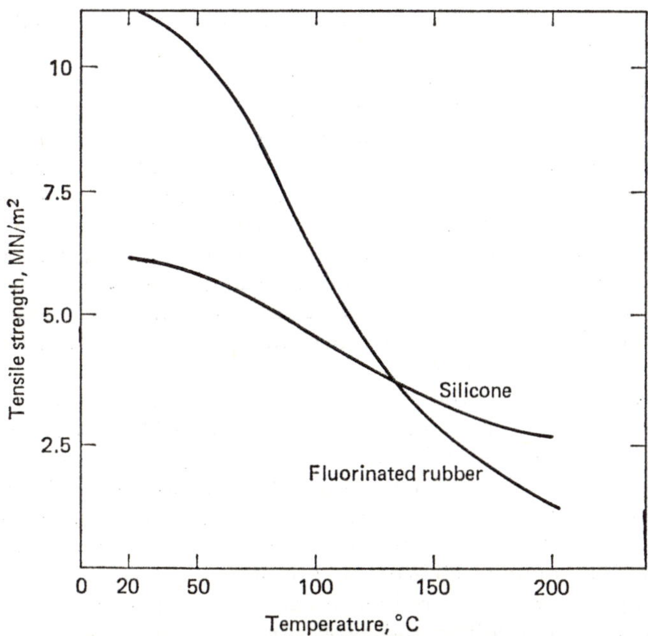

Fig. 2.3.2   Effect of temperature on strength

The damage which occurs to a seal due to extrusion into clearances between metal parts is certainly influenced by strength, and a material of high strength can be expected to suffer less from this cause.

The tensile strengths given in Fig. 2.3.1 are those measured at normal temperatures. At higher temperatures strength falls progressively, and in the case of rubbers used at high temperatures the drop in strength may prove to be a limiting factor.

In Fig. 2.3.2 typical effects of temperature on the strengths of Viton and silicone rubbers are shown. At temperatures above 150 °C the strength of silicone rubber exceeds that of Viton in spite of the two-fold superiority of the latter at room temperature and the behaviour of Viton may be taken as typical of the other organic rubbers.

Elastic recovery

The capacity to recover or tend to recover from deformation is an essential requirement of a seal material for it is this tendency which maintains a loading on the surfaces to be sealed and which enables the seal to follow the movements of the surface resulting from dimensional variations and distortion due to pressure changes. The factors to be considered here are all aspects of the elasticity of rubber, and stress relaxation, compression set and rebound resilience would appear to be the most important. Stress relaxation — the decrease in stress when rubber is held at constant strain over a period of time — may well be a critical factor in the case of static seals working at constant pressure. In the case of moving seals or of those operating over a fluctuating pressure range, however, the ability of the seal material to recover and so follow the inevitable dimensional changes of the surface to be sealed becomes of importance. The ability of rubber to recover after deformation is determined as compression set, the residual deformation after removal of the deforming force, and a low value of compression set indicates a high degree of recovery. Compression set is normally measured after a period of one or more days and at elevated temperature, the latter, either as a means of accelerating the test or of simulating service conditions.

It is more usual to determine compression set rather than stress relaxation because of the more simple test procedure and although the relationship between the two is not precise, agreement is sufficiently close for most purposes of evaluation. It is in practice, therefore, to aim at a low level of compression set in seal materials.

Rebound resilience, although not a factor usually considered in the development of rubbers for sealing purposes, may well be of some significance in that it indicates speed of recovery, unlike compression set which is concerned largely with degree of recovery. In the case of a piston seal, for instance, operating over a wide and rapidly fluctuating pressure range where rapid variations in the bore to be sealed are likely, one would anticipate that a high value of rebound resilience would be advantageous.

Both compression set and resilience can be varied over a wide range by differences in compounding and processing, but the ratings (high values indicate property at most advantageous) in Table 2.3.2 are considered to give an indication of the best that can be achieved with the types of rubbers shown. These ratings will apply up to about 100 °C, but above this temperature silicon rubber would be expected to show a marked superiority.

Table 2.3.2 **Comparative ratings of resilience and compression set**

Rubber	Compression set	Resilience
Nitrile	8	5
Natural	7	9
Silicone	6	7
Neoprene	5	7
Viton	5	5
Polyurethane	5	6
Polyacrylic	4	4
Butyl	4	2

High rating — property at most advantageous

Temperature effects

Both high and low temperatures bring about major changes in rubber. In the case of high temperatures the magnitude of the change is time-dependent, being the result of chemical degradation, and involves the progressive and permanent loss of rubberlike characteristics and eventually in brittleness or resinfication.

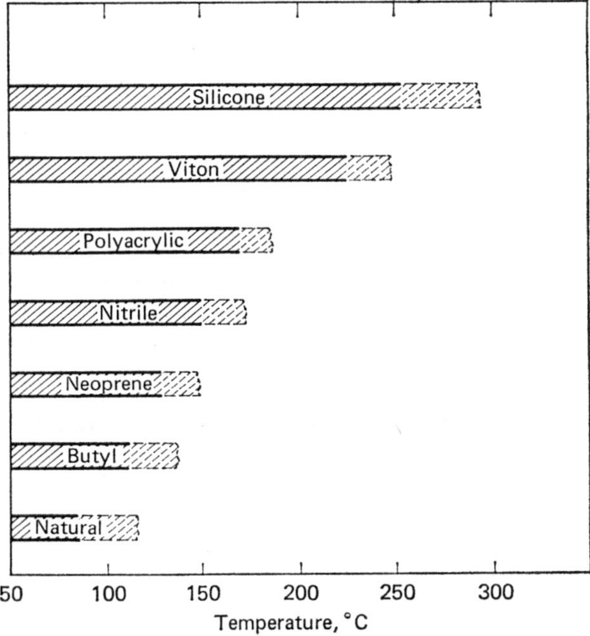

Fig. 2.3.3 **Service temperatures of rubbers**

Low temperatures, on the other hand, produce a physical change as a result of which stiffness increases with reduction in temperature until the brittle state is reached. This change is neither time dependent or permanent. The effects of the two extremes of temperature are considered separately.

Heat resistance

Because of the time dependence of the effect of heat, there is no one maximum service temperature for a given rubber and assessments will vary according to the service required. As an example, nitrile rubber can be considered suitable for use at 180 °C, if a life of 50 or so hours is acceptable, but if several thousand hours' life is required the maximum temperature envelope would have to be some 50 degC lower.

In Fig. 2.3.3 some assessments of operating temperatures are made. The temperature up to the dotted part of the plot is that at which approximately 500 hours' useful life can be expected. Above this, life will fall off rapidly, but at lower temperatures can be expected to approximately double for each 15 degC drop in temperature.

Cold resistance

Although the effects of low temperature on rubber are less complicated than those of high temperatures, the difficulty of relating laboratory test data to service test performance of seals has not been resolved.

When rubber is cooled, not only does its stiffness increase, but both rate and degree of recovery from deformation (resilience and compression set) are adversely affected. Additionally, the pressure conditions during the low-temperature cycle may have a marked effect on the performance of a seal. In the case of a piston seal, for instance, leakage will occur at low temperatures when the hydraulic pressure is increased if the seal material does not recover sufficiently to follow the cylinder distortion. Under constant pressure conditions, however, the same material may maintain a seal down to a temperature at which, by conventional tests it is bone hard and brittle—a difference possibly of 30 degC or more. In this situation conflicting claims for the low temperature performance of materials are inevitable.

Under normal, fluctuating, conditions of hydraulic pressure, however, one would expect the minimum sealing temperature to be at or about the temperature of zero compression recovery (100 per cent compression set), and this appears to be borne out in practice. The determination of compression recovery at low temperatures is not, however, a convenient test to perform and it has been found at least for a range of nitrile rubbers used for aircraft hydraulic seals, that the temperature of zero recovery corresponds to within a few degrees with the temperature at which rigidity modulus, determined by the somewhat arbitrary method of BS903, Part A13, reaches a value of 68·958 MN/m^2. The relationship may not in fact be haphazard as both values indicate that rubberlike properties are largely absent — see Fig. 2.3.4.

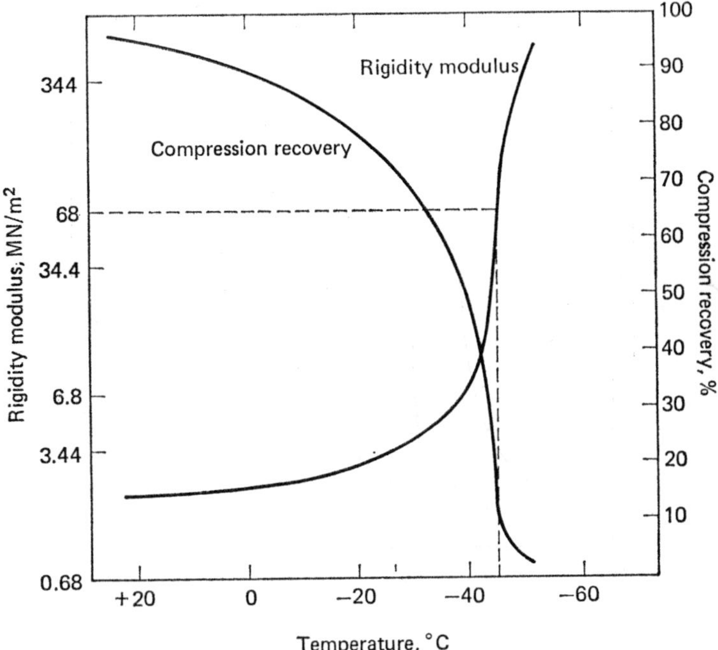

Fig. 2.3.4   Effect of low temperature on the rigidity modulus and compression recovery of a nitrile rubber

Typical changes in stiffness over a low temperature range are shown for a number of rubbers in Fig. 2.3.5. The curve for silicone rubber also shown in Fig. 2.3.5 is that of an 'extreme low temperature' grade which is certainly the best of the commercially available rubbers in this respect. Other types of silicone rubber will show marked stiffening at about 70 °C. Two curves for nitrile rubber are shown, the low swell (high acrylonitrile) type which has poor low temperature characteristics and the high swell (low acrylonitrile) type which indicates about the best which can be achieved with this rubber. Anything between these two extremes can be produced by suitable blending and compounding.

Resistance to fluids

Resistance to fluids on all materials would require a very long chapter and only a general review can be given. For convenience, resistance will be considered under 'Chemical resistance' below.

Chemical resistance

The effect of corrosive chemicals, mineral acids and strong alkalis for instance is to cause progressive surface degradation of rubber, and in some cases, eventual dissolution, rather than swelling.

I

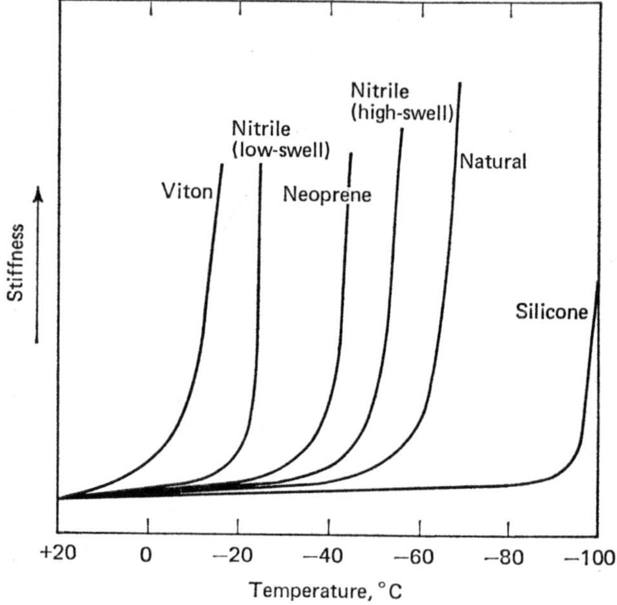

Fig. 2.3.5   Effect of low temperatures on the stiffness of various rubbers — typical

The major advance in chemically resistant rubbers in the last decade has been the introduction of the fluorinated elastomers, notably Viton. The chemical similarity to PTFE has been noted earlier and so this resistance is not unexpected, but unlike PTFE, Viton can truly be classified as a rubber and so its range of application is increased accordingly. Viton will resist all the mineral acids up to high concentrations, including red fuming nitric acid as used for rocket fuels, but not necessarily at elevated temperatures. At lower concentrations or in less corrosive media, butyl is superior to other types of rubber and in view of its low cost compared with Viton it is widely used for less arduous chemical sealing duties.

As far as organic solvents are concerned, no generalisation is possible, but it is of interest that the fluorinated rubbers have enabled us to seal effectively, probably for the first time with rubber-based seals, the aromatic solvents, benzene and toluene, etc. and the chlorinated solvents, trichlorethylene and perchlorethylene.

Oil resistance

Resistance to mineral or petroleum based oils is an extremely important requirement for the great bulk of hydraulic seals and is normally assessed from measurements of volume change after an appropriate period of immersion. In addition, changes in mechanical properties, tensile strength for instance, following immersion may be used in conjunction with volume change to assess suitability.

The swelling of rubber in hydrocarbon oils is a physical phenomenon and, indeed, if the swelling agent is volatile, reversible, the change in mechanical properties bears a fairly definite relationship to volume change. In Fig. 2.3.6 is shown the typical effect of swelling upon the strength of a nitrile rubber and it will be noted that swelling of 20 per cent by volume causes a 50 per cent drop in strength.

Where seals are concerned, the maximum swelling which can be tolerated will vary with the application. However, in the case of dynamic or moving seals it is considered that between 15 per cent and 20 per cent swelling (by volume) is the practical limit. Above this, the dimensional change coupled with the marked decrease in strength and wear resistance which accompanies swelling will give an unacceptably short life. In Fig. 2.3.7 typical time/swelling curves are shown for a number of rubbers in two classes of oil.

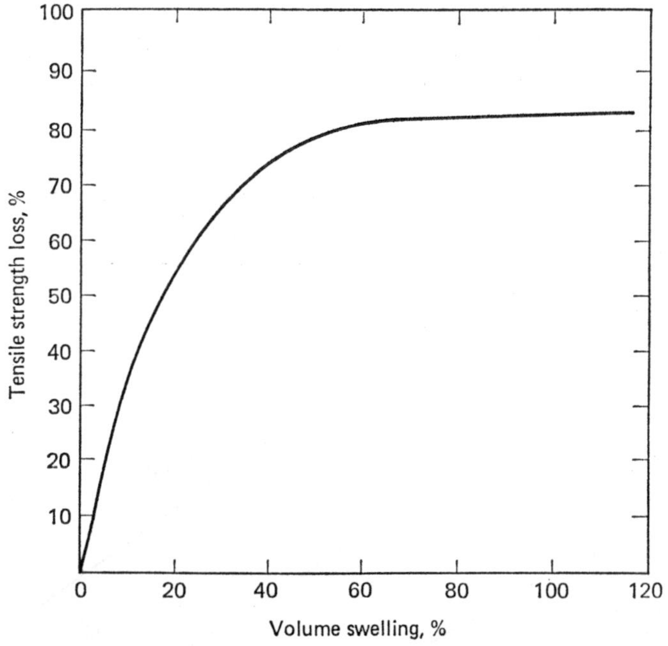

Fig. 2.3.6   Typical effect of swelling on strength of nitrile rubber

The swelling of natural and butyl rubber is such that they are quite unacceptable as seal material for use in mineral oils.

In the case of the four other rubbers shown in Fig. 2.3.7, there is a marked difference in swelling behaviour according to the type of oil. As a generalisation, the lower viscosity oils, as are used as hydraulic fluids, will normally have a greater swelling effect on rubbers of all types than the more viscous lubricating oils. This arises from the higher aromatic content of the low-viscosity oils, and it is the aromatic component of mineral oil which is largely responsible for the swelling of the so-called oil-resistant rubbers.

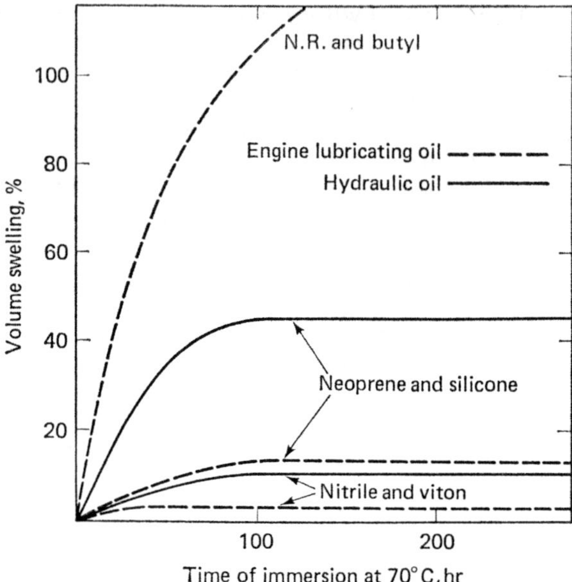

Fig. 2.3.7   Effect of type of oil on the swelling of rubbers

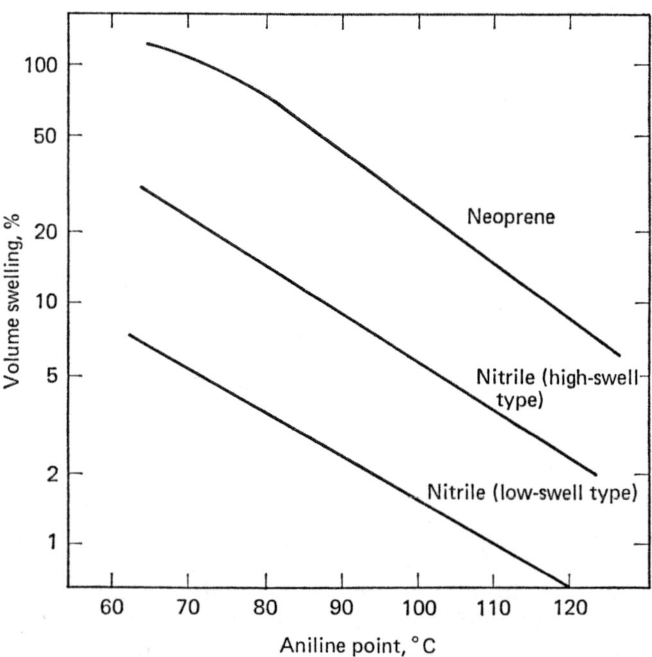

Fig. 2.3.8   Effect of the aniline point of oil on the swelling of neoprene and nitrile rubber at 70 °C

The effect of type of oil is particularly apparent in the case of neoprene and silicone rubbers and these can only be used with confidence in contact with oils of low swelling power. Nitrile and Viton on the other hand can be used with a much wider range of oil types.

In the case of the more oil-resistant rubbers, nitrile particularly, it is possible for shrinkage rather than swelling to occur and this shrinkage arises from the extraction of oil soluble materials which may be incorporated in the rubber, to improve low temperature properties for instance. Shrinkage may be expected particularly when rubbers designed for use in hydraulic oils at low temperatures are used in contact with oils of low swelling power such as engine lubricating oil.

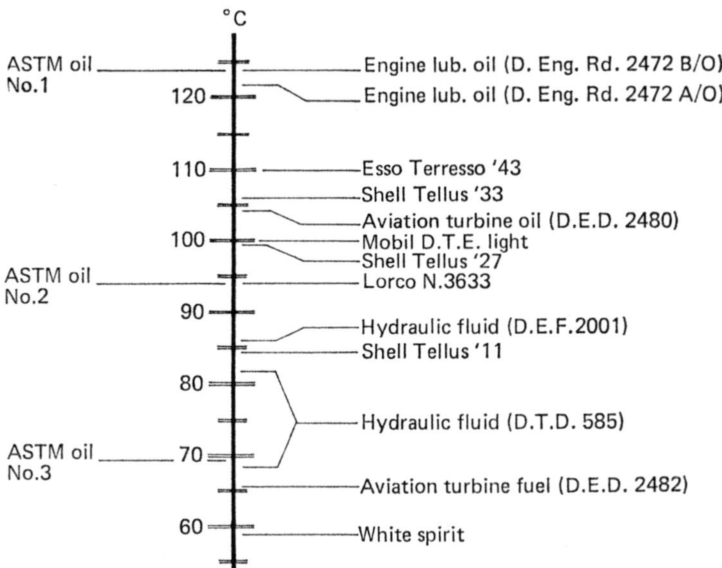

Fig. 2.3.9   Typical aniline points of some British and American oils

The viscosity of an oil gives no more than a rough indication of probable swelling characteristics, and a more reliable and precise relationship has been between the aniline point and swelling power. The aniline point (A.Pt), the temperature at which equal volumes of aniline and the oil are miscible, is inversely related to aromatic content and hence, is similarly related to swelling power. The aniline point may, therefore, be regarded as a reliable index to swelling characteristics; a low A.Pt. oil invariably giving rise to a greater swelling in a given rubber than a higher A.Pt. oil.

It has been found that a logarithmic plot of volume swelling against A.Pt. is approximately linear and typical curves of neoprene and nitrile rubbers are shown in Fig. 2.3.8 and typical aniline points of a number of commercial and specification oils are given in Fig. 2.3.9.

In general, one can say that the swelling of neoprene is likely to be excessive in oils of A.Pt. less than 100 °C, but that nitrile rubbers can be so compounded to work in oils of much lower A.Pt. — down to below 50 °C. For a particular nitrile rubber however, the useful range of A.Pt. over which satisfactory working can be expected is probably not more than 25–30 °C. In oils of higher A.Pt., shrinkage may occur and, in lower A.Pt. oils, swelling could be excessive.

It is important that both seal users and hydraulic equipment manufacturers should appreciate that grades of oil cannot safely be changed without reference to the effect on seals, and the aniline point is suggested as a convenient and reliable means of characterising swelling power.

The selection of a rubber material for a given duty is a complex problem involving, in many cases, compromise in which one property is, in part, sacrificed to achieve another. For the successful solution of this problem the co-operation of the rubber technologist, the seal designer and the equipment designer is essential and an appreciation of the possibilities and limitations of seal materials by all concerned is desirable.

It is the author's hope that this chapter will promote a better understanding of some of the factors involved in the development and use of rubber seal materials.

Chapter 2.4   **Potting shells for connectors**

Moulds are no problem in the vacuum-forming of shells from polypropylene; the low temperatures and pressures obtaining permit the use of wood, plastic, etc., for connectors, but aluminium is best.

Many materials make good vacuum-formed potting shells, but after initial tests, polypropylene is found to be the most versatile, it has a combination of properties which make it almost ideal. It is tough and strong, yet easy to cut and trim. It is rigid enough to hold its shape during the filling and curing of the potting compound, yet the slight degree of flexibility allows easy removal of the shell after curing. It vacuum-forms extremely well, reproducing all surface detail of the pattern, and can be readily deep-drawn with a minimum of webbing. Shrinkage is uniform, non-directional and predictable. While not crystal clear, it is transparent enough to allow visual observation of the filling operation, which helps to eliminate voids and bubbles. Polypropylene has the characteristic non-stick property, common to all olefins and fluorocarbons, which eliminates any need for parting agents; it is chemically inert, not reacting with, or affected by, potting compounds or solvents.

In cases where a large mass of highly exothermic epoxy is required, the resultant temperatures can exceed the softening temperature of polypropylene. In these cases the potting shells can be vacuum-formed from polycarbonate, or polysulphone.

Tooling

Vacuum-forming tooling or patterns, can be made of almost any material. The moderate temperatures and low pressures of the process allow the use of patterns made of wood, plastic, aluminium, rubber, plaster, etc. For potting shell patterns, almost exclusive use of aluminium is adopted. Aluminium has unlimited life, excellent heat transfer and is very easy to machine.

Where a slight undercut is required, or no draft allowed, silicone rubber is used. The flexibility of the rubber allows the pattern to be easily removed from the formed part by the use of compressed air.

Design techniques

Vacuum forming can be most easily accomplished if the patterns have been properly designed. There are several requirements for all vacuum-forming patterns.

Draft angle

A draft angle should be allowed. A minimum of two degrees is the best practice, with greater angles if the design permits.

Vacuum bleed holes

These should be drilled in the pattern at all locations that will trap air when the plastic sheet is formed over the pattern. Large flat surfaces on the top of the pattern, inside corners, and the base of flanges all require bleed holes. For polypropylene, 1·270 mm diameter holes are ample. The bleed holes if very long, can be counterbored from the base for easier drilling. The counterbored hole is also handy for using compressed air to blow off the finished part. Generous corner radii are recommended for stronger parts and better appearance.

Shrinkage allowance

All vacuum-forming patterns must be slightly larger than the desired size of the finished part. *Polypropylene* has a shrinkage of 0·18 mm when formed over a male aluminium pattern; female patterns have slightly more shrinkage.

Two-piece patterns

As a vacuum-forming pattern must be removed from the finished part without damage to the part, two-piece moulds are frequently required for angled or undercut potting shells.

Trimming considerations

Trimming of the finished part can be the most expensive step in the vacuum-forming process, and much thought should be given to pattern design for this operation. Trimming should ideally be accomplished by shearing, this is the fastest, and cheapest method. The trimming location should not be required to hold a dimensional tolerance, eliminating fixturing during trimming.

Vacuum-forming

Heating cycle

The heating cycle is the most important phase of vacuum-forming. Accurate timers and a uniform source of heat are required for good work. Most common plastics require about 9 kW/m² of power for fast-forming cycles. With this amount of power, the heating time is approximately 1 sec per 0·0254 mm of thickness, subject to the following variables:

1. Colour of sheet stock.
2. Thickness.
3. Type of plastic.

4 Distance of plastic from heater.
5 Output of heater.

Cycle times are determined by trial and error. With some experience, the time can be determined by the use of only one sheet. With the machine on manual setting, the heater is placed over the plastic and timed with a stopwatch to determine the ideal point. Most plastics will expand and wrinkle first, then smooth out as they become hotter, flatten to a sheet, and then sag uniformly. The forming is accomplished at the start of the sagging stage. This is the point when the sheet is most elastic and webbing is minimised. Highly plasticised materials will 'blush' or appear milky if forming is attempted before the proper temperature is reached, and almost all plastics will blister if the heating time is excessive. Polycarbonate and polysulphone can be quite accurately controlled by the amount of sag. These high-temperature-forming materials require considerably more heat than most plastics. The sheet plastic should never be allowed to sag so far as to touch the pattern before the vacuum is applied. They will cool and harden immediately and unsightly wrinkles will result. If a considerable sag is required to obtain the proper temperature, e.g. with polycarbonate, the frame must be raised higher from the pattern, or the sheet may be supported outside of the pattern area to prevent touching.

Vacuum

In almost all cases it is the best practice to apply the full vacuum immediately after the heating cycle. In most machines, the sheet frame drops simultaneously with the application of vacuum. This assures immediate and uniform cooling of the formed part. The vacuum should be maintained until the cooling cycle is completed. Control of the degree of vacuum is desirable in some cases and most machines have the capability for presetting the vacuum to 127 to 2 mmHg. Polyethylene and polypropylene in particular are very weak at the forming temperature and immediate application of high vacuum will perforate the sheet before the forming is complete.

When forming these materials, the heat should be removed several seconds prior to the application of vacuum: it should be remembered that the degree of vacuum will determine inside corner radii and amount of detail obtained from the pattern surface.

Cooling cycle

The cooling cycle is not critical, but should be long enough for the plastic sheet to cool below its heat distortion point. This is approximately equal to the heating cycle. High-temperature materials such as polycarbonate and polysulphone cool very rapidly and can be removed from the pattern in 10 to 20 sec.

Heating of patterns

For the best results, the patterns should be heated to a temperature somewhat below the heat distortion point of the plastic. While this increases the cooling cycle, slowing down the production rate, it markedly improves the final part. Warping of flat sections is reduced, thin-out is reduced and closer dimensional tolerances can be maintained. When continuous production is maintained, no external heating of the patterns is usually required, as the heat transfer from the hot plastic sheets is sufficient. When very heavy gauge sheet is being formed, some cooling may be required for fast production cycles. Webbing is the characteristic formation of wrinkles on vacuum formed parts. The amount of webbing is a function of several variables including type of plastic, depth of draw, corner radii on patterns, heating cycle, and general pattern geometry. Webbing can be reduced or eliminated by reducing the heating cycle, large corner radii, ring assist forming techniques, or using anti-webbing 'bumps'.

Webbing is caused by an excess of plastic on corners, i.e. the plastic sheet is not kept in tension during the vacuum-forming operation. The condition is aggravated by excessive heating of the sheet, and a large amount of sag just prior to the forming cycle. Certain types of plastics tend to remain in tension during the heating cycle with little or no sag; these plastics — ABS, high-impact styrene and vacuum-forming grades of polycarbonate primarily — will normally exhibit much less webbing than polyethylene, polypropylene or standard-grade polycarbonate.

For vacuum-formed potting shells, the question of webbing is usually academic, because male patterns are generally used, and any webbing which occurs serves to stiffen the part.

Separation of part from pattern

If ample draft angle is present on the pattern, part removal is very easy. However, some designs will require vertical side walls or very deep draws, and the shrinkage of the plastic tends to make the pattern removal more difficult. Usually a blast of compressed air into the vacuum bleed holes in the bottom of the pattern will suffice to remove the part. In other cases, a tapped hole in the bottom of the pattern allows a handle to be attached to the pattern after forming. For long production runs, the pattern should be secured to the vacuum platen on the machine, and a built-in compressed air 'blow-off' utilised for part removal.

Use of potting shells

The simplest method of using the potting shell is to design it to press-fit over the back of the connector. If properly designed, with an interference fit of 0·0254 to 0·0508 mm, the shell will fit very snugly and no sealing will be required prior to encapsulating. This type of shell, however, must be assembled onto the wire bundle prior to attachment of the connector to the wire ends.

In other cases a split type of potting shell can be used. The two halves of the shell are aligned and secured with several small screws or clamps.

Fill and bleed holes

The potting shells should be drilled with fill and bleed holes in locations that minimise entrapment of air. One advantage of polypropylene shells is that the transparency is such that the filling operation is visible.

Temperature limitations

It is recommended that polypropylene, the most satisfactory material for general use, should be kept below 72 °C during encapsulation and cure. This ensures that the shell retains most of its original rigidity and strength, and good dimensional repeatability results. Therefore, highly exothermic materials cannot be used in large moulds.

Most exotherm problems can be overcome by either using multiple pour techniques, or by allowing gelation to take place at room temperature before even curing. If 72 °C is not exceeded, the polypropylene shells can be re-used several times, provided the design allows the removal of the shell without cutting.

Assessment and limitations

Vacuum-formed potting shells have been demonstrated to have the following advantages:

1. Low cost.
2. A saving of many man-hours over matched-metal mould techniques.
3. Design flexibility and ease of incorporating design changes.
4. Potting shells can be made in quantity, from one pattern, allowing many harnesses to be encapsulated at one time. For a comparable production rate with metal moulds, several sets would be required, greatly increasing tooling costs.
5. No mould release is required when polypropylene is used.
6. Filling can be observed visually, to eliminate trapped air.
7. Harnesses can be encapsulated in place during assembly, after passing final electrical checkout.

The disadvantages are:

1. Somewhat greater dimensional tolerances are required. In most cases 0·762 mm is required for the economical use of plastic potting shells.
2. Flatness is difficult to maintain, due to the slight flexibility of the polypropylene.
3. Variations in surface appearance are noted from batch to batch. Due to surface 'mark-off' during vacuum-forming a consistent high-gloss finish is difficult to maintain.

Chapter 2.5 Fire-retardant thermoplastic and thermosetting materials

Each year, in plastics technology, one sees many advances useful to electronic packaging and production. Increased emphasis has been placed on fire-retardant products, improvements have been made in basic plastics, and elastomers and their formed products have also undergone extensive development. This chapter deals with some of the more significant of these.

Fire-retardant plastics

This is an area which will continue to receive much attention. Tragedies caused by fire have resulted in those with responsibilities for fire safety making increased demands for fire-retardant materials in end-product designs. The electronic plastics industry has responded vigorously in this respect in the development of improved products. However, to keep abreast, every electronic engineer should acquaint himself with the tenets of flame-retardancy, the compromises required of other factors in using fire-retardant plastics and the materials available to him.

The burning of a piece of plastic is a rather complex physical phenomenon. The rate of burning depends on the size, shape and temperature of the sample, as well as the amount of ventilation. Furthermore, the method of ignition, type of gas, height of flame and other factors in the environment such as the proximity, temperature and emissivity of other objects are important. For the purpose of defining flammability of plastic materials the mechanics of sustained burning might be simply described as follows. Vapour is generated thermally from the condensed phase and diffuses into the ambient air until a flammable composition is produced. At this point, combustion occurs. Heat then feeds back from the combustion zone to the condensed phase to generate additional gaseous fuel which sustains the flame. When a flame-retardant is added to a resin mix and the latter is subjected to a flame, the gases produced absorb heat without contributing energy and the flame temperature is reduced to a point too low to maintain the chain of reactions needed for combustion.

Fire-retardancy in plastics is achieved by two primary routes:

1 The blending of fire-retardant additives into plastics.
2 Selecting plastics which are inherently fire-retardant.

Examples of inherently fire-retardant plastics are polyimides, fluorocarbons, polyvinyl chloride and polyvinylidene fluoride. These examples are not sufficient in scope to meet all needs, hence, the use of fire-retardant additives is widespread. Their use has drawbacks in electronics and frequently requires design compromises. Fire-retardant modifications are usually more expensive than the use of conventional plastics, and are often more costly so process because moulding cycles are necessarily slowed. Further, additives may make the final composition less safe to mould and some processors are, therefore, reluctant to mould fire-retardant plastics.

Besides added cost, there is often another serious drawback, especially for electronics. Specifically, desirable properties may be seriously compromised to achieve flame retardancy. Service temperatures may be dangerously low. Permittivities and dissipation factors are usually increased by common additives, often showing dramatic increases above 100 deg C. To reduce density, casting resins can be mixed with hollow glass spheres. These spheres are claimed to be lightweight, non-absorbent, non-flammable with a low heat capacity and thermal conductivity.

Although it would seem wise, it is not always possible to use plastics which are inherently fire-retardant. Rigid polyvinyl chloride is fire-retardant but usually too low in flexibility. The addition of plasticisers, required for flexibility, can yield a material which is not fire-retardent owing to reduction of the chlorine content in the plasticised vinyl material. However, it is possible to react or blend polyvinyl chloride with other plastics, such as polypropylene to produce a self-extinguishing end-material.

A new type of lightweight rigid urethane foam with high flame-resistance has been developed. The new foam shows great promise as an industrial fire protection medium, especially for fires fed by volatile fuels and other highly inflammable material used in industry. This is achieved to increase compressive strength by adding glass and carbon fibres, making the material useful for structural and mechanical cushioning, as well as for fire prevention.

Basic plastics and elastomers

Advances in basic plastics and elastomers continue to be made with regularity; not only has much work been done in developing fire-retardant or self-extinguishing properties, but much work has been done in developing plastics with improved performance, and improved processing capabilities. Processing problems can and do frequently hinder the use of advanced engineering plastics. Often, for instance, higher thermal performance of plastics is accompanied by poor moulding properties. This, coupled with fire-retarding factors has recently been a major development area.

Thermoplastics and glass-reinforced thermoplastics

These materials, amenable to conventional injection moulding, have allowed thermoplastics to gain on thermosetting compounds. More recently, however, much development has been underway on injection mouldable thermosetting plastics, such as cross-linkable polyolfin compounds and phenolics. In addition, there has been growth in liquid moulding, wherein liquid thermosetting plastics are moulded more rapidly and economically and at lower pressures than powdered or pelleted thermosetting materials. There have also been other recent thermosetting developments, based on polybutadienes.

Further, the widely used epoxy-glass laminate has yielded improvements in thermal stability.

Cross-linkable polyolefins

A new family of these compounds has been recently developed, intended primarily for injection moulding. They are claimed to be suitable also for use in compression and transfer moulding. Only a few compounds have been thus far made available. However, a broad range of compounds is expected.

The bonding of parts with adhesive offers advantages, but there are limitations. For effective use, a choice must be made from the various classes of adhesives, and then the best individual formulation must be identified.

Adhesive bonding offers many design advantages. Holes are not required for an adhesive bond, so that the high stress concentrations around holes are eliminated and the stress is uniformly distributed along the bonded area. In addition, there are no projections or irregularities on the surface above and below the bond. The stress-distributing characteristics and inherent flexibility of adhesives provide bonds of superior fatigue resistance. Warping and burning associated with welding and brazing are eliminated because almost all adhesives are cured below 204 °C, and many adhesives cure at room temperature. Adhesives do not conduct electricity and this may prevent galvanic corrosion when dissimilar metals are bonded.

Frequently, adhesives perform multiple functions. In addition to providing physical strength, an adhesive may function as a sealant, an electrical insulator, a thermal barrier, or a gap filler.

Design limitations

For optimum design the advantage of adhesives must be weighed against certain limitations. The operating-temperature range is the most serious limitation to bonding. There are organic adhesives that perform well between 51 and 80 °C, and a few will withstand short excursions to 260 °C. The strength of organic adhesives at elevated temperatures is time-dependent. If the operating temperature is lower, the useful life of the adhesive is increased. However, it is generally agreed that oxidation rather than thermal destruction is the primary mechanism of adhesive degradation.

The choice of an adhesive depends on the manner in which it is to be stressed; rigid adhesives perform best when the adhesive is stressed in shear. Conversely, flexible adhesives must not be loaded in sheer, but perform best when subject to peeling or cleavage stress. Since it is rarely possible to isolate loading conditions a compromise between flexible and rigid adhesives must be made. Flexible adhesives are more susceptable to creep under stress than rigid adhesives and a compromise is required to achieve the best creep performance.

Production limitations

Several problems must be considered when bonding operations are projected. Automatic mixing, metering and dispensing equipment may be required if two-component adhesives are considered. Some adhesives require a short drying period at an elevated temperature to remove the solvent before assembly and cure; other adhesives must be cured under pressure. This requires either simple clamping devices or, in extreme cases, hydraulic presses. Adhesives are composed of materials that may present personnel hazards including dermatitis, in which case the necessary precautions must be considered

Adhesion

Adhesion is the result of molecular forces that exist between materials. Mechanical interlocking due to surface roughness provides a very small portion of the strength in an adhesive bond.

The surface of a solid material is never truly smooth and rarely has the same composition as the bulk of the material. The surface layers of solids are a maze of peaks and valleys, composed of a mixture of the bulk material, absorbed gasses, and airborne impurities. In order to function properly, the adhesive must penetrate the crevices on the solid's surface and displace the adsorbed gases. This means that dirt and grease must be removed so that a clean surface is presented to the adhesive. The process of establishing continuous contact between an adhesive and the base material is called 'wetting'. After intimate contact the molecular and electrical forces establish the bond, and provide the greatest portion of the strength achieved in an adhesive bond.

Primary valence bonds also contribute to the ultimate bond strength. The theoretical strength possible as a result of these forces is never achieved because of incomplete wetting and internal stresses present in the adhesive. These stresses occur as a result of shrinkage during the curing process and differences in coefficients of thermal expansion between the adhesive and the base material.

Adhesive materials

Adhesives may be classified according to either function or composition. The functional classification includes structural and non-structural adhesives.

Structural adhesives are materials of high strength and integrity that bond structures which support a substantial load; non-structural adhesives are not required to support great loads but simply keep lightweight materials in place. The composition classification describes adhesives as thermosetting, thermoplastic elastomeric or combinations of these.

Another distinction possessed by adhesives is the manner in which they cure or harden. Some adhesives cure by simply losing solvent while others harden as a result of a chemical reaction that provides a highly cross-linked molecular structure. In others, solvent evaporation and chemical cross-linking occur simultaneously.

For easier assimulation of this subject I propose to classify adhesives according to composition.

Thermosetting adhesives

Thermosetting adhesives are materials that cannot be heated and softened repeatedly after the initial cure. Adhesives of this sort may cure at room or elevated temperature. Some thermosetting adhesives are in a solvent medium to facilitate application while others are solventless liquids. Still others are supplied as solids. Substantial pressure may be required with some thermosets, and others will provide strong bonds with contact pressure.

Epoxy

Epoxy-resin adhesives are probably the most versatile bonding materials available. These adhesives result in highly cross-linked molecular structures with excellent tensile-shear strength, but poor peel or cleavage strength. Low cure shrinkage and high resistance to creep under prolonged stress are characteristic of epoxy adhesives. Their resistance to moisture and solvents is outstanding.

Epoxy adhesives exhibit a high degree of specific adhesion to glass, metals, and thermosetting plastic laminates. However, adhesion to untreated polyethylene, fluorocarbons, silicones, highly plasticised thermoplastics, and butyl rubber is very poor. Epoxy adhesives are supplied as 100 per cent solids or diluted with a solvent. The viscosity ranges from sprayable solvent systems through thixotropic pastes to solid rods and powders. Also, epoxy adhesives are available as unsupported films or in combination with glass-cloth and synthetic-cloth fibre substrates.

The wide utilisation of epoxy-resin adhesives is the result of a vast number of variations that can be made. An epoxy adhesive is primarily composed of an epoxy resin and a curing agent. Secondary ingredients include reactive diluents to adjust viscosity; mineral extenders to lower cost, adjust viscosity, and modify the coefficient of thermal expansion; fibrous fillers to provide viscosity adjustment and improve cohesive strength; and other resins to provide toughness, impact strength, and high peel strength and otherwise modify the cohesive and adhesive properties.

The epoxy resins available are the products of the reaction of bisphenol A and epichlorohydrin, the peroxidation of polyolefins or polybutadiene with peracetic acid, or the reaction of novalac resin and epichlorohydrin.

There are many polymers in each type of epoxy resin. Epoxy resins derived from mixtures of bisphenol A and epichlorohydrin are most common. The peroxidised polyfins have not been used as extensively as the resins just described. However, these materials are rather new. The cured unmodified resins are hard, brittle, and relatively unsuitable for adhesive bonding. Useful adhesives and sealants have been made by reacting the peroxidised polyolefins with long-chain essential oils such as oiticica oil.

Peroxidised polybutadiene resins are also quite new. These materials are unique because they can be cured not only through the epoxide group with anhydride curing agents but also at the vinyl linkage with peroxide catalysts. This combination provides adhesives with relatively high thermal resistance.

Epoxidised novalacs are either viscous semi-solids or low-melting-point solids; these resins are usually cured with acid anhydrides, and adhesive

K

bonds made with them have high thermal resistance, good tensile-shear strength and poor peel strength.

Epoxy curing agents
The epoxy resins derived from bisphenol A and epichlorohydrin are those most commonly used in epoxy adhesives. There are three types of curing agents for these materials:

1 Polyfunctional primary and secondary amines provide room-temperature-curing adhesive systems. The adhesive is supplied as two components, the resin and the hardener, because the mixed materials will cure in a short time at room temperature.

2 Anhydrides of organic acids are used to cure epoxy resins at elevated temperatures. Longer pot lives, often in excess of 24 hr can be expected with anhydride curing agents. Epoxy adhesives cured with anhydrides generally have better bond strength at elevated temperatures. Tertiary amines may be used to accelerate the cure of anhydride-hardened adhesives.

3 Latent curing agents provide extremely long shelf lives to single-component epoxy adhesives. Shelf lives in excess of six months can be expected under proper storage conditions. Boron trifluoride complexes are one type of latent hardener that cures the adhesive when it is exposed to heat. Molecular sieves loaded with primary and secondary amines cure epoxy resins when the amine is driven from the molecular sieve with either heat or moisture. Dicyandiamide is a solid, latent curing agent that is inert until the adhesive is heated to 175 °C.

Some of the curing agents available for epoxy resins are described in Table 2.6.1 and Table 2.6.2. describes the bond strengths achieved with various epoxy-resin curing agents. The values are somewhat lower than can be expected of an adhesive containing fillers, diluter, and other modifiers.

Fillers perform several functions in epoxy adhesives. First, they decrease the exotherm from the cure cycle thus reducing the thermal stress induced in the adhesive bond. As illustrated in Fig. 2.6.1 fillers help to match the coefficient of thermal expansion of the adhesive to that of the base material. Viscosity is adjusted with the addition of fillers, and cure shrinkage is decreased. Chemical resistance and electrical properties may be adjusted with the use of the proper filler. The effect of various fillers on the tensile-shear strength of epoxy adhesives is shown in Fig. 2.6.2.

Some fillers will either inhibit or accelerate the cure of epoxy adhesives. This phenomenon can be highly specific. Consequently the combination of filler and curing agent should be investigated. For example, hydrated fillers, i.e. aluminium oxide trihydride, will accelerate the cure rate of anhydride curing agents. Liquid modifiers for epoxy resins may be used to lower viscosity, improve flexibility or toughness, or improve adhesion. Usually any advantage gained is to the detriment of another property. For example, most diluters tend to lower the thermal stability of adhesives. Some liquid modifiers perform several functions polyamide resin and polysulfide rubber function as both flexibilisers and curing agents. Triphenyl phosphate lowers viscosity,

Table 2.6.1 **Characteristics of curing agents with epoxy resins in adhesive formulations**

Curing agent	Physical form	Amount requirement*	Cure temperature, °C	Pot life, 23 °C†	Completion of cure conditions	Maximum temperature of use, °C
Triethylenetriamine	Liquid	11–13	21–135	30 min	7 days at 24 °C	71
Diethylenetriamine	Liquid	10–12	21–93	30 min	7 days at 24 °C	71
Diethylaminopropylamine	Liquid	6–8	29–149	5 hr	30 min at 24 °C	85
m-Phenylenediamine	Solid	12–14	65–204	8 hr	1 hr at 85 °C	148
Methlenedianiline	Solid	26–30	65–204	8 hr	2 hr at 162 °C	148
Boron trifluoride monoethylamine	Solid	1–4	135–204	6 months	3 hr at 165 °C	162
Methylnadic anhydride	Liquid	80–100	121–37	5 days	3 hr at 160 °C	162
Triethylamine	Liquid	11–13	21–93	30 min	7 days at 24 °C	82
Polyamides:						
Amine value 80–90	Semi-solid	30–70	21–149	5 hr	5 days at 24 °C	‡
Amine value 210–230	Liquid	30–70	21–149	5 hr	5 days at 24 °C	‡
Amine value 290–320	Liquid	30–70	21–149	5 hr	5 days at 24 °C	‡

* Per 100 parts by weight; for an epoxy resin with an epoxide equivalent of 180–190

† 500 per batch; with a bisphenol A-epichlorohydrin derived epoxy resin with an epoxide equivalent of 180–190

‡ Highly dependant on concentration

Table 2.6.2 **Influence of curing agent on the bond strength obtained with various base materials**

Curing agent	Amount*	Cure cycle, hr at °C	Polyester glass-mat laminate	Tensile strength, MN/m²				
				Polyester glass-cloth laminate	Cold-rolled steel	Aluminium	Brass	Copper
Triethylamine	6	24 at 23 ; 4 at 65	12·75	14·47	16·9	12·48	12·16	3·82
Trimethylamine	6	24 at 23 ; 4 at 65	7·27	9·99	9·54	10·6	10·5	12·0
Triethylenetetramine	12	24 at 23 ; 4 at 65	7·9	11·2	9·8	11·5	11·2	9·1
Pyrrolidine	5	24 at 23 ; 4 at 65	8·6	11·6	8·9	11·2	11·25	9·79
Polyamidamine	35–65	24 at 23 ; 4 at 65	8·27	9·9	16·13	21·5	13·8	12·9
m-Phenylenediamine	12·5	4 at 167	5·3	4·4	14·8	15·5	14·8	11·3
Diethylenetriamine	11·0	24 at 23 ; 4 at 65	6·9	7·7	9·3	9·7	7·8	8·5
Boron trifluoride monoethylamine	3·0	3 at 190	—	—	11·9	12·9	10·5	11·2
Dicyaniamide		4 at 176	3·6	2·97	18·4	19·2	18·1	17·5
Methylnadic anhydride	85	6 at 176	4·1	5·2	15·7	14·9	13·4	12·6

* Per 100 parts by weight of resin

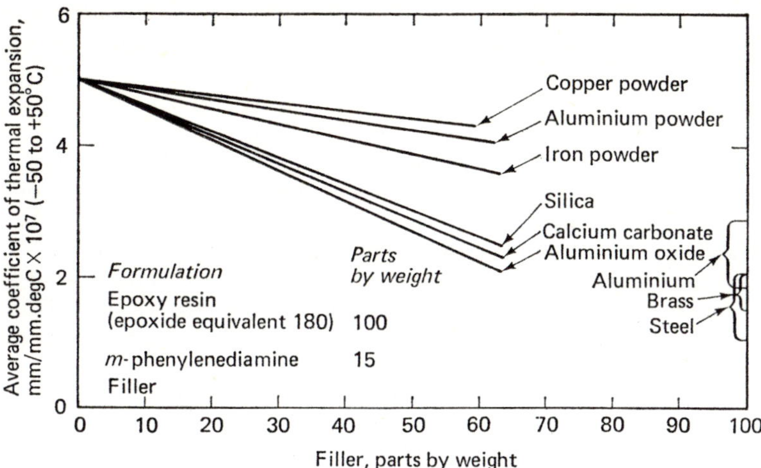

Fig. 2.6.1   Coefficients of thermal expansion of filler epoxy resins compared with those of the common metals

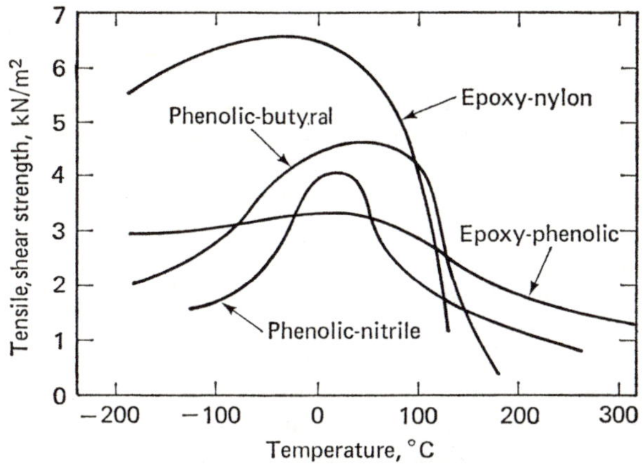

Fig. 2.6.2   Effect of mineral extenders on the tensile shear strength of epoxy adhesives: base material — aluminium

Table 2.6.3   **Summary of liquid modifiers for epoxy-resin adhesives**

Modifier	Function
Butylglycidyl ether	Lower viscosity, permits higher filler loadings
Phenylglycidyl ether	Lower viscosity, permits higher filler loadings
Triphenyl phosphite	Lower viscosity, improves adhesion, accelerates cure
Polyamide resins	Flexibiliser, curing agent
Polysulphide rubber	Flexibiliser, curing agent
Polyester resins	Lower viscosity (sometimes), improves flexibility
Vinylcyclohexane dioxide	Lower viscosity
Polyisobutylene	Tackifier

improves adhesion to aluminium, and accelerates cure. A summary of liquid modifiers for epoxy-resin adhesives is presented in Table 2.6.3.

Polyurethane

Polyurethanes are the reaction product of a hydroxylated polyester, poly-ethers, or other polyols and a polyfunctional isocyanate. These materials have excellent wetting ability, and the highly reactive isocyanate groups combine readily with many functional groups. The isocyanate group reacts with surface moisture and hydrated oxide layers on metals. The result is a clean metal surface that permits intimate adhesive contact. The highly polar hydroxylated polyesters, polyethers and polyols facilitate wetting. The urea groups in the polyurethane molecule react with the functional groups on the surface of polymeric base materials.

Polyurethane adhesives are generally quite tough and somewhat flexible. Consequently, the bond has excellent peel strength and good tensile-shear strength. Excellent bonds are formed with polyethylene terephthalate, poly-methylmethacrylate, chlorinated polyether, polycarbonate, cellulose acetate, polyvinylchloride, polyester and phenolic resins. Polyurethane adhesives have been used extensively for bonding plastics and elastomers to aluminium and steel.

These adhesives are generally supplied as two-component systems without any inert fillers. Solvents or reactive monomers may be used to adjust viscosity. Fillers are avoided because it is difficult to obtain completely dry ones and they have a tendency to pick up atmospheric moisture.

Polyurethanes must be stored under dry conditions. The isocyanate group will react with atmospheric moisture and lower the reactivity and effective-ness of the adhesive. A proper handling procedure must be established and adequate ventilation provided.

The polyurethane adhesives have excellent tensile-shear properties at 100 °C. The tensile-shear strength decreases somewhat at extremely low temperatures. The effect of temperature on the peel strength and tensile-shear strength of two polyurethane adhesives is shown in Fig. 2.6.3. The primary difference between these two adhesives is the hydroxylated poly-ester; the same polyfunctional isocyanate was used in both cases. For simplicity let us categorise the adhesive according to their hardness at room temperature; the softer of the two adhesives had a Shore A hardness of 90. Mylar film 0·050 mm thick was bonded to 1·5 mm thick aluminium metal.

Elastomer adhesives

Elastomeric or rubber-based adhesives usually have excellent peel strength and low tensile-shear strengths. These adhesives are used down to −45·6° C, but soften and often decompose at −65 °C. Elastomeric adhesives are resistant to impact loadings and generally have excellent fatigue character-istics within the temperature limits outlined above. However, they will cold-flow or creep under load. The basic types of elastomeric adhesives are natural or reclaimed rubber, neoprene rubber, nitrile rubber, polyiso-butylene rubber and butyl rubber.

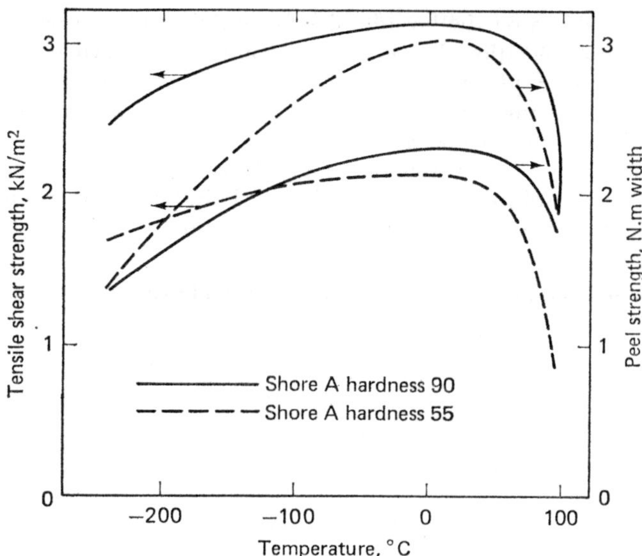

Fig. 2.6.3   Influence of temperature and hardness on the tensile shear strength and peel strength of polyurethane adhesives. Polyethylene teraphthalate was bonded to 1·6 mm thick aluminium

These adhesives may be supplied as solvent solutions, latex cements, distersions and pressure-sensitive tapes. The first three forms involve either dissolving (solvent solution) or dispersing solid particles of the rubber in a liquid, either an organic solvent or water. Tapes are made by placing the adhesive on a backing material such as rubber, vinyl, canvas, or cotton cloth. In addition to the solvents used, elastomeric adhesives may include fillers, antioxidents, tackifiers, pigments, plasticisers, wetting agents, curing agents and combinations of other resins and elastomers.

Numerous formulations are available. To make the situation even more complex, each formulation was often devised to fulfil a specific set of requirements. Table 2.6.4 indicates how and where these adhesives are used.

Thermoplastic adhesives

Thermoplastic materials can be repeatedly heated until they flow; they can then be cooled to the solid state without adversely affecting their properties, unless plasticisers, tackifiers or other modifiers are lost during the heating cycle. Thermoplastic adhesives are useful in the temperature range of −28 to +65 °C. The physical properties vary over a very wide range. Some formulations may be quite tacky and flexible, while others are very rigid. Vinyl adhesives are based on polyvinyl formal, or polyvinyl butyral.

These adhesives may be supplied as solvent solutions, water-based emulsions and hot melts. Usually not one, but a combination, of these polymers is used in a particular adhesive formulation. Consequently, defining properties and applications becomes quite complex. All of these adhesives are useful for bonding plastic films. The hot melt systems have been used extensively for sealing paper and plastic films. The solvent solutions and water-based emulsions are useful in bonding porous materials such as wood and plastic foams. The water-based emulsions have been used to bond materials that are soluble in organic solvents. Polystyrene foam is an example of a material that is soluble in many organic solvents.

Table 2.6.4   Properties and applications of elastomeric adhesives

Adhesive	Type	Cure	Applications	Advantages	Limitations
Reclaimed rubber	Solvent system, latex cement, dispersion, pressure-sensitive tape, viscous mastic	They may lose solvent or remain permanently tacky	Bonding paper, rubber, plastic and ceramic tile, plastic films, fibrous sound insulation and weather-stripping; also used for the adhesive on surgical and electrical tape	Low cost, applied very easily with roller coating, spraying, dipping, or brushing, gains strength very rapidly after joining, excellent moisture and water resistance	Becomes quite brit with age, poor resis ance to organic solvents
Natural rubber	Solvent system, latex cement, dispersion, pressure-sensitive tape	They may lose solvent or remain permanently tacky	Same as reclaimed rubber; also used for bonding leather and rubber sides to shoes	Excellent resilience, moisture and water resistance	Becomes quite brit with age, poor resi ance to organic solvents, does not bond well to metal
Neoprene rubber	Solvent system	Adhesive provides a useful bond after the solvent is removed; a better bond can be obtained if the adhesive is cured at room temperature with appropriate accelerators or heat	Bonding weather stripping and fibrous sound-proofing materials to metal; used extensively in industry; bonding synthetic fibres	Good strength to 65 °C fair resistance to creep	Poor storage life, high cost, small amounts of hydrochloric acid evolve during ageing that may cause corrosi in closed systems, poor resistance to sunlight
Nitrile rubber	Solvent system, latex cement, mastic, tape	They lose solvent; however, curing agents are used to promote crosslinking when maximum thermal, and oil resistance is required	Bonding plastic films to metals, and fibrous materials such as wood and fabrics to aluminium, brass, and steel; also, bonding nylon to nylon and other materials	Most stable synthetic rubber adhesive, excellent oil resistance, easily modified by addition of thermosetting resins	Does not bond we natural rubber or butyl rubber
Polyisobutylene	Solvent system	Solvent evaporation	Bonding rubber to itself and plastic materials; also, bonding polyethylene terephthalate film to itself, aluminium foil and other plastic films	Good ageing characteristics	Attacked by hydro carbons, poor ther resistance
Butyl	Solvent system, latex cement	Solvent evaporation and chemical crosslinking with curing agents and heat	Bonding rubber to itself and metals; forms good bonding with most plastic films such as polyethylene terephthalate and polyvinylidene chloride	Excellent ageing characteristics; chemically crosslinked materials have good thermal properties	Metals should be treated with an ap priate primer befor bonding; attacked hydrocarbons

Adhesive alloys

Adhesive alloys are made by combining two thermosetting resins or a thermosetting resin with either an elastomer or a thermoplastic material. These adhesives utilise the most useful properties of each material. However, the adhesive alloy usually exhibits properties between the maximum and minimum properties of the two materials. Thermosetting resins, being decidedly polar, provide a highly cross-linked, three-dimensional molecular structure with excellent adhesive strength. Impact, bending and peeling strength is provided by adding elastomers to thermoplastic materials. Obviously the degree to which a material will influence the properties of an adhesive depends on the amount of the material added to the formulation.

Adhesive alloys are structural materials. They are expected to support substantial loads with a minimum of creep. Frequently, these adhesives are exposed to elevated temperatures.

The most common adhesive alloys are phenolic epoxy, phenolic nitrile, phenolic butyral, phenolic formal and nylon epoxy. Phenolic resins have excellent thermal stability, but are quite brittle and unusable as adhesives. Hence the phenolic resins are combined with other materials to improve physical properties and handling characteristics. The tensile-shear properties of these alloys are illustrated in Fig. 2.6.4. Additional physical properties and applications are listed in Table 2.6.5.

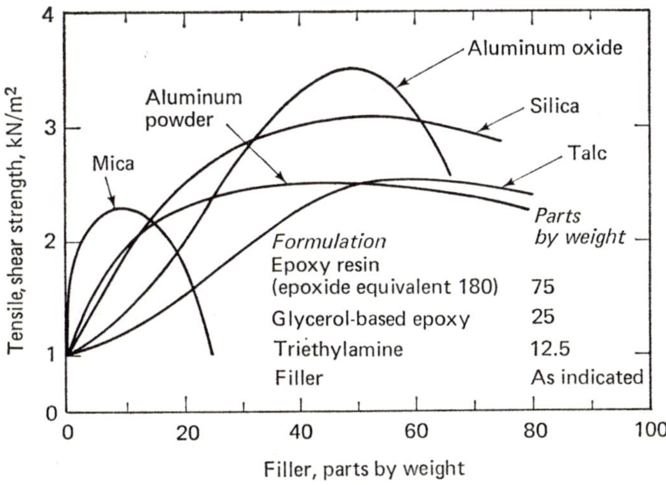

Fig. 2.6.4   Effect of temperature on the tensile shear strength of adhesive alloys: base material — aluminium

Silicone

Silicone resins are thermosetting materials. The silicon atom in the backbone of the silicone-resin molecule differentiates these resins from the organic resins previously discussed. Silicone adhesives have excellent heat resistance. Unfortunately, the initial strength is usually considerably less than that of organic materials such as epoxies and phenolic resin alloys.

Silicone adhesives are supplied as solutions, as a glass-cloth tape, on polytetrafluoroethylene tape, and as a self-sealing tape. All of these products are used in applications that call for superior heat resistance. The materials achieve maximum electrical, physical, and thermal properties only after being cured at elevated temperatures. Some of these materials require pressure during the cure cycle.

Room-temperature-vulcanising silicone adhesives form flexible bonds that withstand temperature in excess of 305 °C for several days. Their tensile shear strength is low, 2·75–3·45 MN/m², but their peel strength is relatively high (10–15 N/m width). Vulcanising silicone rubber cures in 24 hr at room temperature. Not all silicone rubber cures in this fashion, so that the RTV type (room-temperature-vulcanising) must be specified. Acetic acid is released during the cure cycle; consequently, corrosion of metals such as copper and brass may be a problem.

Table 2.6.5   Physical properties and applications of adhesive alloys

Adhesive	Type	Peel strength, N/m width	Applications	Advantages	Limitations
Phenolic butyral	Solvent solutions, unsupported and supported film	1·5–3·5	Bonding copper sheet to printed circuit board, metal sheets to each other, and metal and plastic skins to honeycomb core	Good impact resistance; fair creep and fatigue resistance; good solvent, oil, and fuel resistance	Solvent entrapment m degrade properties; u ful in temperature ran of 37–148 °C
Phenolic formal	Solvent solutions, unsupported and supported film	0·5–2·0	Same as phenolic-butyral	Better thermal properties than phenolic butyral; good solvent, oil, and fuel resistance	Solvent entrapment m degrade properties; u ful in temperature ran of 37 to 176 °C peel and impact strengths lower than phenolic-butyral
Phenolic nitrile	Solvent solutions, unsupported and supported film	2·5–4·0	Bonding brake linings, clutches, facings, metal to metal sheet, plastics to each other and to glass	Good solvent, oil, and fuel resistance; excellent resistance to creep and fatigue	Solvent entrapment m degrade properties; u ful in temperature ran of 37–121 °C, does not fillet well, hence h limited use in bonding honeycomb construc tions
Phenolic epoxy	Supported film	0·1–0·4	Bonding aluminium skin panels in high-performance aircraft	Excellent creep resistance; retains about 50 per cent of room-temperature strength at useful for short periods up to 370 °C	Limited shelf life
Nylon epoxy	Solvent system, unsupported film	2·5–3·0	Bonding aluminium skins to honeycomb core	Excellent filleting characteristics; excellent tensile-shear strength at cryogenic temperatures	Limited shelf life; lim strength beyond 100 °C

Note: Heat and pressure are used for cure

Electrical applications

Adhesives are used in the electrical industry for a variety of reasons. Consider a group of transistors, resistors and other components mounted on an aluminium heat sink. All of the components can be held in place with straps. However, metal-filled epoxy adhesives not only fasten the components to the heat sink but provide a more efficient means of conducting heat from the components. The cylindrical components only make point contact with the heat sink. The adhesive may be filleted around the sides of the components to provide a larger bonding area and a greater heat-conducting surface.

Phenolic nitrile adhesives can be used to bond motor-stator laminations together. The adhesive is applied to both faces of each lamination. Roller coating equipment is used to apply the adhesive to small laminations, and paint-spraying techniques are used to apply the adhesive to large laminations. The proper number of motor-stator laminations are stacked and then cured with heat and pressure.

Adhesive bonding is useful in this application because the cure temperatures are so low (149 °C) that the warping obtained when the stator laminations are welded is eliminated. The noise and vibration level is reduced because the voids between the laminations are eliminated and the operating temperature of the stator core is decreased because the air voids between the laminations are replaced by a more heat-conductive adhesive.

An adhesive should not be required to function as interlaminar insulation because the adhesive film may be damaged slightly during the manufacturing operation. Slight damage will not detract appreciably from the adhesive bond but may cause an electrical short-circuit between adjacent laminations. Very thin glue lines (0·0127 mm) are required so that the adhesive does not add significantly to the thickness of the stator core. Also, the adhesive should have sufficient 'hot' strength to permit handling immediately after the cure cycle is complete.

Chapter 2.7 **Tubing and sleeving materials**

Today, the term flexible tubing is used to describe extrusions made from rubber, plastic, or elastomeric material without fabric reinforcement. Sleeving describes a flexible tubular product braided from such materials as cotton, rayon, fibre glass, nylon, asbestos, etc. When the braided sleeving has been treated by impregnation, using such materials as oleoresinous varnish, acrylic resin, epoxy or polyester resin, polyurethane resins or rubber, silicone resin or rubber polytetrafluoroethylene resin, polymide resin, etc., it is known as coated sleeving. Specialty products such as heat-shrinkable and slide-fastenable tubing (jacketing with a zipper closure) will be covered separately.

Typical applications

Flexible tubings may be used for insulations such as:

1. On lead wires.
2. Magnet wires.
3. Components.
4. Terminations, harnessing or cabling of wires and cables.
5. Provide protection against mechanical or thermal abuse.
6. Identification through the use of transparencies or with colours or imprinting.

Selection factors

Several factors under the broad headings of mechanical, electrical, thermal, chemical, and moisture must be considered when evaluating a flexible tubing or sleeving for a particular application. Some of the more important properties are also mentioned in this chapter. However, it should be remembered that many properties are permanently or temporarily affected by heat, cold, moisture, radiation and other environmental conditions, it would be impossible to discuss all the ramifications for each factor being evaluated.

Dielectric breakdown

This is a measure of the ability of tubing or sleeving to withstand electrical stress without failure. To be meaningful, values should always be accompanied by the temperature and relative humidity conditions under which the tests were conducted.

Dielectric constant (permittivity)

This is a measure of the ability of a material to store electrical potential energy under the influence of an electrical field. Other electrical properties which may be considered are dissipation factor, insulation resistance and corona resistance.

Radiation resistance

This is becoming increasingly important as more designs requiring extreme reliability are sent into space. Outgassing, or the loss of volatile ingredients, can also be an important factor in space-oriented systems where 'fogging' of experiments would be detrimental. It is measured by the reduction in weight by vapourisation after heating. The effect of such volatiles on surrounding materials should also be considered.

Compatability

Materials within an insulation system are often important, in some cases the tubing or sleeving may be degraded by certain solvents, varnishes, insulating compounds, oils and greases and other chemicals.

Effects of temperature

Both high- and low-temperature effects should be evaluated carefully. Upper and lower continuous temperatures, as well as intermittent service temperatures, must be established. In the case of thermoplastics the melting-point should be determined. Low-temperature flexibility and crazing properties are important in some applications. Ageing is a general term used in reference to changes which may or may not take place over a long period of time. Care must be exercised when extrapolating results of accelerated ageing when using temperatures beyond normal operating levels as they do not necessarily correlate with normal ageing. Loss of flexibility, reduction in ultimate elongation and weight loss are used to measure ageing.

Mechanical effects

Abrasion resistance, tear strength, tensile strength and elongation are very important factors. Tubings and sleevings can be subjected to severe mechanical forces not only during service but also during assembly. Bond strength concerns only coated sleevings and refers to the ability of the coating to adhere to the base braid despite fabrication and service abuse.

Such properties as moisture absorption and adsorption wicking of fabrics, hydrolytic stability and effect of humidity on other properties are considered under the general heading of moisture resistance.

Today's complex and compact packages, where space is at a premium, make wall thickness an important factor. Care must be exercised to investigate fully the cold flow and cut-through properties in order to avoid possible

Table 2.7.1

Properties	Silicone rubber	Vinylidene fluoride	PCTFE	FEP	PTFE	Low-temperature vinyl	General-purpose vinyl	UL 105 °C vinyl	Poly-ethylene	Poly-carbonate
Specific gravity	—	1·76	2·1	2·14–2·17	2·1–2·2	1·7	1·26–1·38	1·26–1·75	0·95	1·2
Tensile strength, MN/m^2	6·2–8·2	48·0	31·0–37·0	13·7–21·3	13·7–72·2	1·24–21·3	15·8–27·5	15·8–25·5	10·3	55·0–62·2
Elongation, %	300–500	300	120–250	200–300	100–400	200–450	275–450	300–400	400	60–100
Water absorption per 24 hr, %	0·5	0·04	0·01	0·01	0·01	0·15–0·5	0·11–0·5	0·07–0·4	0·01	0·35
Softening/service temperature, °C	200–220	150	200	200	250–260	+60 to +80	+65 to +90	−12 to −60	−60	+135
Dielectric strength, V/m	400–660	—	620	800–1,400	500–2,000	250–1,200	700–1,100	700–1,100	800	400
Power factor (50 Hz)	—	—	—	0·001	0·0003	0·9	0·3–0·7	0·3–0·1	0·01	0·0009
Permittivity (1 MHz)	2·97	—	—	—	2·0–2·2	—	3·2–5	—	2·3–2·5	—
Voltage resistivity, Ω/cm	10^{15}	—	10^{17}	—	10^{12}–10^{17}	10^4–10^{10}	10^{10}–10^{14}	10^{12}–10^{14}	—	10^{17}
Low-temperature brittle point, °C	−70 to −85	−62	—	−55	−55 to +90	—	—	—	—	—

breakdown. Cold flow means that over a period of time and under pressure —
the material may be reduced in cross section at the point of pressure. Actual
penetration may result, especially if elevated temperatures are involved.

Pushback

This ability refers to that property which permits the material to be receded
on itself, thus decreasing its length temporarily. This characteristic is
desirable when insulated connections must be exposed for soldering.

Other factors which may be considered include fungus resistance, colour
stability, conformability, and inertness.

Flexible tubing

As defined above, flexible extruded tubings are those which are extruded
from rubber, plastic or elastomeric materials which have no fabric reinforcement. Typical properties are shown in Table 2.7.1.

Probably the most popular extruded tubing is polyvinyl chloride (PVC).
This is due in a large measure to its versatility, availability in many sizes and
low cost. It is available from different manufacturers in many types and
grades. However, they are generally graded to standard specifications for
non-rigid PVC tubing:

Grade A	General purpose
Grade B	Low temperature
Grade C	High temperature
Grade AFR	General purpose, fungus-resistant
Grade BFR	Low temperature, fungus-resistant
Grade CFR	High temperature, fungus-resistant

The high-temperature grades of PVC tubings are usually designed for continuous operation at 105 °C. Many of these are suitable for appliances and
switchgear. The low-temperature grades are designed for use in aircraft and
missiles, and similar applications where low temperatures are likely to be
encountered.

Polyethylene

This tubing is also used quite widely. It combines good electrical and
mechanical characteristics with excellent flexibility. It is generally rated at
105 °C.

Polyurethane

This tubing has excellent physical properties being one of the toughest
materials available today. It has good flexibility over a wide temperature
range −55 to 125 °C, together with excellent chemical resistance which
is so important in many cabling applications.

Silicone rubber

This type of tubing is extremely flexible and has a continuous operating temperature rating of 200 °C. It exhibits excellent ageing properties and a maintainance of outstanding electrical properties under conditions of high humidity. Its low-temperature brittle point is -70 °C or better and it has good resistance to many chemicals. Chemical compatibility can be improved by selecting a fluorosilicone tubing without sacrificing the other properties associated with silicone elastomers.

Vinylidene fluoride

PVF_2 is a tough, self-extinguishing tubing, exhibiting excellent cut-through resistance and a serviceable temperature range of -62 to 150 °C. It is resistant to most chemicals and solvents and is recommended for aerospace, hermetic and other critical applications.

Polytetrafluoroethylene

PTFE offers the highest heat resistance of all the extruded tubings -250 °C continuously. It is chemically inert, affected only by molten alkali metals. It has outstanding resistance in electrical properties to fungus. Its moisture absorption is considered to be zero. Although the material takes much abuse, there is a limitation: long-term cold flow could be a problem. It is available to fit standard wire sizes with wall thicknesses of standard-thin and super-thin. It is particularly recommended for use in extremely compact designs where soldering iron temperatures will not affect it during assembly and rework.

Braided sleeving

Braided or woven sleeving is available in both untreated and coated constructions. Base braids and various types of coated sleevings are separately mentioned.

Base braids

Commonly used fabrics for untreated braided sleeving include glass fibre, rayon, cotton and asbestos. Other materials used on a limited basis are fused quartz, polyamide, and ceramic fibres. In some cases base braids are lightly coated to prevent fraying and improve abrasion resistance. Uncoated or lightly treated sleevings are used for mechanical protection, harnessing, separation, identification, tying and lashing as a good base for varnish treatments in electrical units.

This type of sleeving is not designed to provide electrical insulation properties beyond that of an equivalent thickness of air. Flexibility is exceptionally high in the untreated types of sleeving. Extreme 'pushback' ability and 'stretch' are also important.

Braided glass-fibre sleevings

These are available in both flat and round form in nearly any size desired. Properties of glass-fibre sleevings include high tensile strength, non-flammability, high heat resistance and excellent resistance to fungus, moisture and chemicals. The commonly used type of glass fibre maintains its physical properties to 204 °C and retains about half its strength up to 371 °C.

Asbestos

These braids are characterised by excellent heat resistance. They are available in size from 1·6 mm internal diameter in wall thicknesses of 0·4 to 3 mm.

Braided cotton and rayon sleevings

They are characterised by good flexibility, high absorption ability, moderate strength and abrasion resistance. They are flammable and have limited heat resistance. Glass-fibre sleevings have largely replaced the use of cotton and rayon sleevings.

Ceramic fibre sleeving

This type provides a positive space factor insulation capable of continuous service up to 1,260 °C. It is available in diameters up to 25 mm with a minimum wall thickness of 0·8 mm.

Polyamide sleevings

These have exceptional heat resistance and retain over 60 per cent of their original room temperature strength after exposure to 250 °C for 500 hr. It is very difficult to ignite and is self-extinguishing when the flame is removed. Dimensional stability, resistance to chemicals and radiation are good to excellent.

Coated sleevings

The flexibility of coated or treated sleevings is directly related to the flexibility or elasticity of the treatment. Dielectric strength, flexibility, flammability, ageing and other properties vary with the coating material. In order for the sleeving to be used in electrical insulation applications the coating must be continuous.

Coated electrical sleeving types are classified according to the following:

1 Organic base materials, such as cotton, rayon, nylon or other organic fibres, which have been impregnated or coated with an organic substance which by experience or accepted tests can be shown to be capable of operation at 105 °C.

2 Inorganic base materials, such as glass fibre, which has been impregnated, and coated with a thermosetting organic substance which by experience or accepted tests can be shown to be capable of operation at 130 °C.

L

3 Inorganic base materials, such as glass fibre, which has been impregnated or coated with a thermoplastic material, such as polyvinyl chloride plastic compound for operation at 130 °C.

4 Inorganic base materials, such as glass fibre, which has been impregnated or coated with a silicone resin or other resinous materials for 200 °C.

5 Inorganic base materials, such as glass fibre, impregnated, coated or impregnated and coated with an insulating material, such as some epoxies, polyesters or polyurethanes for operation at 155 °C.

6 Inorganic base materials, such as glass fibre, impregnated or coated with a silicone elastomer or other materials at 200 °C.

Common types of coated sleeving

1 Vinyl-coated glass-fibre sleeving offers maximum resistance to cut-through and abrasion with extreme flexibility. It has superior pushback ability, good oil and chemical resistance, and is serviceable from −60 to 130 °C. It is a type 3 material.

2 Silicone-resin-coated glass-fibre types have excellent dielectric stability over a wide temperature range. It is self-extinguishing and is fungus resistant. It has excellent radiation resistance and is capable of operating at 200 °C. Type 4 sleeving.

3 Polyamide-varnish-coated glass fibre is a physically tough sleeving having high heat resistance but limited flexibility. It is non-flammable and has outstanding resistance to all chemicals and oils, except concentrated acids and alkalies. Rated at 250 °C continuous service it will probably obtain a Type 4 rating.

4 Silicone-rubber-coated glass-fibre sleevings have excellent heat resistance, flexibility, pushback and resilience. They are slow burning, generally inert to fungi and have excellent corona resistance. The physical strength of this coating varies between manufacturers. They are rate for 200 °C and belong in Type 5.

5 Epoxy-coated glass-fibre sleevings have good flexibility, very high resistance to abrasion and cut-through. It is flexible at −60 °C, exhibits good performance in most solvents including fuels and is a Type 6 material.

6 Polyurethane-resin and rubber-coated glass-fibre sleeving has excellent abrasion and cut-through resistance, good heat stability and chemical resistance. The coatings have excellent adhesion to the braid, with the rubber offering considerably more flexibility. It is a Type 6 sleeving rated for 155 °C.

7 Other types include glass-fibre sleevings coated with polyester resin, acrylic varnish or resin and acid polytetrafluoroethylene (PTFE) resin.

Heat-shrinkable tubings

Heat-shrinkable tubings possess the unique property of 'elastic memory', with the application of heat they recover to a predetermined dimension. They come in a wide range of materials, sizes, properties, shrink temperatures and recovery ratios. Polyolefin, polyvinylchloride (PVC), polyvinylidene fluoride (PVF$_2$), polytetrafluoroethylene (PTFE), polyester and fluorinated ethylene propylene (FEP) modified polymers are used as well as neoprene, silicon and butyl elastomers. Physical properties vary with the material and are available in very flexible to rigid grades. Most products recover to 50 per cent of the diameter supplied, although shrink ratios as high as 10 to 1 are possible.

Most materials will recover in the range of 100 to 175 °C. For fast shrinkage and high production rates, temperatures of 250 to 350 °C can be used. This has the added advantage of exposing the object being covered to heating for a minimum length of time, keeping it relatively cool during the shrinking process. Shrinking can be accomplished through the use of hot-air blowers, conductive heating tool, infrared radiation units, dielectric and induction heating devices, ovens, and heat transfer fluids.

Two types of shrinking mechanism are used to cause the recovery action. Some polymers can be processed so that the molecules are 'frozen-in-place' while in a strained condition. After heating at a specified temperature for a minimum time, the strains are relaxed and the material seeks its original dimensions. A second category includes polymers which as a result of their cross-linking qualities will no longer meet, but will exhibit perfect elasticity above their crystalline melting-point. They can now be heated, expanded and cooled to 'freeze' them in the expanded state. Upon heating them again they return to their original dimensions. Longitudinal shrinkage is kept to a minimum with this mechanism as well as precise control of other recovered dimensions.

Typical properties of different materials are shown in Table 2.7.2.

Special constructions

One unique construction available is a selectively cross-linked, dual-wall polyolefin tubing. The outer wall is cross-linked which shrinks with heat but does not melt. The non-cross-linked inner wall melts during heating and due to the shrinking action of the outer wall is forced into voids. This provides a moisture resistant encapsulation system for such things as wire splices and components.

Where space is at a premium a polyester film tubing is available having a recovered wall thickness of 0·58–0·63 mm. It has a thermally welded, non-adhesive seam. Another manufacturer offers a spirally wound polyester film tubing using a special adhesive as a bonding agent, this is available in wall thickness down to 0·025 mm.

Slide-fastenable tubing

Jacketing with a zipper closure was developed to provide a tubing which could enclose, insulate, protect, shield and identify wires, cable bundles and other products. It features a plastic zipper-track that permits installation and re-entry for inspection or modification that is not available with other types of tubings and sleevings. If desired the track can be sealed permanently with

Table 2.7.2 **Typical properties of heat-shrinkable tubing**

Properties	Irradiated polyolefin					Irradiated PVF$_2$	Flexible PVC	Flexible Irradiated PVC	Semi-rigid irradiated PVC	Neoprene rubber	Silicone rubber	Butyl rubber	Polyester film	PTFE
	Flexible opaque	Flexible clear	Semi-rigid opaque	Semi-rigid clear	Dual wall									
Tensile strength, MN/m^2	17·2	17·2	20·6	20·6	13·7	55·0	20·6	20·6	34·4	13·1	6·2	11·0	137·0	31·0
Ultimate elongation, %	400	400	400	400	400	300	300	300	250	220	300	350	—	250
Brittleness temperature, °C	−60	−85	−60	−90	—	−73	−20	−20	−20	−40	−75	—	−60	−90
Hardness	98A	90A	—	—	—	—	85A	—	—	85A	70A	80A	—	—
Specific gravity	1·3	0·93	1·3	0·95	0·94	1·76	1·4	1·35	1·4	1·4	1·2	1·2	—	2·2
Water absorption, %	0·05	0·01	0·05	0·01	0·1	0·1	—	0·6	0·6	0·5	0·5	0·1	—	0·01
Dielectric strength, V/m	1,300	1,300	1,300	1,300	1,100	1,500	750	750	900	300	300	130	>3,500	1,200
Voltage resistivity, Ω/cm	10^{15}	10^{17}	10^{15}	10^{17}	10^{16}	—	10^{12}	10^{12}	>10^{13}	10^{11}	10^{15}	10^{12}	10^{14}	10^{18}
Permittivity	2·7	2·3	2·7	2·4	2·4	—	5·4	—	—	—	3·3	—	3·4	2·1
Power factor	0·003	0·0003	0·003	0·0003	0·0005	—	0·12	—	—	—	—	—	0·003	0·0002
Fungus resistance	Inert	Inert	Inert	Inert	Inert	Inert	Inert	Inert	Inert	Inert	Inert	Inert	Inert	Inert
Fuel and oil resistance	Exc.	Exc.	Exc.	Exc.	Exc.	Exc.	—	RExc.	Exc.	Good	Fair	Fair	Exc.	Exc.
Hydraulic fluid resistance	Exc.	Exc.	Exc.	Exc.	Exc.	Exc.	—	Exc.	Exc.	Fair	Poor	Good	—	Exc.
Solvent resistance	Good	Good	Good	Good	Good	Exc.	—	Exc.	Exc.	Fair	Fair	Fair	Exc.	Exc.
Acid and alkali resistance	Exc.	Exc.	Exc.	Exc.	Exc.	Exc.	—	Exc.	Exc.	Good	Good	Good	Good	Exc.
Flammability	Self-Exting.	Burns Slowly	Self-Exting.	Burns Slowly	—	Non-Burning	Self-Exting.	Self-Exting.	Self-Exting.	Self-Exting.	Self-Exting.	Burns Slowly	**	Non-Burning

the application of a solvent which chemically fuses the plastic and provides water-tight protection.

It is available in a variety of materials including polyvinyl chloride (PVC) polyethylene, vinyl-impregnated nylon cloth, PVC or PTFE-impregnated glass cloth and aluminised asbestos, designed to meet specific temperature, abrasion and insulation requirements.

To meet specific RFI or magnetic shielding needs, the jacketing material can be fabricated with inner linings of woven copper-clad wire mesh, aluminium and nickel foil. A grounding braid of tinned copper can be attached to the inside of the overlap.

Chapter 2.8   **Flexible cables, printing wiring, printed circuits**

The urgent need to save weight and space and to obtain the utmost reliability in electrical and electronic components has led to the development of flexible flat cables, flexible printed wiring and flexible printed circuits. Their success has resulted in their adoption for many other electronic applications utilising high-density wiring where similar attributes may not be so essential but are certainly highly desirable.

Though the three products have many features in common and all can be time, labour and money saving, it is by considering them separately that their similarities and differences become apparent.

Flexible flat cables

Flexible flat cables consist of a number of conductors of thin, ribbon-like, flat strips of metal — usually copper embedded side by side and parallel to each other between films of insulating material, usually a plastic. The conductors may vary in number from two to fifty or more, and their width and spacing are usually uniform to suit standard connectors. The conductors may vary in width or spacing and be of different metals and alloys.

Flexible flat cable, like conventional cable, is usually used to connect and inter-connect separate or remote component packages. Its flexibility, combined with toughness, is of special use when the components have to move in relation to each other or when the cable is installed in confined spaces. Flexible flat cable is usually supplied cut to length, with or without terminations but can be supplied in lengths of up to 30 m, to be cut as required. All the conductors are parallel to and at a constant distance from each other throughout. It is usually used with special connectors at each end in the same way as conventional wiring.

Advantages

Many of the advantages of flexible flat cable result directly from the shape of the conductor, which presents a greater surface area than round conductors of the same cross-sectional area and are more efficient in terms of electrical performance.

Greater current capacity is provided by a flat conductor than by a round conductor because heat can be dissipated more quickly, especially as the plastic film can be relatively thin and still provide adequate insulation.

Reduction in weight compared with circular cable is obtained directly as a result of the reduction of relative volumes of material and the use of a thinner and lighter installation. As a result of improved heat losses, the current-carrying capacity of the conductors may be uprated so that conductors of smaller area may be used. Further over-all reduction in weight is possible because in installation the cables can be attached by the application of adhesives instead of by the use of solid and relatively heavy fasteners and fittings. It is claimed that weight reductions of up to 85 per cent can be achieved over that of conventional cable of a similar current-carrying capacity.

Economy in space is obtained primarily by the fact that the cable is a thin flat strip. Only small clearance is necessary for the cable and no allowance as regards space need be made to fix or accommodate clamps, brackets, nuts or bolts or other fastening devices. The total space required may be only a quarter of that of a conventional cable.

Consistent electrical characteristics are possible because the conductor spacing is fixed. Capacitive cross talk and inductive interference coupling are therefore consistent, and in some cases can be calculated. Effective shielding is simple.

Easy and rapid installation is a feature that is not only valuable in itself, but contributes to the reduction in overall costs. The cable folds and bends readily in one plane, conforms to the mounting area and fastens easily. Visible, parallel conductors in a fixed position within the dielectric simplify coding, inspection and circuit tracing. The cable can bend, flex, fold or slide into available spaces, and so adapt itself to existing layouts of components.

Great strength and flexibility is obtained because apart from the inherent strength of the tough, abrasion resistant insulation; the tensile load is borne by the conductors and insulation together. Flexing does not impose a severe strain on the conductors as they lie between the films of insulation on the central axis of the cable. The insulation has a high-resistance to mechanical damage and to chemical attack.

Materials

Conductors in flexible flat cables are usually of high conductivity rolled copper strip of rectangular section, annealed dead soft, the shape provides maximum flexibility in one plane with good heat dissipation so that higher current ratings are possible. For plug-in terminations, conductors may be plated with nickel or precious metal to ensure low resistance contact, long life and to give protection against heat and corrosion. Other materials may also be used, including some with plated or oxidised surfaces. For heating and instrument cables, conductors of special alloy are produced.

Insulating film materials are available in a wide variety of materials with high insulating and thermal characteristics, and good abrasion resistance. A range of materials can be used, the choice being dependent on the particular environment conditions. The prices of these materials generally increase according to the severity of the environment in which they will have to work. Most materials have a high degree of transparency.

When choosing flat cable, additional factors to be considered include the ambient and maximum temperatures that are likely to be encountered, electrical parameters, flexibility requirements and any special environmental conditions likely to be encountered.

Polyvinyl chloride (PVC) is widely used but is limited by its sensitivity to heat, as it becomes brittle at relatively low temperatures and deteriorates at temperatures above 90 °C, although special compounds have been formulated to increase the temperature range.

Polyesters have good heat-conducting characteristics, have little shrink, stretch or cold flow below 100 °C and have good dielectric properties. They also have high tensile strength and high bond strength to copper which is valuable when maximum flexural life is required.

Fluorinated ethylene propylene (FEP) can withstand temperatures up to 200 °C and can be used at temperatures up to 225 °C when reinforced with glass. It has excellent electrical properties and good stability and resistance to abrasion.

Polychlorotrifluoroethylene (PCTFE) can withstand temperatures up to 150 °C and exhibits good electrical properties. With high-temperature installations the effect of heat on the conductors, particularly at terminations, must be taken into account.

Screening

Full screening or shielding is easily accomplished by sealing into the cable on both sides of the main conductors a metal foil of aluminium or copper, which may be earthed by conductors in contact with them. Semi-screened cables have a foil screen on one side only.

Capacitance

As the conductors are positioned edge-to-edge and the thickness is small, the capacitance is lower than that of relative conventional multi-core cables. The cables can be used on balanced or unbalanced circuits.

For unscreened types of a typical cable, the capacitance is 16 to 25 pF/m for balanced operations depending on the size of the cable, with polyester insulation as with pvc insulation the capacitance increases by about 60 per cent but is still less than that of conventional circular cables. The capacitance is increased by the incorporation of a screen; a single screen increases the capacitance compared with that of unscreened versions by a factor of between ten and fifteen times and a double screen by a factor of between 20 and 30.

Retractable performs

Retractable performs can be obtained to maintain connections between components which move in relation to one another. These may be either of the double coil or the concertina type. They are extensible under tension and retract on release, ensuring long life and maximum space saving. When a component may have to be moved some distance from another, the perform may require extra support.

Flexible printed wiring

Whereas flexible flat cable is generally used to connect separate component packages, flexible printed wiring is used more often for inter-connection within a particular component wiring system. As such, it is usually designed for one specific application, and the conductors may be parallel to each other for considerable distances but may change in direction at any point, according to the required design.

Whilst only one conductive layer is used, conductors can be arranged to cross each other by simply folding 'spurs' made within the pattern. As each layer is thin (typically 0·3 mm) more than one layer may be accommodated.

To make full use of flexible wiring the equipment should be designed with the flexible printed wiring concept in mind to ensure that the finished design is compatible with economic tooling, i.e. avoiding awkward shapes and long branches; this will ensure the use of the minimum area of flexible material and allow relatively simple assembly leading to significant cost reductions.

As in the case of flexible flat cables the system is, in effect, a sandwich of wiring between plastic film, but the method of manufacture is different, flexible printed wiring being produced by a photomechanical, or silk-screen printing process, or by die-stamping.

Production

Etching process

In one method of production a master drawing of the wiring pattern is produced, exact in every detail according to the requirements of the final product. The size may be up to ten times that of the final product, but the pattern of the required size is obtained photographically.

The material in which the wiring is produced is formed of a sheet of conducting material, usually high purity copper, 0·05, 0·077 or 0·10 mm thick, as required, which is processed chemically to produce a suitable oxidised surface, the sheet of metal is then heat-bonded under pressure to a plastic film, the oxidised surface of the foil permitting a higher bond strength. The wiring pattern of the correct size is then printed either photomechanically or by using a silk screen on to the processed copper film using a material that resists standard etching solutions. The copper film is then etched, so that the pattern of the wiring appears after which the plastic film 'cover coat' is heat-bonded under pressure to the laminate thus completely enclosing the wiring. Various parts of the wiring such as solder pads can be left uncovered and these areas can be plated. The complete laminate is then cut to the required size and the whole is then ready for assembly into a particular piece of equipment.

Master drawing

The accuracy of the final circuit depends on the precision and clarity of the original artwork. Therefore, great care should be taken in preparing them, and a number of points must be considered. The material on which the master drawing is prepared should be dimensionally stable, especially when the dimensions are critical or when it is necessary to store the drawing for some time pending further use.

Die-stamping

In another method of production the conductor pattern of copper or aluminium is die-stamped on to one of a variety of base materials, including polyester, PVC, and glass cloth. The conductors can be of differing thicknesses, and therefore, of different current capacity, when required. These composite circuits are bonded to a common insulating base and encapsulated with an additional layer of pre-pierced film, which can be tinted for rapid identification of the conductor surfaces. This method is claimed to be the more economical for long runs.

Materials

The foils from which the etched conductors are made are usually of high-purity electrolytic copper (the thickness usually being given in terms of its weight per square foot — 1-oz, 2-oz, 3-oz). The minimum width of the conductor should finally be 0·725 mm and the maximum width is determined by the stiffness that is permissible in the wiring, the design of the termination used and the permissible temperature rise. For example, a conductor 6·35 mm wide of 85·047 g (2-oz) copper is adequate normally to carry 10A, with a rise in temperature of the conductor of 25 °C above the ambient.

Other alloys used for the foils for the conductor are brass, bronze, beryllium copper and pure nickel. There are a number of plastic films used as insulating material, the type chosen depending, of course, on the condition of the application; one of the chief factors is that the extremes of surface temperature, it must be remembered that the dielectric has to withstand the heat of soldering.

Nearly any type of metal foil, such as copper, nickel and aluminium can be used for die-stamping process.

Screening

Flexible printed wiring can be screened in the same way as flexible flat cables; by the use of a grounded shield, bonded to one or both sides of the wiring.

Multilayer circuits

To achieve the required circuit density, multilayer circuits may be used. In order to retain maximum flexibility, the layers should not be bonded together, screening can be interleaved to relieve any possible interference between the layers.

Flexible printed circuits

The terms 'flexible printed wiring' and 'flexible printed circuits' are sometimes used as if they were interchangeable. The fact is, of course, that most so-called printed circuits consist only of wiring and that a true printed circuit includes not only the basic wiring but all circuit components too, so that when the term 'printed circuit' is applied, it is usually that 'wiring circuit' is really meant.

Some components can be produced on flexible film by printed wiring techniques; these include inductors in the flat plain form, resistance elements etched from resistive metals of different characteristics and parallel plate condensers formed between the conductor faces of the clad laminate; many components produced by these methods can be incorporated with the basic wiring pattern, but it is sometimes difficult to produce them in one technique.

Connectors

It is true to say that though flexible flat cable and flexible printed wiring were introduced a number of years ago, their adoption was relatively slow because there appeared to be no versatile, reliable and easily applied connectors. That situation has been largely remedied now and a wide variety of connectors is available.

From the point of view of the production of these products, the range of facilities offered is wide. As mentioned, flexible flat cables are usually supplied wound on reels, varying in number, width and thickness of the conductors, and up to 30 m long. Flexible printed wiring being usually made for a specific application, necessitates firstly the preparation of a master drawing. This can be prepared by the purchaser and handed to the supplier of the printed wiring, or the supplier will have it drawn according to the instructions of the purchaser and then produce the printed wiring. For users of large runs of flexible printed wiring, complete systems, are provided, consisting of equipment for exposure, development, etching and laminating the product.

Section Three

Chapter 3.1   **Dielectric materials**

For many years it was adequate, in the classification of materials, to specify only their thermal capabilities. The materials available were natural and simple. But with the advent of synthetic materials and the tailoring of properties by molecular structuring, by compounding, and by adding stabilisers and antioxidants, this classification system was no longer sufficient.

The most recent method of classification that has evolved 'life-by-test', permits the assignment of temperature indices to a material. These indices depend on the material's use and the deterioration in properties that govern its suitability for a given application. The governing parameters may be chemical, physical, electrical, or such combinations as softening, embrittlement, dielectric strength, insulation resistance, loss, etc., and are chosen to be specific for each application. For example, coil wound with insulated magnet wire are normally rated at 105 °C in air: in oil the same coil might be used at temperatures as high as 175 °C. Cross-linked polyolefin wire might be rated at 120 °C for a useful life of 20,000 hr, for use in a missile it might be rated at 120 °C for 20,000 hr during shipping or storage, and at 400 °C for 5 min during flight.

In using dielectrics, the engineer has many types of materials with which to work: gases (including vacuum), liquids, and solids. Seldom is one particular material used alone. Consequently, the designer must know or be able to predict the effects produced on combinations by environment (temperature, contaminants, moisture, radiation), ageing, wear, and shifting of dielectric parameters. To work effectively and economically in this field requires a conceptual understanding of the mechanisms and forces involved. For this purpose of discussing these basic concepts, this chapter is divided into three sections: conductivity, capacitance and loss, and high voltage.

Conductivity

In any material, resistance increases with the length l or the conduction path and decreases with the cross-sectional area A. The expression for resistance is

$$R = pl/A$$

where p is volume resistivity (or resistivity) and is most commonly expressed

$$p = RA/l \quad \Omega.\text{cm}$$

For a two dimensional system, resistance decreases with the width P of the conduction path instead of the cross-sectional area and increases with length g. The proportionality constant σ, where $\sigma = RP/g$, is surface resistivity and is most commonly expressed in ohms. Physically, σ is the resistance between electrodes on opposite sides of a square and independent of the size of that square.

For engineering purposes, the surface includes the material lying along the surface adsorbed moisture and gases, contaminants, etc.

Resistivity is a convenient measure for classifying materials as conductors, insulators, or semiconductors, since it is independent of material dimensions. Thus, materials with resistivities between 10^{-7} and 1 are generally classified as conductors, those with resistivities between 1 and 10^6 are semiconductors, and those with resistivities above 10^6 are insulators. Often materials are more conveniently described in terms of conductivity. In such cases, the units are the mho per centimetre for volume conductivity and the mho for surface conductivity.

The principle charge carriers in most conductors (metals) are the outer orbital electrons, which move freely through the lattice structure of the material. Non-conductors do not have these free orbital electrons, and current flow is the result of slow chemical changes.

Dielectric-absorption characteristics

Changes in an insulating material that are caused by an applied electric field are called polarisation and concern the displacement of various charges from their initial position. The changes occur slowly requiring microseconds to weeks or months to be significant. As these changes occur, the resistivity of the material increases. Their time dependence is analogous to the flow of viscous liquids. The analogy is usually referred to as the viscoelastic theory of matter and is useful in explaining material behaviour under electrical and physical stresses. However, the mechanisms of these changes are often extremely complex, and speculation about their nature is supported by minimum experimental verification and anology. The effects of these changes are usually referred to as dielectric absorption.

As volume resistivity of a material increases with time in an applied electric field, it approaches an asymptotic value that is determined by the nature and condition of the material and the environment (including temperature and moisture content). Because the change is slow, the ultimate resistivity is seldom if ever reported for a material. Established test methods usually stipulate the time of electrification as $(60 \pm 5 \text{ sec})$. During this interval, change in resistivity may be as little as one decade for low-loss materials or as high as five, seven, or more decades. But a material with a high 1-min resistivity may reach a much lower ultimate resistivity than a material with a much lower 1-min resistivity and a larger susceptance to dielectric absorption effects. (Fig. 3.1.1). Thus, it is impossible to draw any conclusions about the relative resistance of two materials in a specific component or system from a knowledge of short-time resistivity.

Another method of investigating dielectric absorption is the voltage-recovery technique. Essentially, this method makes use of the energy stored in the viscoelastic stresses produced by the applied electric field.

The stresses are relieved when the electrodes are momentarily short-

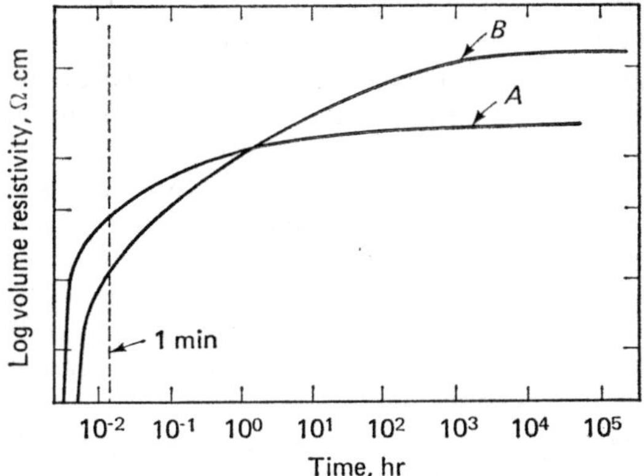

Fig. 3.1.1   Variation of resistivity with time for two materials: *A* — with high short-term volume resistivity and low dielectric absorption susceptance; *B* — with lower short-term volume resistivity and higher dielectric absorption susceptance

circuited and the stored energy appears at the electrodes. The electrode voltage after short-circuiting is a measure of the dielectric absorption of the material.

Environmental effects on conductivity

Our discussion so far has assumed constant environmental conditions. However, variations in temperature and humidity have a considerable effect on conductivity.

Temperature

An increase in temperature increases available energy, facilitates formation of charge carriers, and causes a general increase in conductivity of insulating materials. Changes in material structure over short temperature spans frequently cause reversals in the conductivity — temperature relationship, producing a maximum and a minimum. However, if a phase change, e.g. solid to liquid, also occurs within the same temperature span, only a portion of this curve may be evident.

Humidity

Moisture content in a material resulting from immersion in water or exposure to a higher relative humidity, increases charge mobility and results in a general decrease in resistivity, also enhances ionisation and charge formation.

When these changes in resistivity are observed, the test method must be able to separate changes in resistivity caused by dielectric absorption which may occur simultaneously and may mask the effect of moisture.

Radiation

Ionising radiation produces charge carriers in a material and, at high intensities, changes insulators into conductors. Recovery time is measured by the half-life of the charge carriers and may range from milliseconds to

M

hours or days. After recovery from the transient-radiation-electronics effects (often called TREE), the material remains permanently degraded. The amount of degradation is measured by the integrated-dosage-radiation intensity times the exposure time or the area under the intensity–time curve. The integrated-dosage degration is usually reported in the literature. The transient effects, which are very difficult to measure are usually many orders of magnitude more severe than the final recovered effects: they must be considered when designing systems for nuclear environments. An often neglected radiation effect is that of the by-products of the interaction of ionising radiation and dielectric. For example, at very mild dosage (10^2 to 10^3 r), wire insulation containing halogen produces gaseous halogen acid which attacks the surface of the wire.

Silver-plated wire becomes covered with a black silver-halide coating of poor conductivity and high contact resistance.

Resistance measurements

The electrical conductors are the electrodes when the insulation resistance of systems, machines and components is being measured. In cable systems, the insulation resistance of each conductor is measured with respect to all others and to ground, the latter consisting of the outer shield or metallic sheath of the cable or a tank of water in which the cable is immersed. In polyphase apparatus, each phase is tested with respect to all others and to ground.

For insulating materials, taper pins set into reamed holes for close contact can be used to simulate electrodes to measure surface and volume resistance and, for laminates, the interlaminar resistance. Washer electrodes, placed under screws, simulate industrial circuit mounting boards for resistance measuring purposes. Conductive comb patterns on surfaces simulate electronic circuit boards and permit evaluation of conformal coatings used to reduce the effects of surface-adsorbed and condensed moisture. Bars placed across films and tapes are a convenient way of testing thin and often narrow films and strips.

Each of these methods provides test values under given temperature and other environmental conditions. The values are combinations of surface and volume resistances but there is no indication of the dominant resistance.

For materials research and development and for quality control, the measurement of volume and surface resistivities is indispensible. Guarded or three-electrode configurations are usually used to define the specimen dimensions precisely and to eliminate undesired leakage currents. The familiar bulls-eye patterns on flat and cylindrical specimens are universally used. Volume resistivity measurements are most accurate and reproducible when the air gap between the guard and the measuring electrode is as small as possible without permitting undue conductivity between them.

Capacitance and loss

The total current that flows through any dielectric is equal to the vector sum of the capacitive current I_c and a loss current I_l (Fig. 3.1.2). The loss current,

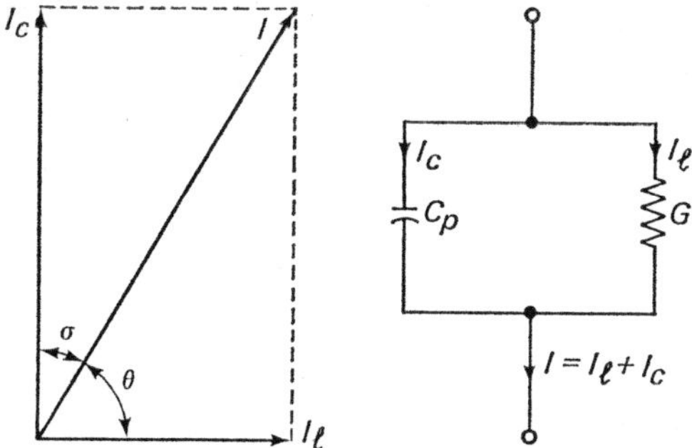

Fig. 3.1.2   A dielectric material can be represented as an ideal capacitor in parallel with a conductance. The total current through the dielectric is the vector sum of the capacitative current I_c and loss current I_l. Power factor and dissipation current are related to the angular differences between total current and its component parts

a result of all the energy-absorption mechanisms, includes the conduction current and the induced 'molecular friction' (the viscoelastic absorption mechanisms from the molecular motions induced by the field). The captive current, related to the geometric capacitance of the dielectric, is the d.c. inrush current that exponentially falls to zero. The geometric capacitance C_o is the capacitance of a vacuum of fixed volume V_o. If we designate C_x as the capacitance of a dielectric material of volume V_o, the dielectric constant K' of the material at a given temperature and frequency is equal to C_x/C_o. Thus the dielectric constant of a vacuum is 1 and that of all other materials is greater than 1 (a vacuum is considered an engineering material).

Polarisation

When atoms are placed in an electric field, the electrons are displaced toward the anode from their unstressed equilibrium position, and the nucleus is displaced toward the cathode to a lesser extent because of its much higher mass. This off-centre displacement of positive and negative charges is called an induced dipole because the charges return to their previous equilibrium positions upon removal of the field. The mechanism is considered to be perfectly elastic, returning to the system all of the displacement energy, and is known as electronic polarisation (Fig. 3.1.3). Atomic polarisation occurs when the applied field alters the separation (bond length) between two atoms or the bond angle among three atoms.

Random charges caused by cosmic radiation or thermal deterioration, or left over from the formation process, constitute what is known as the space charge. Polarisation of the space charge occurs when an applied field causes a displacement of the charges and if their mobility permits a drift in the direction of the field. In a material, dipoles that are not perfectly aligned with the applied electric field have a torque applied to them. The angular displacement resulting from the torque is called orientation polarisation.

The response time of orientation polarisation is slow because a large mass

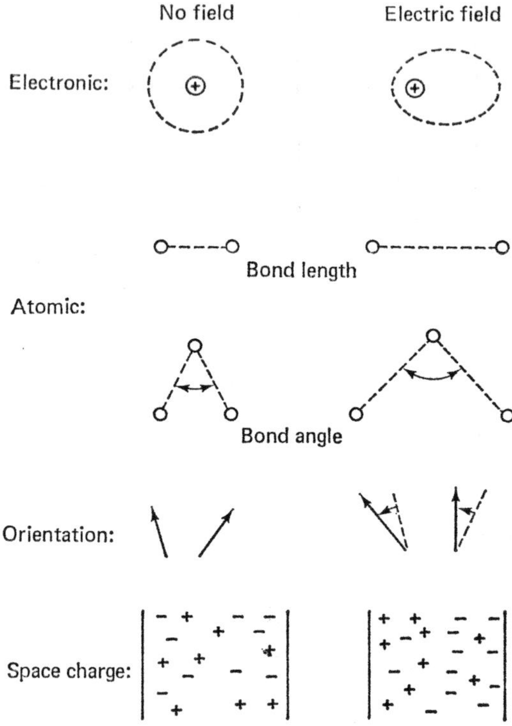

Fig. 3.1.3   Types of polarisation

is involved, and because the rotating dipole molecule is impeded by neighbouring molecules. In general, the response time for any polarisation mechanism depends on the material and the temperature, and may be less than a microsecond or as much as a second or even years. It must be emphasised that the polarisation mechanisms of which we speak are only very small fractional responses. However, they are the basis for the time dependence of dielectric absorption and the variation of dielectric constant and loss of frequency.

Frequency effects

When a steady or very low-frequency potential is applied to a capacitor filled with a polar material having a high concentration of dipoles, and there is sufficient time for the dipoles to respond to the electric field, the dielectric constant is high. As the frequency is increased the high-mass dipoles reach a point of resonance where maximum energy is absorbed from the field. Upon further increase in frequency these dipoles can no longer interact with the electric field, and their contribution to the dielectric constant is lost. Each successive loss of contribution is shown by a peak in the dissipation factor and a decrease in the dielectric constant.

Temperature effects

As temperature increases, thermal expansion occurs and the greater energy and molecular mobility alters dipole interaction with the electric field. The general effect is a decrease in dielectric constant and an increase in loss. (In limited temperature ranges, the opposite effect frequently occurs). Changes in energy, volume, mobility, and ordered structure are shown by inflection points in the temperature curves for dielectric constant and loss.

First-order transition changes (such as the phase change in the melting of a crystal) are shown as sharp and discontinuous inflection points. Second-order transitions (such as change in crystal structure from brittle to ductile) frequently show rounded inflection points attributed to the long time required for complete transformation. Another cause is hysteresis: loss values depend upon whether the measurements were made with increasing or decreasing temperature. Over short temperature ranges the variation of dielectric constant and loss can be adjusted to produce temperature curves with minimum and maximum values. The manipulation of these property values is a specialised study and is used, for example, to develop capacitors with positive, negative, and zero temperature coefficients over practical temperature ranges.

Moisture effects

Water, which has a high permittivity (about 80), readily encourages ionisation that in some materials can change the dissipation factor as much as three orders of magnitude. Moisture can also change the volume resistivity of a material by several orders of magnitude: in extreme cases, an insulator can take on semiconducting characteristics. For example, a rubber compound at equilibrium with 50 per cent relative humidity has had its volume resistivity reduced from 10 to 2,000 Ω cm after water immersion.

High-voltage stresses in a dielectric may induce electromechanical interactions which appear as an increase in dielectric constant. Internal ionisation which appears as an increase in dissipation factor (loss of power-factor tip-up), can occur simultaneously. Further increases in voltage stress increase dielectric constant and loss up to the point of dielectric breakdown.

Permittivity and dissipation-factor measurements

The various techniques for measuring permittivity, dielectric constant and loss, or dissipation factor can be divided into those that use lumped-parameter or distributed parameter circuits (Fig. 3.1.4). The former are not used in very high-frequency measurements because of stray capacitance, lead impedance, and other proximity effects. The upper frequency limit for lumped-parameter circuits is usually between 10 and 20 Hz.

Null methods

Resistive-arm, inductive-arm, and transformer-ratio capacitance bridges and parallel-T networks use null methods of measure capacitance and loss of

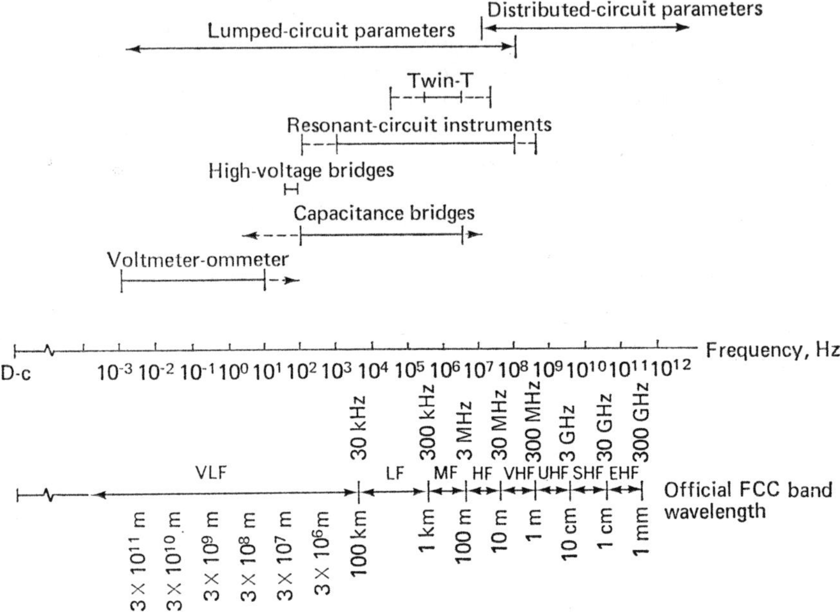

Fig. 3.1.4   Dielectric measurement spectra

dielectrics. Accuracy depends on the resolution of the balance arms and the detection sensitivity of the null-balance indicator.

Optimum performance is achieved when the capacitance in the bridge arms balancing the total specimen and stray capacitances is as small as possible. In the lower portion of the applicable band three-terminal measurements (guarded electrodes) are preferred: when surface conductivity of the specimen is important the guarded electrodes are essential.

At the upper end of the band, two-terminal measurements are generally preferred because of circuit complications caused by guarded electrodes. However, with the transformer ratio bridge, three-terminal measurements are suitable over the entire frequency band.

Resonant-circuit methods

Resonant-circuit instruments, such as the Q-meter, the R–X meter and the admittance meter, consist of precisely calibrated capacitors in series or in parallel with inductors. Resonance is indicated by a high-impedance VTVM. The loss in the circuit is measured by the width of the resonance curve. The ability to tune the resonance tank circuit varies directly with the square of the Q and inversely with the capacitance. Hence, it is desirable to use a high Q with instrument capacitance no larger than necessary to balance the capacitance of the specimen and electrodes. Inductors with a Q of 500 to 1,000 have become available in recent years: their use is essential to realise fully the benefits offered by resonant-circuit techniques. These techniques do not permit the use of a guard terminal: hence two-terminal electrodes are used. (Fortunately, surface conductivity of the test specimen is of negligible importance in the frequency band compatible with resonant-circuit instruments).

Generally, fringing effects (distortion and extension of the electrical field at the edge of the specimen) control the realisable accuracy of dielectric constant by permittivity measurements. Accuracy may be increased ten-fold by immersing the electrodes and specimen in a low-loss dielectric liquid of similar dielectric constant to reduce fringing (and specimen dimension) and contact errors to a negligible quantity.

High-frequency (above 3 kHz) methods of measuring dielectric characteristics include the use of transmission lines, wave guides, or resonant cavities and are not standardised. Essentially, the dielectric constant of the specimen is derived from the change of some resonant mode when the specimen is placed in the path of the wave: the dielectric loss is related to a change in the width of the resonant curve.

Electrode systems

Electrode systems are geometrically constructed so that the measurement of capacitance and circuit loss may be related by calculation to the dielectric constant and dissipation factor of the insulant. Plane-parallel and cylindrical or coaxial electrodes are the configurations most frequently encountered.

A most useful tool for understanding electrode systems in dielectric measurements is the equivalent circuit. Figure 3.1.5a is the equivalent electrical circuit of an ideal loss-free capacitance with perfect electrode contact. In practice, minor surface irregularities and dust particles prevent perfect contact between the electrode and the dielectric, introducing a small series capacitance, Fig. 3.1.5b, that lowers the apparent measured capacitance.

Better surface contact is achieved with contact electrodes such as conductive paint, vacuum-deposited metal, sprayed liquid metal, fused metal, or pasted foil. Although these electrodes reduce the capacitance error, they introduce impedances that raise the apparent measured dissipation factor. The dissipation factor error increases with the square root of the frequency.

One method of eliminating capacitance and dissipation-factor errors is to measure dissipation factor before the contact electrodes are applied and dielectric constant after they are applied. A better method is to keep one or both of the electrodes from contacting the specimen as in the air-gap or liquid-displacement techniques. The techniques introduce another capacitance, in series with the specimen, which can be controlled and whose effect can be known with a high degree of certainty (Fig. 3.1.5c).

When the dielectric constant of the separating medium nearly matches that of the specimen, an additional advantage is achieved: ringing of the electric field is not changed by insertion of the specimen, and the accuracy of the permittivity measurement is significantly increased. The equivalent circuit for electrodes leads at low frequencies (Fig. 3.1.6) includes typical electrode-contact impedances from Fig. 3.1.5. At the upper frequencies in which lumped parameters can be used, the lead impedance increases,

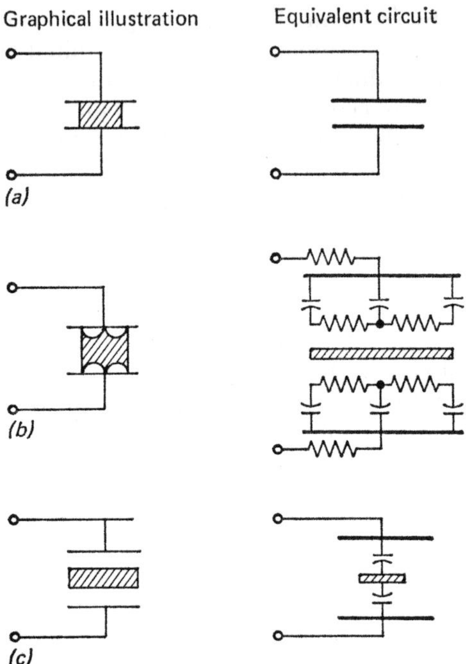

Fig. 3.1.5   Electrode equivalent circuit in frequency range 0–10⁸ Hz: (a) perfect electrode contact; (b) distributed, poorly defined contact impedances; (c) non-contact electrodes with lumped, easily defined contact impedances

determining the maximum frequency of measurement by lumped-parameter techniques.

High voltage

The mechanisms of electrical conduction in any material are associated with its chemical structure, and they follow the laws of kinetics and thermodynamics. Each reaction that produces or removes charge carriers requires a minimum amount of energy transfer. This minimum, or threshold level, is called activation energy. As energy is added to a system, the rate at which existing reactions occur increases and new reactions begin to occur. The latter provide additional energy which may be, for example, in the form of heat or electromagnetic energy. The heat is either applied to the system or generated within the system by the electric field.

Voltage effects

Consider the behaviour of a typical insulating material as the electrical stress (or voltage) increases from zero to breakdown (Fig. 3.1.7). At low stress levels the conduction current increases at a rate determined by the increased number of charge carriers that are either injected at the electrodes or created within the material by the electric field. These charge carriers are accelerated toward the electrodes by the field but have insufficient energy before colliding to generate new carriers.

At higher voltage stress, the concentration of charge carriers begins to increase at a greater rate. This portion of the current-voltage curve is called

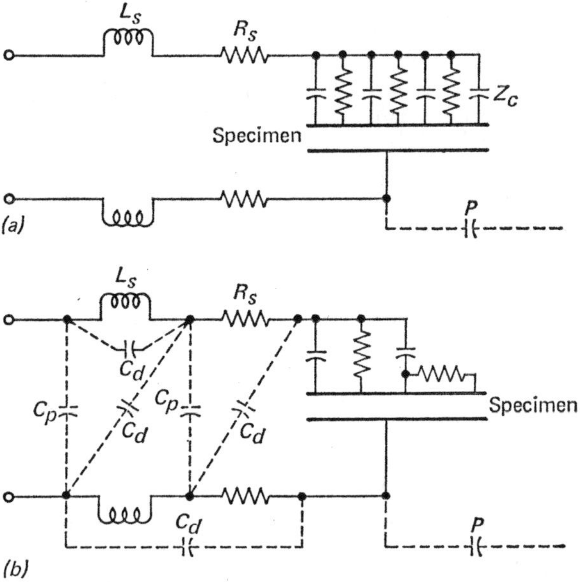

Fig. 3.1.6   Electrode-lead equivalent circuits in frequency range of 0–10⁸ Hz: (a) low-frequency lumped parameters; (b) high-frequency distributed parameters; L_s — series (lead) inductance; R_s — series (lead) resistance; C_p — parallel lead capacitance; C_d — distributed capacitances; Z_c — electrode contact impedances (RCL); P — proximity effects

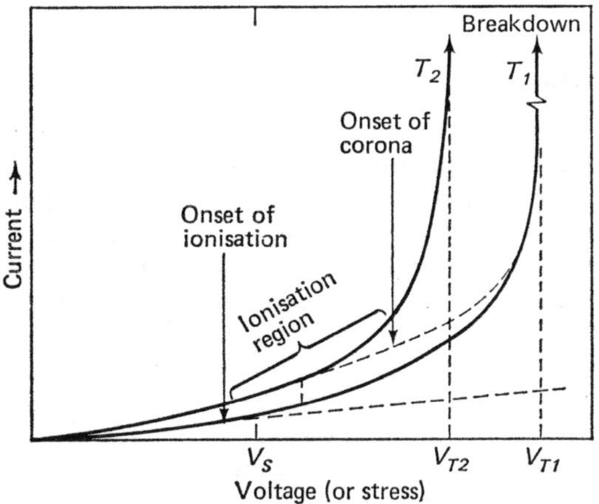

Fig. 3.1.7   Conduction current in an insulator up to the dielectric breakdown voltage at temperatures T_1 and T_2, where $T_2 > T_1$. V_s is safe operating voltage

the ionisation region (it is also called 'loss tip-up' or 'dissipation-factor tip-up') and is also evidenced by an increase in the dissipation factor. As the voltage is increased beyond the ionisation region, an increasing number of charge carriers attain sufficient energy to multiply: current increases rapidly and corona occurs.

Internal heating also increases, causing a higher corona level: this generates more heat, so that eventually thermal or corona runaway occurs and the dielectric breaks down. The corona region is sensitive to the rate of voltage increase. At slow rates of voltage increase, there is more time for heating of the material and consequently a lower breakdown voltage.

In carefully controlled laboratory tests, the magnitude of the temperature rise can be estimated. With a voltage decrease, the conduction current decreases (but at a slower rate than that at which it increased), intersecting higher temperature curves at each point. This intersection with a higher temperature curve occurs because of the heating of the dielectric by the internal corona and ionisation. The point at which corona current is no longer detectable is known as the corona extinction voltage (CEV). If sufficient time has elapsed for cooling to take place, a second increase in voltage causes the current to retrace the original curve, leaving an open-loop area.

Factors affecting dielectric strength

The usual voltage-breakdown test provides data which are influenced to varying extents by the environment, the dissipation factor, the temperature coefficient, the thermal conductivity, and the geometry (dimensions and design of the electrodes) of the dielectric. In general, the dielectric strength decreases with increasing temperature and as an exponential function with increasing thickness of the test specimen. High-frequency dielectric strength is lower than that at power frequencies, an effect usually attributed to increased dielectric heating of the material and high corona levels.

Breakdown tests

If we eliminate internal-heating effects, coronas, the surrounding medium, the geometry and electrode influences, we obtain the ultimate dielectric strength of the material itself at a given temperature. This is the intrinsic dielectric strength. It is measured at the point where extremely strong fields accelerating electrons through the material immediately produce the characteristic catastrophic failure. In simple structured materials, experimental results closely duplicate theoretically calculated values (based on bond strengths and lattice energies) of intrinsic dielectric strength. However, in organic structures, the competing reactions for electron generation, ionising collisions, and recombinations, and the effects of dipole-relaxation times are too complex to permit theoretical calculations to approximate experimental data.

Impulse tests for intrinsic voltage breakdown use contact electrodes to eliminate corona formation in the surrounding medium and fast-rise-time impulse voltages to minimise dielectric heating. These tests are used principally for theoretical research on molecular structure and to determine the effects of switching transients and lightning on power generators and transmission apparatus (the test for the latter is called the 'basic insulation test').

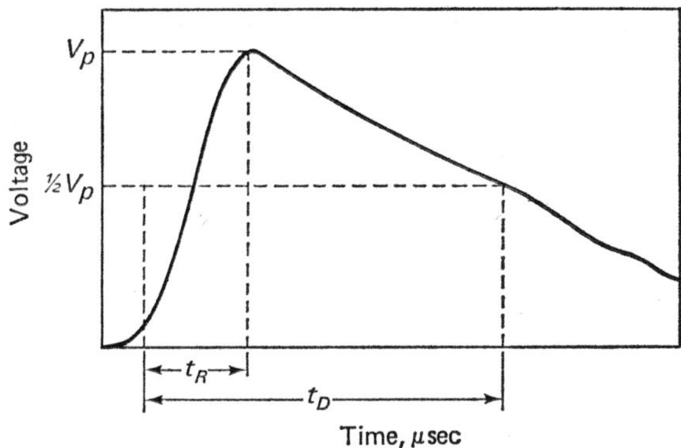

Fig. 3.1.8   Shape of impulse test waveform is specified by t_R/t_D

The shape of the impulse or surge voltage (Fig. 3.1.8) is given by specifying in microseconds the rise time to peak voltage and the time of decay to one-half the peak voltage. For example, the 1/50 curve has a 1 μsec rise time t_R and a 50 μsec decay time t_D.

In most tests on commercial materials, breakdown is caused by electric discharges that produce high local field stresses (Fig. 3.1.9). With solid materials corona discharges occur in the ambient medium and produce failure just beyond the edge of the electrode.

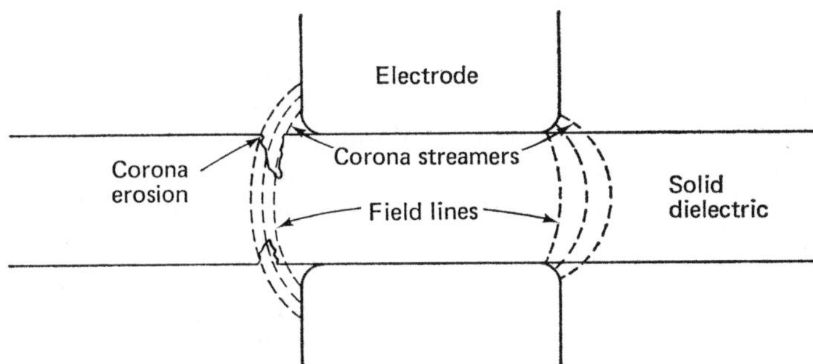

Fig. 3.1.9   Corona discharge produces high field stresses in solid dielectric and eventually cause failure

Two types of commercial tests are the short-term test, in which failure occurs in minutes and which is important for quality control and materials specification, and the long-term test, in which failure occurs in 20 to 30 years and which determines the voltage endurance and useful life of electrical apparatus. The conditions to be used in a number of these tests (the electrodes, surrounding media, etc.). The difference in voltages of short-term and long-term tests varies widely from material to material. For some the ratio may be 2 : 1, for others it is 50 : 1 or greater.

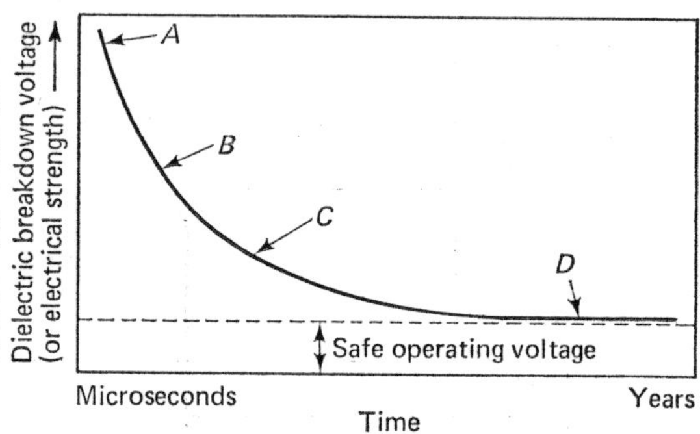

Fig. 3.1.10   With a constant applied voltage, the dielectric breakdown voltage decreases with time. Predominant failure mechanism during time of electrification are : A — electromechanical ; B — corona streamers in ambient media ; C — thermal heating ; D — corona-generated erosion, oxidation, and tracking

Figure 3.1.10 shows the dependence of dielectric strength of dielectric withstand voltage on the time of electrification for a given voltage-stress level. For extremely short periods (impulse conditions) the mechanism of failure is electromechanical and is associated with the molecular structure. Values of dielectric strength under these conditions begin to approach the intrinsic strength of the insulation. For periods of 10 to 1,000 sec, corona at the electrode edges and thermal heating become the major failure mechanisms. Values of dielectric strength may be 1/2 to 1/50 of those obtained under impulse conditions. As time is increased thermal-failure mechanisms come into control. For low voltage-stress levels applied for long periods, internal temperature rise is reduced by a lower rate of dielectric heating, allowing time for heat loss to the ambient surroundings. This permits corona to deteriorate the insulation by erosion, cutting, tracking, and oxidation (especially by formation of ozone, a more powerful oxidising agent than oxygen). Voltage-stress levels below the asymptote of the dielectric strength curve are the long-term safe operating voltages. This asymptotic value of dielectric withstand voltage corresponds to the value V_s in Fig. 3.1.7.

In practice, allowance must be made for shift in ionisation and corona levels of the material system due to ageing, environmental conditions, mechanical abuse, and chemical deterioration.

This brings us to a very important conclusion: dielectric-strength values obtained by short-term tests are generally of little or no value for the design of electrical insulating systems and are better suited for material specification purposes.

For the effective design of insulating systems, a knowledge of the corona extinction level (CEV) of the electrical insulating system and the conductors which are a functional part of that system is essential: the variation of the CEV with operating conditions and ageing is of paramount importance in predicting reliability.

Corona measurements

Corona is any electrically detectable, field-intensified ionisation that does not immediately result in complete breakdown of the insulation and conductor system. Corona is a low-power arc, whereas a true voltage breakdown is a high-power arc; in terms of equivalent circuits, corona is the current which flows through an equivalent high resistance, whereas in a voltage breakdown the current flows through an equivalent short circuit.

Corona degradation progressing into eventual catastrophic failure proceeds by several mechanisms. Corona generates heat that raises the temperature of the insulation, thereby increasing the dissipation factor, lowering the volume resistivity and dielectric strength, and accelerating the rate of deterioration. Corona generates ozone which is a more powerful oxidising agent than oxygen. The corona-current ions produce a cutting action on insulation surfaces and in voids. This supported by increased oxidation, leads to carbonisation, tracking and, finally, dielectric breakdown.

Applications

Corona measurements are a most useful tool for designing insulating systems and can inexpensively and quickly provide valuable information about the probable life of a particular design. They overcome most of the deficiencies inherent in dielectric strength and over-voltage tests. The principle areas where corona measurements can be used advantageously are:

1   APPARATUS DESIGN. In prototype stages, design weaknesses in the system that would lead to eventual long term failure can be located quickly and measures taken to correct them.

2   QUALITY CONTROL. Corona testing can locate production deficiencies such as voids in moulded sections and embedded parts, cracks, and conductive paths. Insulation-resistance testing is another useful tool in this area, especially when used in combination with high-humidity exposure or water immersion.

3   MAINTENANCE. Corona tests, in conjunction with insulation-resistance tests, can be used in the scheduling of motors, transformers, and other gear for cleaning to prevent down time.

4   MATERIALS INVESTIGATIONS. Corona tests (especially corona endurance tests at various intensity levels) provide an important key for the development of improved materials, compounds, molecular structures, and additives (stabilisers, antioxidants etc.).

Breakdown in fluids

Although I have covered breakdown phenomena as related to solid insulations, the discussion also applies generally to liquids and gases with the differences described below.

Voltage breakdown in a liquid is influenced by contaminants and moisture. An applied electric field tends to sweep contaminants into the area of highest stress. The increasing concentration of contaminants at the point of highest potential gradient eventually produces a 'bridge' across which arcing can occur. Arcing produces decomposition products, which are themselves contaminants, and raises the local temperature, producing higher losses, higher conductivity, and as a result, further deterioration. When voltage stress is removed, the contaminants can diffuse out of the area of breakdown and, if the total volume of liquid is great enough can become sufficiently dilute to allow the system to operate at least for a time.

Voltage breakdown in gases is influenced to a great extent by electrode separation and pressure. It is generally stated as Paschen's law: the potential necessary to cause sparkover in a gas depends on the product of the gas pressure and the spark length in a uniform field (Fig. 3.1.11). Voltage-breakdown test results are influenced by the surface condition of the electrode, the concentration of field stress at the surface (a function of radius of curvature), and the composition of the electrode.

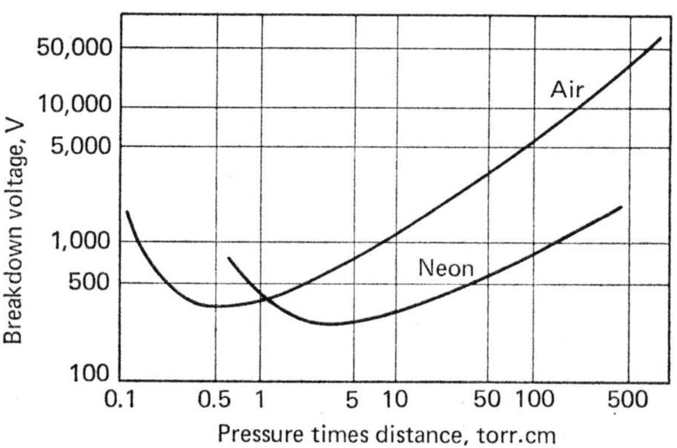

Fig. 3.1.11   Voltage breakdown in air and neon depends on the distance between the parallel electrode plates and on the gas pressure

A measure of the ease with which electrons are released from a material and reactions that occur during gas discharge are highly complex and involve ionisation, recombination, and elastic collisions of electrons. External energy sources such as ultraviolet light produces ions which lower the breakdown point. Charge carriers are either generated within the gas or injected by the electrodes.

Insulation between conductors with air as a parallel insulator (Fig. 3.1.1 poses a special situation, as it is complex in its field configuration and highly subject to environment. Solid surfaces in contact with air normally

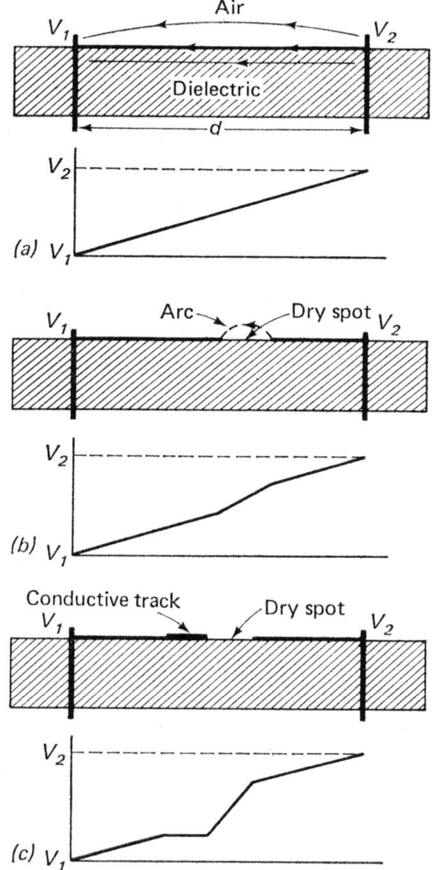

Fig. 3.1.12   (*a*) Conduction paths for leakage current through air, dielectric, and air-solid interface; (*b*) increased current caused by arcing across dry spot generates heat, and dry spot expands (eventually, dry-spot size prevents sustained arcing and dielectric cools); (*c*) repeated arcing and resultant expansion of dry spot produces decomposition products which form a conductive path (track) along the surface of the dielectric

have an absorbed layer of moisture on the surface. Water, which has a high dielectric constant, readily ionises impurities, producing a semiconducting film between the conductor surfaces. Leakage current in the moisture film generates heat, and eventually the moisture evaporates in spots, leaving dry un-ionised surfaces with relatively good insulating properties. If the potential difference across a spot exceeds the breakdown strength of the air across the width of the dry spot, arcing occurs. Large amounts of heat generated by the arc further increase the width of the dry spot, allowing the arc to creep across the insulation toward the conductors. If the potential difference is insufficient to sustain the arc across the full gap, the arc is extinguished and the system cools. Upon reformation of the film of moisture, the arcing process is repeated. Eventually, the arcing causes the insulation to deteriorate to the point where a conductive carbon track is formed, thereby completing the failure of the insulating system. Materials which, by themselves do not produce carbonous products on pyrolysis (for example, ceramics and silicons) are termed non-tracking and are usually chosen for high-voltage applications. In dusty locations, carbon tracks may be formed by the repeated pyrolysis under action of the arc of the dust particles, thereby causing system failures.

Early methods for determining arc resistance used a low-power intermittent arc and measured the life of a material in seconds to failure. The electrodes are often inverted so that the arc is in close contact with the material under test. (The test is only suited for materials which do not melt under the heat of the arc.) The methods are of limited value in that they fail to rate materials in order of their survival under arcing conditions in switches and apparatus. Later methods using synthetic dust and water spray are more useful.

Dielectric materials in capacitors

Dielectrics are essentially materials in which the application of an electric field gives rise to no net flow of electric charge but only to a displacement of charge, i.e. polarisation occurs. This fact implies that dielectrics are capable of storing charge and it is this property which is exploited in capacitive devices. The capacitance of a parallel plate capacitor is given by the following relationship:

$$C = kA/4\pi t$$

where k is the permittivity of the dielectric, A the surface area of the electrodes and t the thickness of the dielectric. For a given dielectric material, capacitance is increased by either increasing the surface area or decreasing the thickness. Both these factors are usually controllable, and are important in capacitor design.

When an ideal capacitor is subjected to an alternating potential the current leads the voltage by a phase angle of $\pi/2$, i.e. by a quarter of a cycle. Under these circumstances no power is absorbed by the dielectric and the capacitor is said to have zero loss. However, a real dielectric always possesses some loss, although this may be negligibly small in many practical capacitors. Losses may occur for the following reasons.

1 Migration of charge through the dielectric (leakage current).
2 Dielectric absorption.
3 Ohmic resistance of the capacitor electrodes and leads.

Loss may be defined either in terms of the cosine of the angle between the voltage and the current (power factor) or the tangent of the angle by which the current lags behind that in the ideal dielectric (dissipation factor).

Provided the power factor does not exceed about 0·1, the two expressions give essentially similar values. Percentage dissipation factor, i.e. 100 tan δ is widely used in commercial practice and is used throughout this chapter.

Capacitance is normally measured by comparing the impedance of the unknown capacitor connected to either a resistance in parallel C_p, or a resistance in series C_s. If the dissipation factor is high these values are not identical and are related by the following equation:

$$C_p = C_s/l \tan \delta^2$$

Similarly dielectric loss may be considered to occur either in a parallel resistance R_p or series resistance R_s. The dissipation factor is then given by:

$$\tan \delta = (2\pi f C_p R_p)^{-1} 2\pi f C_s R_s$$

where f is the frequency of the measuring signal. It should be emphasised that with practical capacitors, neither capacitance or dissipation factor are invariable but depend on frequency, voltage and temperature of measurement.

Capacitor dielectrics may be conveniently classified according to thickness as follows: thin $\langle 1\ \mu m$, thick$\rangle 1\ \mu m$. As a consequence of the rapid growth in microelectronics in recent years, considerable research activity has been concentrated in thin-film areas.

Advances in the techniques of deposition have been made using a wide variety of methods, but especially those involving vacuum processes. The thick dielectrics belong, in general, to an older technology and progress in this field is related essentially to an improvement in quality.

Thin-film dielectrics are primarily used in capacitors intended for low voltage applications, often in the region of 10 to 35 V. If the unit is to be one element in a thin film microcircuit, it is essential that the processing conditions should be compatible with those for applying other passive or active devices on the same substrate. In most cases the dielectrics are inorganic oxides and usually have either an amorphous or microcrystalline structure. Although normally possessing a high intrinsic breakdown strength 1 to 10 MV/cm, because of the extreme thinness of the layer, the electrical properties are often dominated by the presence of pin-holes or other structural defects. Capacitors having thick dielectrics are generally used in conventional circuitry and are available for use at voltages up to several thousand volts. Thick dielectrics are often organic polymers but also include a number of inorganic materials. Breakdown strengths are normally in the region of 0·01 to 1·0 MV/cm.

Thin-film dielectrics

Anodised oxides

Capacitors with dielectrics prepared by anodisation were the earliest thin-film types. They are used in both micro and conventional circuitry, these dielectrics are formed at the surface of certain metals, the so-called 'valve' metals, during anodic polarisation in a liquid electrolyte. The oxide layer so produced is characterised by a voltage-dependent thickness, a mainly amorphous structure and a high breakdown strength. Since some of the electrolyte is incorporated in the oxide during growth, the physical and electrical properties show some variation according to the solution used. Typical properties of a number of the anode oxides are shown in Table 3.1.1.

In spite of numerous studies of alternative metals only anodized aluminium

N

Table 3.1.1   **Typical properties of anodic oxides in liquid electrolytes**

Metal oxide	Thickness relationship, Å/V	Permittivity	Field at limiting thickness, MV/cm
Aluminium	13·5	8	7
Tantalum	16	27	6
Niobium	24 to 30	41	4
Zirconium	25	20 to 30	4
Titanium	50*	100*	2*
Silicon	5	3·5	20

* Anodised in a fused salt electrolyte. Markedly different values are obtained when formed in conventional liquid electrolytes

or tantalum are at present employed commercially. As used in conventional circuits the electrolytic capacitor consists essentially of a metal coated with an anodic dielectric separated from a second metal, not normally anodized, by a liquid electrolyte. When a wound construction is used, paper spacers which readily absorb the electrolyte are interposed between foil electrodes. Unless constructed with two components or two anodes in opposition, these capacitors may only be used under conditions in which the anodized electrode remains essentially positive, since high currents are passed under reverse polarization.

The reason for the rectifying behaviour of the metal-oxide–electrolyte system has not been satisfactorily explained. However, two general theories have been advanced. The first, taking into account the large number of physical defects in the oxide supposes that these features become filled with oxygen during anodic polarisation, so insulating the underlying metal from the electrolyte. On reverse bias, the hydrogen which is liberated may either dissolve in the metal substrate, thus failing to 'seal' the fissures, or, alternatively become incorporated in the oxide layer, resulting in a considerable increase in conductivity. The second theory regards the rectification as originating from the presence of a *pn* or a *pin* junction within the oxide. This is said to arise from a consideration of the mechanism of formation, which involves movement of metal and oxygen ions in opposite directions. In consequence, an oxide layer containing excess metal *p*-type is formed adjacent to the metal and a layer containing excess oxygen *n*-type is formed adjacent to the electrolyte.

The capacitance of an electrolytic capacitor is enhanced by using highly etched foils or, additionally in the case of tantalum, porous bodies prepared from vacuum-sintered powder. The presence of the electrolyte ensures the self-healing of areas of dielectric damaged during construction. Accordingly, electrolytic capacitors are usually very reliable in operation. This is particularly so with the tantalum type, the oxide of which is less chemically reactive and, therefore, more stable in the presence of the electrolyte, than aluminium oxide. For example, exceptionally high capacitance, low voltage, tantalum units are available whose dielectrics are only a few tens of Angstrom units thick. In the absence of an electrolyte, such capacitors would be impractical. However, because of the presence of the rather poorly conducting liquid, wet electrolytic capacitors often have high dissipation factors 1 to 15 per cent at 1 kHz and also a restricted operating range in both frequency and temperature. A development which has improved the frequency and temperature characteristics of the electrolytic capacitor is the substitution of the liquid by a semiconducting oxide. In practice, this innovation has largely been confined to porous tantalum types impregnated with manganese dioxide, the counter electrode being usually applied in the form of a dispersion of graphite. These units, known as solid electrolytics, are still polar but will sustain rather higher reverse voltages than the wet type.

Owing to the thinness of the dielectric layer, electrolytic capacitors are notable for their large capacitance values and comparatively small size. Thus commercial units may range in value from a few to several thousand microfarads. The wet aluminium types are widely used in radio and television circuits for smoothing, decoupling and by-pass applications. The non-polar capacitor is suitable for intermittent use in a.c. systems, e.g. motor starting. The wet tantalum types are generally used when higher reliability and lower leakage current are important: solid units have a superior performance at higher frequency and lower temperature.

In existing microelectronic practice, only anodised tantalum or evaporated silicon monoxide dielectrics are used to any degree. Thin films of sputtered tantalum, because of their relatively high purity and surface smoothness compared with bulk material, permit the formation of dielectric layers of such improved quality that evaporated metal counter electrodes may be applied directly without the need of an intervening layer, either of electrolyte or manganese dioxide. These units are very largely non-polar and perform well over a wide range of temperature and frequency. Dissipation factors are of the order of 0·5 to 1·0 per cent at a frequency of 1 kHz for units rated at 0·1 μF/cm^2. Resistor elements in the circuit are prepared either from sputtered tantalum patterns, which are anodised to trim to the desired value, or from sputtered tantalum nitride. A recent development is the anodisation of aluminium or tantalum in an oxygen glow discharge. Using this technique, capacitors may be prepared entirely in the vacuum system, a factor of some importance in assessing production capabilities. The gaseous process also makes possible the anodisation of certain materials, e.g. a lanthanum-titanium alloy which may be difficult in a liquid electrolyte. A summary of recent work on thin-film anodised capacitors is given in Table 3.1.2.

Vacuum-deposited films

The impetus behind the use of vacuum techniques for preparing dielectric films has been the need for high-quality capacitors in microelectronic applications, e.g. active filters and coupling and decoupling in linear amplifiers. The systems studied have generally embraced the valve metal oxides and a few organic polymers. In practice a parallel plate geometry is employed, the dielectric being interleaved between thin vacuum deposited metal electrodes. The thickness of the metal electrodes is usually of the order of 1,000 to 4,000 Å and represents a compromise. Thick layers are desirable to maintain the dissipation factor at low values, particularly at high frequency:

Table 3.1.2   **Summary of capacitor characteristics and applications**

Type	Permittivity		Characteristics	Applications
Electrolytic:				
(a) wet (Al and Ta)	Al 8 Ta 27		Polar, large capacitance-volume ratio. Limited temperature and frequency range. Ta more expensive.	Smoothing, decoupling, by-pass, etc. Ta chosen for higher reliability and lower leakage current. Non-polar Al types suitable for intermittent use in a.c. circuits.
(b) solid (Ta)			Generally as for wet, improved performance at higher frequencies and overwider temperature range.	As above.
(c) thin film (Ta)			Largely non-polar, high capacitance per unit area.	Microelectronic circuits.
Evaporated silicon monoxide	5		Lower capacitance than Ta types but superior dissipation factor.	
Paper	Approx. 5		Cheap, generally good performance, insulation resistance falls rapidly with increasing temperature.	General purpose: power factor correction, blocking, audio and high frequency by-pass.
Plastic:				
(a) polystyrene	2·3		High insulation resistance, low dielectric absorption and dissipation factor.	Charge storage, filter circuits.
(b) melinex	3·1		Capable of working at higher temperatures than paper or polystyrene.	Replacement for paper in higher temperature systems.
Mica	2·5 to 7		High stability and low dissipation factor.	Filter circuits. Standard capacitors.
Ceramic:				
(a) low permittivity	5 to 20		Low dissipation factor.	May be used instead of mica in some circuits.
(b) medium permittivity	20 to 200		Moderately high capacitance-volume ratios. Controlled temperature coefficients of capacitance.	LC resonance circuits.
(c) high permittivity	1,000 to 25,000		Large capacitance-volume ratio approaching that of electrolytics. Permittivity markedly voltage and temperature sensitive.	By-pass, blocking and smoothing circuits.
Glaze	10 to 10,000		Relatively cheap, printed, RC networks. Encapsulation not required.	Microelectronic circuits.

on the other hand, an essential feature of thin film capacitors is self-healing at defects in the dielectric brought about by evaporation of the electrode above the conducting site and, for this action, relatively thin metal layers are required.

Owing to the possible need for capacitance values in the region of $1 \cdot 0 \, \mu\text{F/cm}^2$, or higher, a few ferroelectric materials have also been investigated. However, high effective permittivities have rarely been achieved, partly because of the difficulty in obtaining crystalline structures by vacuum processes and partly because of the intervention of a space charge effect known as field penetration, at the electrodes. This effect leads to the formation, under the influence of an electric field, of insulating layers which may have capacitance values of the order of $1 \, \mu\text{F/cm}^2$, in the surface regions of an electrode at a dielectric-metal interface. Since the layers are in series with the primary dielectric, an over-all reduction in capacitance is obtained. The effect is apparent with metal–dielectric–metal systems of very thin anodic oxides but is not found when the counter electrode is replaced by a liquid electrolyte. Where relatively high permittivity values have been achieved with vacuum deposited dielectrics, dissipation factors are high and breakdown strengths are reduced.

In general, it is often found that highest breakdown strengths are achieved only with essentially amorphous films. The reason is probably associated with the concomitant absence both of a well defined electron band structure and also of grain boundaries.

Evaporation

Silicon monoxide, prepared by vacuum evaporation of the bulk material from a heated boat, is the most important of the vacuum deposited layers at the present time. Silicon monoxide can have a relatively low dissipation factor, of the order of 0·1 per cent at 1 kHz for capacitors rated at 0·01 $\mu F/$cm^2, and is fully compatible with evaporated nickel–chromium or silicon monoxide–chromium resistors. However, the dielectric properties depend very much on such factors as the partial pressure and nature of the atmosphere during evaporation, rate of deposition, temperature of substrate and treatment after deposition, especially annealing at elevated temperatures. The actual composition of the dielectric layer is uncertain but probably lies between silicon monoxide and dioxide. Evaporation of silicon dioxide, which is difficult owing to its high boiling-point, probably gives essentially the same material when the deposition is effected from an electrically heated boat.

However, if the evaporation is performed by electron bombardment the dielectric layers are close to silicon dioxide in composition. The dissipation factor of the dioxide is higher than that of the monoxide but the breakdown strengths of the two materials appear to be similar.

Various attempts have been made to evaporate high-permittivity dielectrics of certain ferroelectric materials, especially barium titanate. Direct evaporation of barium titanate leads to a partial separation of the constituent oxides, resulting in the deposition of a film rich in barium oxide. This may be avoided by flash evaporation of small grains of the bulk material: however, to ensure stoichiometric composition it is necessary to evaporate in a partial oxygen atmosphere, otherwise a reduced and conductive titanate is formed. An alternative approach is to co-evaporate barium and titanate oxides using electron beam heating, both methods have given films with permittivities of 500 to 1,330 but high dissipation factors, approximately 15 per cent at a frequency of 1 kHz, and low breakdown strength, 0·2 MV/cm, are also found. The low-permittivity layers appear to be amorphous but those having higher values show the ferroelectric perovskite structure.

Reactive sputtering

The term sputtering is applied to a process in which atoms are ejected from a metal cathode as a result of bombardment by positive gas ions, usually in a glow discharge. When accomplished in an atmosphere of oxygen or an argon–oxygen mixture, metal oxides are deposited. The method has distinct advantages since the high temperatures associated with evaporation are avoided and it is clearly most useful for materials having intrinsically high melting points. The mechanism of oxidation has not been fully explained, however, it seems likely that oxidation occurs on the substrate after deposition in an atmosphere of low oxygen content, and at the surface of the cathode prior to emission in an atmosphere of high oxygen content. The discharge is normally maintained by a d.c. potential but there are certain advantages, notably in improving the purity of the deposit, in using asymmetric a.c. field. More recently RF fields have been used to sputter from insulating electrodes, e.g. glass and silica.

Most of the simple oxide dielectrics, e.g. tantalum, zirconium, titanium, aluminium and silicon, have been prepared by reactive sputtering. The processes are generally slow, especially with aluminium oxide, and the deposits, for reasons not yet explained, show a marked tendency to be inter-penetrated with pinholes. However, intrinsic breakdown strengths can be high, in the region of several megavolts per centimetre. Dissipation factors do not, in general seem to be very different from those obtained by anodisation. The preparation of materials possibly possessing ferroelectric properties may be investigated by reactive sputtering either from multi-component cathodes, e.g. lead–titanium or lead–titanium–zirconium, or from cathodes of conductive ferroelectric materials. A preferred method, which gives greater control over composition, is to use a split-sector cathode in which the current through the separated metal components can be individually controlled.

Only moderately high permittivities of 100 to 200 have been obtained in these studies and, as with the evaporated ferroelectrics, breakdown strengths were quite low — 0·2 MV/cm. A study of the reactively sputtered lead–titanium system has indicated a low dielectric constant of 20 in a region of composition corresponding to 55 to 85 per cent lead content. Within this region, the films appear to be quite amorphous and non-ferroelectric. However, they are free of pinholes and have breakdown strengths of 2 to 3 MV/cm.

Catalysed deposition

A relatively new but rapidly expanding field of study is concerned with the catalytic deposition of dielectric layers from the vapour phase at solid surfaces. Often an external stimulation of the gas; by a glow discharge or by irradiation with either ultraviolet or an electron beam, is required. However, reaction may also occur at hot or cold surfaces in the absence of excitation. Many of these processes have not yet been fully assessed but it is probable that they will play a considerable part in future thin film developments.

When excitation is accomplished by a glow discharge, the plasma is usually maintained with an RF field, so eliminating the need for internal electrodes. Silica may be deposited at a pressure of about 0·1 torr using a mixture of silane with either nitrous oxide or carbon dioxide. These films may be deposited at rates as high as 2 to 10 μm/hr and so the process may also have a useful application to encapsulating microcircuits. Dielectric layers of silicon nitride have been prepared from mixtures of silane and ammonia and of silicon carbide from silane and methane. Additionally, several typical organic polymers such as polystrene have been deposited by this method and also, rather unusually, polymers of benzene or methane.

Ultraviolet or electron-beam excitation has largely been applied to preparing organic polymers. The ultraviolet method has been used for the polymerisation of 1,3-butadiene or methyl methacrylate and the electron-beam method for diffusion pump oil, siloaxanes, hydrocarbons and epoxides.

Silica may be deposited without excitation from tetra-ethyl silicate at a pressure of approximately 1 torr on surfaces heated to 650 °C. Good results including relatively high breakdown strengths and low dissipation factors, have been obtained but owing to the high temperature required this method is not generally compatible with microelectronic techniques. Similarly, aluminium oxide has been prepared by passing aluminium chloride, hydrogen and carbon dioxide over either molybdenum or nickel strips heated to 1,000 °C. The deposited film appears to consist of crystals of α-alumina dispersed in an amorphous phase. Layers of thickness in excess of 1 μm are reported to have a dissipation factor of 0·1 per cent at a frequency of 1 MHz and a breakdown strength of several kilovolts per centimetre.

An interesting organic polymer, poly-p-xylylene, with a permittivity of 2·65, is formed by heating xylene under reduced pressure to approximately 550 °C and then cooling to 50 °C. The polymer, which is deposited directly the vapour comes into contact with metal substrates, can be prepared in thicknesses varying from 100 Å to 300 μm. Wound capacitors made from aluminium foil electrodes coated with several micrometres of poly-p-xylylene will operate at temperatures up to 170 °C. Dissipation factors are stated to be less than 0·1 per cent at frequencies between 10^2 to 10^6 Hz and breakdown strength appears to be about 3 MV/cm.

Thermal oxidation

Few satisfactory oxide dielectric can be prepared by thermal oxidation of the metal in an appropriate atmosphere of air, oxygen, steam or a mixture of these. Typically metals like tantalum or aluminium form only very thin amorphous oxides, e.g. 50 to 100 Å at temperatures up to about 500 °C. At higher temperatures, mixed amorphous-crystalline oxides possessing inferior dielectric properties are formed. There is also a tendency, very marked with tantalum and niobium, for the metal to partly reduce the oxide at the metal–oxide interface, leading to the establishment of a semiconducting layer having undesirable field-dependent properties. However, satisfactory dielectrics possessing high breakdown strengths have been prepared at temperatures in the region of 900 to 1,200 °C on silicon.

Capacitors prepared from these films are used in silicon integrated microcircuits in which all devices, both active and passive are made on a single slice of silicon.

An improved technique which has recently been described involves a low pressure (1·0 torr) oxygen plasma, excited by microwave radiation. Silicon disks supported in the plasma form oxide films up to a limiting thickness of 4,400 Å at temperatures stated to be less than 300 °C. Thicker films may be prepared at an anode by applying a small d.c. potential (30 to 90 V) between two disks; under these conditions the cathode remains substantially free of film. Although the mechanism of film formation has been explained in terms of thermal oxidation, it is possible that this process is essentially another form of gaseous anodisation. However, the method is capable of producing quite thick films of high quality with breakdown strengths of 10 MV/cm.

Thick dielectrics

Paper

Paper capacitors are typically prepared in a tubular form by rolling strips of paper between thin metal foil electrodes. The intrinsic permittivity of the papers used is in the region of 5·6 to 6·6. The wound coil is vacuum-impregnated after drying and prior to encapsulation with a fluid in order to eliminate air pockets. This leads to an increase in effective permittivity and also to an improvement in stability, since plasma breakdown of the gas phase is thereby largely avoided. The choice of impregnant is governed by such factors as circuit application, voltage requirements, power dissipation and operating temperature. The fluids are usually:

1 Hydrocarbons.
2 Mineral oil.
3 Petroleum jelly.
4 Chlorinated hydrocarbons.
5 Chlorinated diphenyls.

Unless a suitable stabiliser is added, the chlorinated liquids are not used in capacitors intended for d.c. operation, since corrosion of the anode can occur as a result of the discharge of chlorine. Additionally, it is essential to have good sealing, otherwise hydrolysis of the impregnant may also lead to corrosion. However, the chlorinated materials can lead to a reduction in costs and give increased capacitance. For high temperatures up to about 200 °C, fluorinated esters may be used; another impregnant sometimes employed is epoxy resin which has the advantage that further encapsulation is not required.

Owing to the presence of impurity inclusions, paper dielectric contains minute short circuiting paths. Consequently, capacitors are normally constructed with several sheets of paper, each having a thickness in the region of 5 to 10 μm, the precise number of layers is determined by the voltage requirement. However, for low voltage requirements up to about 200 V, single sheets of metallised paper are used. Because of the thinness of the electrodes, often approximately 1,000 Å of aluminium, these units are self-healing.

Paper capacitors are available in values from a few hundred picofarads to several microfarads and are relatively cheap. Breakdown strength is in the

region of 2 to 4×10^5 V/cm and dissipation factors are approximately 0·5 to
1·0 per cent at 11 Hz. However, insulation resistance falls rapidly with
increasing temperature, although not having outstanding electrical properties,
paper capacitors are sufficiently good for use in a wide range of applications,
e.g. power factor correction 25 to 400 Hz, filtering of rectified a.c. blocking,
and audio- and high-frequency by-pass.

Plastic

The principle plastic dielectric in present use is polystyrene. Capacitor construction is similar to that used for paper dielectrics, i.e. wound tubular units
containing plastic films interposed between thin metal foils. The plastic is
biaxially stressed during manufacture. After winding, the units are heated
above the relaxation temperature, under which condition the films contract
into a stable form with the partial elimination of air pockets between the
electrodes. Metallised film is used for the lowest voltage applications.

Unfortunately, unlike equivalent paper units, these are not normally self-
healing, therefore high electric current surges, which occurs at conducting
sites in the films, can produce deposites of carbon which lead to the presence
of permanent short circuits. An interesting method of overcoming this effect
is to coat the back of the electrode with an anodising medium, this process
eliminates the carbonaceous material in the form of carbon dioxide.

Polystyrene capacitors are characterised by an insulation resistance which
is higher than that of any other dielectric. The insulation resistance falls
relatively slowly with temperature and additionally, the units show a
markedly low dielectric absorption. Consequently these units are especially
useful in applications requiring a temporary storage of charge followed by
transfer to another part of the circuit, e.g. digital computing or timing
circuits. Dissipation factors are about 0·02 per cent at 1 kHz and are virtually
independent of frequency. Because of this and generally good stability, polystyrene capacitors are tending to replace mica in wave filter circuits and as
standards. However, the dielectric has a relatively low softening-point and
the temperature of operation is limited to 70 °C

Another plastic film in commercial use is polyethylene terephthalate,
which has a permittivity of 3·1. Capacitors employing this dielectric are
capable of working at temperatures up to 130 °C and are cheaper to prepare
than polystyrene. Dissipation factors are about 1·0 per cent at 1 kHz and
these units can generally be used in place of paper capacitors when the
higher temperature of operation is required. New materials now coming into
use, e.g. polypropylene with a permittivity of 2·2 and polytetrafluorethylene
with a permittivity of 2·0, will enable plastic capacitors to operate at substantially higher temperatures.

Mica

Mica, a naturally occurring silicate, exists in a laminated state and is capable
of being cleaved into thin sheets. The permittivity varies with source of origin
but is generally in the region of 2·5 to 7. Capacitors are constructed in the

form of stacked layers, thin sheets of mica, each approximately 25 μm thick, being interleaved with metal foil. An alternative construction, which results in improved stability, is the use of silvered mica sheets. The sheets are often impregnated with wax in order to prevent the intrusion of water and to eliminate air gaps, the latter being of special importance with the foil types.

Mica capacitors are made in values from a few picofarads to several microfarads. Breakdown strengths are about 4×10^5 V/cm and the units have an excellent long-term stability. Dissipation factors, which are almost independent of frequency, are low, about 0·03 per cent at 1 kHz and may substantially reduce if the mica is specially selected and dried. These capacitors are used in wave filter and network applications and also as standards.

Ceramic

Ceramic dielectrics embrace all those which are prepared by a process of thermal sintering of inorganic powders, usually mineral oxides, at temperatures in the region of 1,100 to 1,400 °C. In the green unsintered state, the powders, which are coated with an organic binder, may be pressure moulded or extruded to the shape required of the finished capacitors; disks, plates or tubes. Because many of the compositions are based on titanium dioxide, sintering must be accomplished in air if the dielectrics are to have a high insulation resistance. The dense, sintered bodies are very hard and chemically inert. The electrodes are normally applied by thermal decomposition of a brushed or screen printed silver paint.

Low permittivity, 5 to 20

Low-permittivity compositions are used when minimum dissipation factor is paramount and high capacitance, i.e. not exceeding about 10,000 pF, is not required. Dissipation factors are of the order of 0·1 per cent or less at a frequency of 1 MHz and show a tendency to increase at either higher or lower frequencies, specially prepared alumina can have an exceptionally low loss in the region of 0·003 per cent, which is superior to the best mica capacitors. Breakdown strengths of these dielectrics are 0·5 to 1·0 MV/cm and temperature coefficients are in the region of ± 100 ppm/degC. Low-permittivity ceramic capacitors may be used in place of mica in some circuits.

Medium permittivity, 20 to 200

These dielectrics fulfil a requirement for relatively high quality units which possess rather higher capacitance values than can be conveniently achieved with the low permittivity materials. The two principle compositions used are titania 90 to 100 or calcium titanate 160. Although having generally similar properties to those prepared from low-permittivity dielectrics, these capacitors invariably have markedly higher temperature coefficients.

Titania has a negative temperature coefficient, − 800 ppm/degC. These characteristics have led to the development of a range of units known as temperature compensating capacitors. The basic feature is the addition in varying proportions of either magnesium or zinc oxide or a mixture of the two, with titania, to give a composition after sintering which has a controlled temperature coefficient. These capacitors are used in LC resonance circuits, changes in impedance with temperature of the components being counterbalanced to prevent detuning.

High permittivity, 1,000 to 20,000

High-permittivity dielectrics are largely based upon the ferroelectric material, barium titanate. Ferroelectrics are crystalline materials which possess a spontaneous polarisation that can be reversed in an electric field. The relationship between dielectric displacement and polarisation is not linear, but exhibits hysteresis. The origin of ferroelectricity lies in a distortion of the unit cell of the lattice which results in the presence of a finite electric moment. At a certain temperature the Curie point which coincides with a sharply defined maximum in permittivity, spontaneous polarisation, and hence ferroelectricity, disappears. In the case of barium titanate, the permittivity reaches a value of 10 to 12×10^3 at the Curie point, 120 °C.

The objective in preparing a high-permittivity composition is, firstly to displace the Curie point towards room temperature, and secondly, so to smooth the relation between permittivity and temperature that practical capacitors having relatively low temperature coefficients can be prepared. This may be achieved by adding strontium, calcium, tin and zirconium oxides to the barium titanate prior to sintering. Compositions based on Ba, Sr, Ca, O, Sn, Zr, O_2 have permittivities in the region of 1,000 to 3,000 and temperature coefficients of about $\pm 5,000$ ppm/degC in the range 20 to 100 deg C. Lead oxide additions tend to increase the Curie temperature; however, when combined with compositions of the type Ba, Pb, O, Ti, Sn, Zr, O_2 very high permittivities up to 25,000 may be obtained in the region of room temperature. As may be expected, capacitors prepared from these dielectrics possess a sharp peak in permittivity–temperature relationship and have temperature coefficients of $\pm 20,000$ ppm/degC.

Capacitors with modified barium titanate dielectric are made in values ranging from about 0·001 to 10 μF. The effect of d.c. fields is to reduce the capacitance and to flatten the capacitance–temperature relationship. Breakdown strengths are of the order of 0·1 MV/cm and dissipation factors are between 1 and 2 per cent at 1 kHz. Because of the variability of permittivity with temperature and voltage, and to some extent with time, these capacitors are used in by-pass blocking, or smoothing circuits.

Various attempts have been made to improve further the capacitance–volume factor of the high permittivity capacitors. One method is to stack thin plastic-bound layers of barium titanate with interposed electrode patterns, either of powder or screen printed metal.

Because the subsequently pressed block must be sintered in air, platinum or palladium, electrodes are used. However, if a proportion of manganese dioxide is added to the titanate, sintering may be performed in reducing atmospheres, and in this case a considerable saving in cost may be made by using nickel electrodes. Using this system, packing densities approaching 3 μF/cm for units working at 50 V have been obtained. Dissipation factors may be only 1 per cent between 0 and 100 °C. The breakdown strength of the multilayered dielectric is 0·02 MV/cm. An alternative approach makes use of the highly conductive body which is obtained by heating barium titanate in hydrogen or carbon monoxide. If re-oxidation in air is performed under carefully controlled conditions it is possible to limit the process to the surface layers only, using this technique it is possible to prepare small discs having capacitance values equivalent to several tens of microfarads per centimetre with working voltages of a few volts. However, the capacitance of these units are voltage dependent and dissipation factors may be several per cent at 50 Hz, rising rapidly with increasing frequency. Higher capacitance values can be obtained by the use of either etched or semi-porous disks. Efforts have also been made to reoxidise the reduced ceramic by anodisation, but this method does not appear to be entirely suitable.

A relatively new range of high permittivity dielectrics, which are not ferro-electric, is based on solid solutions of strontium and bismuth titanates, these compositions possess permittivities of about 1,000 and have a high insulation resistance. The change of permittivity with temperature is relatively smooth, hence these materials are suitable for capacitors intended for rather high voltage and temperature applications.

Printed glaze

Capacitors with glass dielectrics, in the form of stacked sheets or either as a screen-printed glaze or enamel, have been manufactured for a number of years. Recently there has been a renewed interest in the glaze technique because of the possibility of using this process to fabricate microcircuits. The glaze consists of a dispersion of glass-forming oxides in a supporting liquid medium which is capable of being printed through silk screens to give well defined patterns. On heating, usually at temperatures in the range of 400 to 600 °C, the glaze melts to form a continuous film. Conducting layers can be prepared by introducing into the dispersion noble metal compounds which are reduced to the metal on heating. Thus, capacitors, inductances and interconnections can all be prepared by this process and it is possible that active devices may also be achieved eventually, furthermore not only is the printed technique essentially cheap but, because of the glaze, no further encapsulation is normally needed.

Glaze dielectrics prepared from mixtures of lead monoxide and silica have a permittivity of about 10, but appreciably higher values of the order of 1,000 to 10,000 have been claimed for zinc oxide–bismuth oxide systems. Commercial capacitors are available with dielectrics having a thickness of several micrometres and may range in value from about 5 to 10,000 pF. Dissipation factors are in the region of 0·2 per cent at a frequency of 100 kHz and breakdown strength appears to be about 10 V/cm.

Most of the present research activity in the field of dielectric materials is devoted to the preparation of thin-film capacitors. These units are primarily intended for incorporation in complex microelectronic circuits, in which a whole range of passive and active devices with their associated interconnection patterns are arranged on a single substrate. It is not evident at the moment which of the many systems under investigation will eventually augment or displace, the silicon monoxide or anodised capacitors in current use. However, as the thin-film technique becomes more widely established in the course of the next decade or so, there is likely to be a gradual decrease in the use of the conventional capacitors employing thick dielectrics, although these will still be required for high voltage applications. The change towards microcircuitry will be marked not only by a reduction in size, but also an increase in reliability and an over-all decrease in costs.

Chapter 3.2   **Connector materials**

There are events occurring at present in the connector field that indicate
that developments will become common place, and to understand these the
following questions are indicated:

1 What is the state of packing densities relating to connectors?
2 What effect will automatic termination have on densities?
3 How important is the trend toward matched-impedance connectors?
4 Will there be replacement of copper by aluminium; if so, what connector
   problems will arise?
5 Are there any strong tendencies towards standardisation?

Connector density

As a result of technical advances in the connector industry over the past few
years, interest has been concentrated in high density connectors. Several
requirements have combined to influence the trend towards greater connector
density; the circuits designed and built today require that more components
be put on a board. The need for increased area efficiency requires that the
packages are smaller, and the result could very well be circuits terminating
on two sides of the board with each more complex than the single sided
board.

One typically new type of large connector has 86 dual-readout contacts —
43 on each side. Now there is a requirement for the same type of connectors
with 180 to 200 contacts per side because of the need for greater densities.
Higher density connectors will introduce difficult problems, for instance, the
thermal efficiency is considered as a limiting factor, difficulty in assembly,
field service, fragility of materials, and closer tolerances.

Automatic termination

The economies of termination with larger and larger contact numbers dictate
the use of automatic methods, but where closer spacings are involved
problems associated with automatic terminations can become quite formid-
able.

The soldering and welding methods are becoming competitive. Automatic
soldering methods such as wave-soldering, are replacing hand-soldering in
many applications; the advantage of these methods, however, is that it is
easier to work within greater contact densities.

Although it cannot be considered a truly high-speed system, in typical computer or telephone applications, solderless wrapping and crimp methods are usually faster than hand-soldering.

Matched-impedance connectors

As the speeds and operating frequencies of computers have increased, shielding and impedance-matching have become increasingly important. With the introduction of more sophisticated technology there will be a growing demand for matched-impedance connectors; where impedance-matching is critical, there are usually other critical functions to be considered, such as frequency, dB losses and propagation factors. The design of these connectors will be influenced greatly by the performance of the circuitry with which their use is intended. The type of board material will have an indirect effect on the design, and the dielectric materials used with the connector will be most important. It is obviously of importance that the length of a terminal be as short as possible and, in some cases, this too could present a significant design problem, particularly when reviewing the distance the contact member must flex during engagement and disengagement with the printed board.

In the past, mechanical characteristics were probably the most important connector considerations. There existed the problem of getting the wires into and out of the black box, and their attachment; also contacts were not replaceable.

With impedance-matching, the electrical characteristics becomes equally important to the system designer. The designer now must consider the voltage standing wave ratio and cross-talk characteristics also other electrical parameters that in previous connector applications were not necessary considerations. The connector now becomes an integral part of the system, and may be considered as a resistor capacitor within that system whereas, in the past, it was frequently considered as hardware.

Replacement of copper by aluminium

One of the possibilities facing the connector industry is the replacement, at least in part, of copper by aluminium. The increased use of aluminium in electrical and electronic applications is a reflection of the rapid rise in the price of copper combined with periodic copper shortages.

In consideration of the possible replacement of copper by aluminium, the packaging technologists must consider the density and cost advantages of aluminium, the handling disadvantages associated with aluminium, also a new set of properties associated with the aluminium–iron alloys, and the connector problems associated with terminating aluminium wire.

The usual method of comparing aluminium with copper is based on considerations involving the weight and cost advantages of aluminium and the conductivity advantages of elemental copper. For example, copper has more than three times the density of aluminium but less than twice the conductivity on a volume basis and less than half the conductivity on a weight basis. The current price of aluminium per pound is substantially less.

The disadvantages associated with the use of aluminium as a replacement for copper must be included in the packaging technologist's calculations. Upon exposure to air, aluminium formes a thin but adherent oxide almost instantaneously, the oxide, difficult to remove, is electrically non-conductive and its presence or absence must be considered in the design of aluminium connectors. Another disadvantage with the use of aluminium is its unique physical properties when pressures are applied; the material's creep or flow properties, dependent on time and temperature, cause it to yield continually to a stress point equal to its ultimate strength. Alternatively copper creep is much smaller and equilibrium is reached after a short time. The factor of stress relaxation, also time dependent, must be considered, although it is not associated with a dimensional change. Stress relaxation, like creep, results in a reduction of the contact pressures in mechanical joints and cause failure of the electrical connection — hence the importance of nanosecond interruption testing.

Decided disadvantages associated with the use of aluminium have prevented its acceptance by the industry as a replacement for copper. Two developments have shown great promise in circumventing these difficulties:

1 The provision of aluminium–iron alloys with greater flexibility, bend strength and tensile strength with approximately the same electrical resistivity as aluminium wire.
2 The formation of copper-clad aluminium with the obvious concomitant connector advantages.

The influence of aluminium on connector technology is dependent upon how far one carries the aluminium. If the aluminium is only in the wire, intended to terminate to a copper contact, the dissimilar metal problems become evident. However, if the aluminium is to terminate to an aluminium contact, the problems disappear and that of fabricating the aluminium contact appears. In either case, it is evident that the circular conductivity of the aluminium will necessitate a change in the back end of the connector to accept larger wire.

The use of aluminium wires and the resultant connector problems will probably come about only when a very definite economic advantage can be shown. It is probable that a small economic advantage will not overcome the inertia of resistance to change. It seems fairly certain that although most connector manufacturers anticipate problems with aluminium they are preparing to deal with this material, particularly in the light of possible future copper shortages.

Standardisation of connectors

Connector standardisation has been the subject of discussion for the past several years, and has been concerned primarily with the area of the new, rapidly moving electronic products. The problem area is still with the military and aerospace users, but maximum effort is being expended to make standardisation a reality. The entire subject of standardisation of connectors, interchangeability and the like, must be considered as a two-edged sword.

The users would benefit greatly by standardisation of connectors. Standardisation of connectors offers, of course, some very definite advantages to the user because it provides him with multiple sources for the same product and, possibly a competitive situation that would significantly lower the costs.

The packaging engineer is advised, before designing his package beyond the point of no return, to assess the market, to determine what is available in the field of connectors, and to attempt, if at all possible, to use that which is available. The manufacturer's position is that if the packaging engineer faced the connector problem before he got deeply involved in his design requirements, he might find what he needs already available. In that sense, standardisation would not only be possible, but would be likely.

Chapter 3.3 Metallised ceramic materials

The joining of electrical conductors to non-conductors has represented a major problem in the electronics industry. Of the non-conductors, glass was the earliest to be used and for many years was satisfactory. However, it became less useful as the frequencies and temperatures at which components had to operate increased. In addition, glass is fragile, and so is not always suitable for use in miniature components; moreover, although it can be worked into a considerable variety of shapes, it cannot generally be held to very close tolerances. Ceramic materials do not have these disadvantages. They possess low dielectric losses over a wide frequency range and can operate at high temperatures. They can be made dense and vacuum-tight, and they can be prepared to close tolerances although not in the wide variety of shapes possible with glass. Ceramics possess a greater strength — an important economic factor — and a greater resistance to thermal shock.

Choice of sealing method

The process of sealing usually begins with the metallising of the ceramic surface, followed by subsequent soldering or brazing of a metal part, sometimes with an intermediate plating of the metallised part. The method chosen will depend on the particular requirements of the operation. The solder chosen should not have an excessive vapour pressure, should have good flow characteristics and should wet the metal but not react with it to any large extent. Further it should be ductile when cold in order that the high thermal expansion characteristic of this type of material should not set up undesirable stresses in the ceramic.

For applications where the metallised ceramic is to be sealed into a metal housing, consideration must be given to the thermal expansion of the metal. This should match that of the ceramic as nearly as possible. However, if a good match is not possible, a metal should be chosen which results in the ceramic being in compression.

Bonding processes and mechanisms

There are three basic types of ceramic-metal bond. There is, first of all, the type produced by the molten ceramic wetting of a solid metal, as exemplified by enamels. This involves the use of low-melting-temperature ceramics and so the bond has a low operating temperature. The second type is produced by the molten metal wetting of a solid ceramic. Examples of this type of bonding are provided by cermets and hard solder seals. The third type of bonding involves solid-state reactions which occur below the melting-point of the metal and ceramic components.

All these are simply methods of obtaining the necessary intimate contact between the ceramic and metallising. The mechanism of bonding is more complex, and can be described briefly under three headings.

Mechanical bonds

Such bonds involve the interlocking of particles and the penetration of one component into the pores and grain boundaries of the other. Van de Waal's forces can also provide bonding if the materials can be brought into sufficiently close mechanical contact.

Chemical bonds

Chemical bonds can result from solvent action which may give rise to new phases at the metal–ceramic interface, or from the formation of a metalloid structure by an active metal due to a valence shift which results in the transition, metal-suboxide–oxide, across the interface. In addition, certain metals, owing to their atomic size and chemical properties, have an affinity for chemical bonding. Such metals are easily oxidised and the oxides formed are usually refractory and compatible with the ceramic. Examples of such metals are titanium, zirconium, iron, cobalt and nickel from the transition elements, chromium, molybdenum and manganese.

Mechanical and chemical bonding

In many cases, a combination of mechanical and chemical bonding is operative. It is always difficult to examine the interface directly, and hence it is not always possible to determine the precise type and nature of bonding present in a given system.

Process details

The following gives an outline of some of the basic rules that should be used in the quantity production of metallised ceramics in order to achieve uniform quality and a high yield.

1. Ceramics should be inspected for uniform density, mechanical strength and freedom from voids.
2. The ceramics should be carefully cleaned by one or more of the following processes: ultrasonic cleaning; air firing at approximately 1,000 °C for 1 hr; boiling in nitric acid, rinsing and boiling in distilled water, rinsing and boiling in acetone and degreasing in trichlorethylene.
3. The metallising paint should be applied to a controlled thickness of $0\cdot0254$–$0\cdot0508$ μm and fired in an atmosphere of wet hydrogen at a temperature and for a time chosen to suit the part.

4 The parts should be inspected for correct film thickness and freedom from cracks.

5 In some cases, a thin coat of nickel (0·00054 mm) should be applied to help subsequent soldering or brazing. This is achieved either by electroplating or hydrogen reduction of an oxide coating and firing in wet hydrogen at approximately 1,050 °C for 20 min.

6 The parts should be inspected for correct film thickness and freedom from cracks.

7 For soldering applications, the metallised parts should be tinned either by a dipping process or on a soldering iron. For brazing applications, the metallised ceramics should be assembled with their appropriate metal parts in suitable fixtures to maintain their alignment. Rings or washers of braze material should be placed at the parts to be joined and the assembly fired in a hydrogen atmosphere to melt the braze.

8 Parts that have been tinned should be inspected for freedom from missing tin, and assemblies inspected for alignment followed by a leak test in a sensitive leak detector.

9 Handling of parts should be done with tweezers or by hands protected with gloves made from an approved material. Assembly of parts should be carried out in a dust-free, clean area.

Preparation of the paint mixture is one of the most critical steps in the metallising process. Since it is generally recognised that small particle size contributes to the formation of a strong bond, metal powders of particle size 1–5 μm are used. These are milled with an organic vehicle usually consisting of nitro-cellulose lacquer in a suitable solvent.

Application of the metallising coating can be done in several different ways according to the size and shape of the particular component. Brush coating has been a widely used technique for many years. However, this requires considerable skill on the part of the operator and careful control of the viscosity of the paint. Spraying forms the basis of a more easily controllable process and silk-screening lends itself to some degree of mechanisation and gives high uniformity.

Ceramic materials

The ceramic materials consist either of steatite (magnesium silicate) or alumina, and these are metallised with nickel. This produces a strong metal–ceramic bond, and the metal surface is readily tinned for soldering.

The components find many applications throughout the electronics industry. Metallised connector blocks are designed to provide anchoring points possessing high insulation resistance, high working flashover voltages and ease of soldering. Metallised rotor shafts are used in the manufacture of tuning and amplifying units, and a wide variety of metallised insulators is produced.

Silver coatings are used in addition to nickel. In general these do not possess the high bond strengths that can be achieved with nickel, and there is a tendency for silver to dissolve in solder on subsequent processing. However, silver is useful in certain applications since the bond is formed by firing in an oxidising rather than reducing atmosphere. For example, ceramics based on barium titanate for piezoelectric and dielectric applications must be fired in an oxidising atmosphere since any reduction would impair their electrical properties.

Chapter 3.4 **Protective metal coating**

Metals such as iron and steel are easily oxidised in the air and can form rust; many other metals can also corrode or tarnish, especially at elevated temperatures. It is not surprising, therefore, that a large number of coatings have been developed to protect metals against the destructive action of environment. In many cases, the coating performs a function in addition to that of protection, such as enhancing the appearance or imparting hardness, but this chapter deals solely with methods of providing a metal coating on the surface of components for their protection and impact and so impart better resistance to wear, friction, impact, and durability. All the processes are mechanical, and therefore, do not involve electricity, as in the case of electroplating.

This chapter is devoted to describing of methods employed to produce coatings of protective layers on the surface of the component pending such treatment as hot-dipping, cementation and metal spraying.

Coating methods

Hot-dipping

The application of a coating of metal by hot-dipping (the immersion of the article or component into a bath of the molten coating material) has been carried out for many years and is widely used because it is simple and effective. The operation is normally restricted to metals with relatively low melting-points such as tin, zinc and aluminium, because if articles are immersed in a bath of molten material at a very high temperature the metal of the articles might be impaired. Even at the temperatures of the molten coating metals a certain amount of alloying takes place between the coating and basis metal, and impairs the adhesion of the coating metal to the basis metal. The outer coating of pure metal is necessary because, on removal of the article from the bath, a certain amount of the coating metal is retained and solidifies on the surface as it rapidly cools.

Hot-tinning

The low melting-point of tin, the ease with which it wets and spreads on other metals and its high resistance to corrosion in a large variety of circumstances are the reasons for the wide use of hot-tinning. Because tin is non-toxic, it is used in the production of the coating of food containers.

Hot-tinned coatings are produced by immersing the prepared work in a bath of molten tin. Work which is too big for dipping may be tinned by wiping. The tin, in stick, powder or molten form, is applied to the prepared and heated surface and wiped smooth, the tinning can be restricted to pre-scribed areas by masking.

The average thickness of coating obtained by hot-tinning is from 0·0027 to 0·018 mm with good uniformity of thickness. If, however, a somewhat uneven distribution can be tolerated, coatings as thick as 0·025 mm or more can be obtained, but the uniformity depends on the physical shape of the article and the skill employed in manipulation after tinning.

Hot-tinning is a rapid process, requiring only a few seconds. Bulky articles can be hot-tinned singly or a few at a time using jigs, and very small parts can be handled in bulk by means of baskets.

Hot-tinning may be applied to low alloy carbon steel, copper, brass, bronze and cast iron. Other iron and copper base alloys may be tinned provided that alloying elements which produce thick surface oxide layers, such as aluminium, chromium and silican are not present in substantial quantities. Nickel and most of its alloys can be hot-tinned. Other metals that may be hot-tinned are lead, zinc, silver, the noble metals, cobalt, molybdenum, tungsten and cadmium. It is more often likely that the tin coating will be necessary as an aid to subsequent soldering operations rather than as a protection.

Galvanising

Hot-dip galvanising is one of the best known and long established of all metal coating processes. A variety of ferrous materials in the form of sheet, strip, rod and bar, as well as many finished products are made corrosion-resistant by coating with zinc. A zinc coating is economical and long lasting, its ultimate life depending on thickness of coating and correct handling of the material in fabrication or use after processing.

Most types of ferrous materials and products can be hot-dip galvanised. The process is particularly suitable for structural steel and for many finished products. The actual galvanising in molten zinc is preceded by cleaning and pre-treatment processes designed to ensure proper bonding and coating uniformity. After dipping, excess metal is drained off, the final coating thickness for galvanised sheets, wire or tube can be controlled by mechanical rolling or wiping.

The appearance of so-called 'spangles' which form on the surface, is related to the composition of the molten metal. Only in sheet galvanising is this deliberately controlled. Small amounts of aluminium are also often added to the bath to improve the brightness of the coating, while larger quantities (0·2 per cent) are added in sheet galvanising to reduce the amount of alloy layer formation.

The coating itself consists of a series of zinc–iron alloy layers with an external high zinc layer. This alloying which takes place makes for a tight bond. Coating thickness is normally 0·076 to 0·10 mm, but is much thinner for galvanising on sheet.

The quality and structure of the zinc coating in galvanising depends on several factors which can be fairly closely controlled. The length of time in the bath and the temperature of the bath determine the rate of growth of the alloy phase, too much alloying can lead to brittleness of the coating. The normal practice is to keep the bath temperature as low as possible and the immersion time short.

The rate of withdrawal from the bath also affects the thickness of the pure zinc outer layer. Finally, the cooling rate has a bearing on the zinc–iron reaction; this is because further alloying can take place after withdrawal if the heat capacity of the piece is large, quenching is often employed to prevent this. Today, in continuous processes such as those for sheet galvanising, temperature, immersion time withdrawal speed and other factors are all automatically controlled to ensure uniformity and proper structure of the coating.

Lead alloy coatings

A lead coating imparts resistance to atmospheric oxidation and also to acid fumes containing sulphur compounds. Steel articles cannot normally be coated with pure lead by hot-dipping since the molten lead does not alloy with iron. However, by adding such metals as tin and antimony, which alloy with both the basic metal and lead, a coating of the lead alloy can be applied. Terne plate is one of the most common lead alloy coatings, containing from 20 to 50 per cent of tin. It forms an excellent bond with steel, and the tin increases the corrosion resistance and hardness of the plate. The antimony or zinc enables the tin content to be reduced and these additions also increase the fluidity of the lead and, to some extent, the hardness of the coating.

Hot-dip aluminising

Hot-dip aluminising is a process similar to hot-dip galvanising in that the work is immersed for a very short period in molten aluminium. This produces an almost pure aluminium coating with a very thin intermetallic layer. The hot-dip aluminising process is applied as a treatment for various steel products. Continuous processes are used for sheet, strip and wire.

Aluminised steel processed in this way as raw material for fabrication purposes is first cleaned and preheated in a reducing atmosphere. The whole process can be closely controlled. Sheet or coil up to 122 cm wide in gauges from 2·4 to 0·4 mm is produced in this way. The finished material can be formed without damage to the coating; it can be brazed or welded by most techniques, although welding will cause some local damage to the coating. The finish of hot-dip aluminised steel produced by a continuous process is bright and attractive in appearance. Above temperatures of about 400 °C the surface will become dark grey due to alloying of the aluminium with the base metal. Where aluminium coatings have to be painted a chromating treatment is desirable, using any of the commercially available solutions, or a primer of the zinc chromate type may be applied. The resistance to corrosion of aluminised steel makes it very suitable for use where minimum maintenance of painted surfaces is important.

Cementation processes

In cementation processes the metal surface of the article to be protected is alloyed with another metal, the resultant coating being hard and corrosion-resistant, but sometimes somewhat brittle. The coating is produced by heating the article in a container with the coating metal in the form of a powder. The coating metal under the influence of heat gradually diffuses into the surface with which it is in contact. In some cases a reducing atmosphere such as hydrogen must be employed, for both oxygen and water must be excluded from the reaction in the majority of cases.

Sherardising

Sherardising is a method of applying a zinc coating by a dry diffusion process, using heat and a zinc powder. It is limited in application to parts that can be packed into a small drum, usually not more than 61 cm in diameter. But for many small parts such as nuts and bolts, springs and screws, tubing, small cast parts, small pressings and other steel articles, which could not be easily galvanised, sherardising is a convenient and economical process. It ensures uniform coatings, even with complex surfaces.

After suitable pre-treatment the parts are placed in the drum, which is heated and made to revolve continuously for several hours. Operating temperature is usually in the region of 350 to 370 °C, which is well below the melting-point of zinc. In some circumstances higher or lower temperatures may be necessary when coating springs so as not to affect the temper. On the other hand, higher temperatures may be needed when using a zinc powder compound with a lower metallic content. Coating thickness applied is usually about 0·012 to 0·015 mm although heavier coatings are possible.

Sherardised finishes have a clean, grey appearance and the process has been used for architectural metal work. Finishes can be oiled or burnished to enhance their appearance.

Chromising

By heating iron or steel articles with powdered chromium the surfaces can be coated with a hard, solid solution of chromium in iron, the process being known as chromising. The process also provides a surface which is highly resistant to corrosion, abrasion and the effects of high temperature.

There are various techniques, but the method most used is a powder cementation process similar to sherardising. After suitable cleaning or pre-treatment the work is packed into a drum with a mixture of powdered chromium and powdered alumina in the ratio of 55 to 45, the alumina being used to prevent the particles of the metallic powder from fusing together in a solid block or mass. The whole is then heated to about 1,400 °C in an atmosphere of pure hydrogen which provides a non-oxidising atmosphere.

Gaseous and molten salt bath chromising processes are also used. The salt bath is made up of chromium chloride and barium chloride. In the gas process a vapour of chromium and aluminium powder is used. A coating produced by chromising can be polished in the same way as can electro-deposited chromium plating, but the surface is not so hard.

Components for chromising must be of low-carbon content up to 0·2 per cent as carbides hinder the diffusion of chromium. Steels with a higher carbon content may be chromised if they have first been decarburised. The process provides similar resistance to steam and saline and ordinary atmospheres as chromium itself, and has a high resistance to 10 per cent nitric acid. The process is most suitable for small compact parts, but is not really suitable for complicated shapes.

Calorising

For increasing the heat resistance of steel parts where working temperatures do not normally exceed about 750 °C, aluminium coating by the calorising process is often used. Above this temperature the use of special heat-resisting material for the whole part may be more desirable.

Calorising involves heating the metal in powdered aluminium in very high temperature — about 800 to 1,000 °C. The result is an impervious outer coating of highly refractory aluminium oxide and a deep aluminium-rich layer thoroughly diffused in the parent metal. The aluminium content of the layer is in the region of 15 to 20 per cent.

Components to be calorised are first grit- or shot-blasted and then placed in rotating drums and heated. An alternative technique for sensitive or very complex parts is 'pack calorising' in which the work is packed in a container with aluminium powder and heated but not rotated.

Some of the applications for the calorising include exhaust manifolds, aero-engine piping, annealing furnace tubes, firebars, retorts, thermometer sheaths and pressed-steel containers. Pots, boxes and trays of calorised mild steel are used extensively for salt, cyanide and lead hardening, case-hardening, carburising and annealing.

Another use is in the protection of steels of nickel-chromium alloy used at elevated temperatures; as is known, sulphur dioxide is generally present in varying degrees of concentration in the atmospheres of industrial heating operations, and owing to the great affinity of nickel for sulphur, the nickel in the alloy combines with the sulphur in the atmosphere and rapid breakdown of the alloy follows. Such alloys may be effectively protected by calorising, as the surface then effectively resists the effect of hot sulphurous gases and the products of combustion of all common fuels.

Peen plating

Peen plating is a mechanical plating process which has a number of advantages, amongst these being good quality finish with economy, and simplicity and compactness of plant. No heat is used and no electrolytes are required, the action being entirely mechanical. Before the workpiece is peen plated it must be thoroughly degreased, etched and precoated. The etching roughens the surface to provide better adhesion, and coating of immersion copper plating protects the cleaned surface. The components are placed in a revolving drum and tumbled with a dust of the coating metal in water together with an impacting material usually steel grit or glass beads. The action of the impacting material is to force the metal dust on to the workpiece so that successive layers of coatings of the metal are built up.

Zinc, brass, cadmium, tin and lead are the most commonly used coating materials for peen plating, but it is claimed that virtually all metals with the exception of stainless steel, titanium and magnesium can be plated in this way. The process is applicable to a wide variety of components, such as small stampings, and many more hardware parts, but it is not suitable for larger components due to the fact that the costs quickly increase with size, also components having steep crevices cannot be dealt with economically.

Chromating

In the chromating process a coating is applied that retards the corrosion of zinc- and cadmium-plated parts and zinc die castings. It is usually applied by immersion although electrolytic methods are used.

The film produced on zinc or cadmium by the various types of chromate bath (generally known as 'passivation' treatments) give considerable protection against corrosion in humid conditions or in salt-water spray. The film thickness is usually of the order of 1 to 2 μm.

There are a number of different proprietary processes for chromating, but all embody hexavalent chromium ions and mineral acids. Chromated surfaces can be sealed with paint for extra protection. Magnesium alloys are similarly protected against marine and other atmospheric corrosion by chromating processes specially developed for the job.

Metal spraying

Metal spraying or metallising is a technique that has been used for many years; it offers many advantages for the application of coatings to improve resistance to wear, abrasion and corrosion.

The principle is the spraying of molten metallic particles from a gun at high velocity on to the work. The coating metal may be in the form of a powder or wire which is melted in the gun by a burning gas and oxygen mixture and then carried in a stream of high-pressure gas or air or by electric arc.

A feature of metal spraying is that the temperature of the spray stream when it reaches the work is not sufficient to cause warping or distortion; in fact even inflammable materials such as fabric can be metal sprayed without heat damage.

A very high rate of deposition of metal is possible with metal-spraying techniques. Preparation of the work surface before metal spraying is important to ensure proper adhesion of particles — metal surfaces are usually grit-blasted. The adhesion between coating and the parent metal is a mechanical one, the particles cohering on impact to form a solid mass of metal. The finish produced is invariably a matt one, but turning, grinding, or polishing of the surface may be carried out.

Metal-spraying techniques have many applications; the compactness and mobility of the equipment used makes it very suitable for spraying large structures *in situ*, and parts that may be difficult or too large to coat by other means can easily be sprayed. A thin zinc-sprayed coating will provide protection equal to a hot-dipped coating; but since the degree of protection is a function of coating thickness, metal-sprayed systems are used for very long-term protection without costly maintenance or where corrosive conditions are particularly severe. A complete protective system in which organic and inorganic sealers and/or colour coats are painted over a sprayed-metal coating provides the most economical long-term corrosion protection available. The porous nature of the coating provides a very suitable base for painting and helps to ensure durability of the paint film.

Light-gauge sheet and metal work which would be distorted by dipping may be zinc sprayed without risk of damage. Sprayed-aluminium coatings on steel can also provide resistance to corrosion at temperatures up to about 500 °C. The coating is applied to a thickness of from 0·102 to 0·0203 mm, although thicker casings up to as much as 0·25 mm are used. Very pure aluminium coatings may now be applied. Heat-treatment is often applied after spraying or when the part is first put into service. This results in the aluminium taking up iron from the base metal and forming an intermetallic layer. Continuous processes for the spraying of aluminium on to steel are in use. The work passes on a conveyor through a grit-blasting stage and cleaning processes, and then passes to multiple spraying nozzles.

Sprayed-aluminium coatings are matt, off-white in colour and like zinc coatings are often used as a corrosion-resistant base for painting. Although not as yet as popular as zinc coatings, they provide even better protection in marine or industrial atmospheres.

In most immersed conditions and particularly in soft water at temperatures above 60 °C, aluminium protection is superior to zinc, although zinc coatings offer good protection generally in immersed conditions.

In addition to the deposition of protective metal coatings, metal spraying has another important application in the reconditioning of worn surfaces, thus worn bearing sections can be built up with the new metal which may be even more wear resistant than the original.

For reclamation purposes many types of metal may be sprayed, including carbon steel, nickel, stainless steel and copper alloys.

Corrosion and its prevention

When two different metals are in contact and are also bridged by water containing an electrolyte, e.g. salt, acid, combustion product, current flows through the electrolyte from the anodic or baser metal to the cathodic or nobler metal. As a result of the ensuing electrolytic action, the cathodic metal tends to be protected but the anodic metal may suffer greater corrosion than when it is connected to a metal similar to itself. In the past, schedules of potentials have been published which have been of great use in drawing the attention of designers to the dangers of bimetallic corrosion. These schedules can however, be misleading as although the potential differences between metals is the prime driving force of the corroding current, the magnitude of the potentials is not always a reliable guide to the amount of corrosion suffered; in particular, statements that any specified differences of potentials are either safe or unsafe may be unreliable.

The chapter provides information about the conditions in which bimetallic contacts are most likely to lead to troublesome corrosion and also about mean of mitigating such effects (see Table 3.4.1).

Table 3.4.1 is not to be regarded as infallible. The information is largely qualitative. Moreover, knowledge and experience of certain combinations is very inadequate and, in any case, it is impossible to tabulate concisely the effects of a wide variety of exposure conditions.

The groups of metals numbered 1 to 10 are arranged as a sequence which is approximately similar to the 'Schedule of potentials against the saturated calomel electrode in sea water at room temperature'. The groups of metals 11 to 16 are not arranged in any particular sequence. Chromium, stainless steel, titanium and aluminium develop highly protective oxide films and behave as relatively noble metals under conditions favourable to the maintenance of such films, but are in some cases less resistant under conditions in which breakdown of the film is possible. It is these metals which cause difficulty in attempting to predict the behaviour of bimetallic contacts from a schedule of potentials.

The symbols in the table record only the acceleration of corrosion resulting from the dissimilarity of the metals in contact: no information is given on the basic corrosion resistance inherent in the particular metal. The degree of acceleration is classified as follows:

A Corrosion of the 'metal considered' is not increased by the 'contact metal'.

B Corrosion of the 'metal considered' may be slightly increased by the 'contact metal'.

C Corrosion of the 'metal considered' may be markedly increased by the 'contact metal'.

D When moisture is present, this combination is inadvisable, even in mild conditions, without adequate protective measures.

The symbol given for each combination of groups is based on direct experience of bimetallic contacts and general electrochemistry than on the schedule of potentials already referred to. This applies particularly to groups 11 and 16.

Where two symbols are given, e.g. B or C, the acceleration is particularly liable to vary with variations in the environmental conditions or the condition of the metal. The designers general experience of his particular service conditions should aid him in the practical interpretation of these symbols.

The electrolyte and its effects

From a practical point of view, moisture from any source, such as condensation, rain or sea-water, often contains an electrolyte such as salt or acid. Bimetallic contacts affect corrosion only while a continuous film or body of

Table 3.4.1 **Degree of corrosion at bimetallic contacts**

A — Corrosion of the 'metal considered' is not increased by the 'contact metal'.

B — Corrosion of the 'metal considered' may be slightly increased by the 'contact metal'.

C — Corrosion of the 'metal considered' may be markedly increased by the 'contact metal'. (Acceleration is likely to occur only when the metal becomes wetted by moisture containing and electrolyte, e.g. salt, acid, combustion products. In ships, acceleration may be expected to occur under in-board conditions, since salinity and condensation are frequently present. Under less severe conditions the acceleration may be slight of negligible.)

D — When moisture is present this combination is inadvisable, even in mild conditions, without adequate protective measures.

Metal considered	Contact metal					
	1	2	3	4	5	6
	Gold, platinum, rhodium, silver	Monel, Inconel, nickel-molybdenum alloys	Cupronickels, silver solder, aluminium-bronze, tin-bronzes, gunmetals	Copper, brasses, 'nickel-silvers'	Nickel	Lead, tin, soft solders
1. Gold, platinum, rhodium, silver	—	A	A	A	A	A
2. Monel, Inconel, nickel-molybdenum alloys	B	—	A	A	A	A
3. Cupronickels, silver solder, aluminium-bronze, tin-bronzes, gunmetals	C	B or C	—	A	A	A
4. Copper, brasses, 'nickel-silvers'	C	B or C	B or C	—	B or C	B or C
5. Nickel	C	B	A	A	—	A
6. Lead, tin, soft solders	C	B or C	B or C	B or C	B	—
7. Steel and cast iron	C	C	C	C	C	C
8. Cadmium	C	C	C	C	C	B
9. Zinc	C	C	C	C	C	B
10. Magnesium and magnesium alloys	D	D	D	D	D	C
11. Stainless steel Austenitic 18/8 Cr/Ni	A	A	A	A	A	A
12. Stainless steel 18/2 Cr/Ni	C	A or C	A or C	A or C	A	A
13. Stainless steel 13% CR	C	C	C	C	B or C	A
14. Chromium	A	A	A	A	A	A
15. Titanium	A	A	A	A	A	A
16. Aluminium and aluminium alloys	D	C	D	D	C	B or C

water joins the different metals so that the small electric current which forms an essential part of the corrosion process can pass through this water, the circuit being completed by the metallic contact. The severity of the effect in a given time is dependent on the conductivity of the water which is itself dependent on the quantity and nature of the electrolyte dissolved in it. If contact through the water is intermittent, the cumulative bimetallic effect is dependent on the total time of contact.

The effects are more severe when the metals are immersed in water,

7 Steel and cast iron	8 Cadmium	9 Zinc	10 Magnesium and magnesium alloys	11 Stainless steels — Austenitic 18/8 Cr/Ni	12 Stainless steels — 18/2 Cr/Ni	13 Stainless steels — 13% CR	14 Chromium	15 Titanium	16 Aluminium and aluminium alloys
A	A	A	A	A	A	A	A	A	A
A	A	A	A	A	A	A	A	A	A
A	A	A	A	B or C	B	A	B or C	B or C	A
A	A	A	A	B or C	B or C	A	B or C	B or C	A
A	A	A	A	B or C	B or C	A	B or C	B or C	A
A or C	A	A or C	A	B or C	B or C	B or C	B or C	B or C	A
—	A	A	A	C	C	C	C	C	B
C	—	A	A	C	C	C	C	C	B
C	B	—	A	C	C	C	C	C	C
D	B or C	B or C	—	C	C	C	C	C	B or C
A	A	A	A	—	A	A	A	A	A
A	A	A	A	A	—	A	A	—	A
A	A	A	A	C	C	—	C	C	A
A	A	A	A	A	A	A	—	A	A
A	A	A	A	A	A	A	A	—	A
B or C	A	A	A	B or C	B or C	B or C	B or C	C	—

spreading for considerable distances from the line of contact, although diminishing in intensity as the distance from the line increases. In atmospheric conditions the effect is usually more localised to the vicinity of the line of contact. A film of moisture condensed from the air will dissolve electrolyte from 'polluted' atmospheres, e.g. salt windborne from the coast or drifting into the interior of a ship or other structure, combustion products, etc., and can then cause appreciable acceleration of corrosion.

If the water contains certain substances which form complexes such a cyanides, citrates and tartrates, the data given may be incorrect, since the effect of the complexes has not been taken into account in the compilation of Table 3.4.1. Thus, if one of the reagents forms 'complexes' containing a given metal, that metal will behave less nobly than the symbol in the table would indicate. The most common reagent of this type is ammonia, which has no action on iron but attacks copper (normally a more noble metal) in the presence of air.

One source of trouble with bimetallic contacts is the use of certain fluxes at soldered joints which leave hygroscopic residues and produce moist conditions at joints. This may be avoided by thorough cleaning after soldering or, preferably by the use of suitable non-corrosive fluxes. It must be stressed that serious acceleration of corrosion can occur when the two metals are not directly joined together.

Condition of the metal

The different alloys within one of the groups in Table 3.4.1 may differ significantly in composition and may, therefore, give rise to bimetallic effects when in contact with each other; for example, the corrosion of one brass can be accelerated by contact with another brass. These effects within the groups have been disregarded in order to minimise the number of groups and to keep the table as simple as possible. Moreover, the behaviour of a given alloy in bimetallic contact is affected by its condition. For example, with the H.15 type of aluminium–copper alloy, the bimetallic effect is dependent on heat treatment conditions, the naturally aged alloy being more noble than the artificially aged alloy. Again these effects are disregarded to avoid undue complication of the table. The more noble character of the stainless steels is dependent upon the protective film on the metal, and the maintenance of this film, is, in turn, dependent upon the continuous presence of oxygen.

Where there is doubt as to the significance of these complex effects in a particular structure or piece of equipment, the designer should carry out tests.

Metallic coatings

Some of the metals in the table, e.g. chromium, nickel, cadmium and zinc, are most commonly used as protective coatings on another metal. The electrochemical behaviour of a metallic coating is generally similar to that of the same metal in the massive state, provided that it is reasonably thick and free from pores; reference should be made to the appropriate group in the table. Where discontinuities exist in a coating, a corrosion cell may be set up which can cause attack on the base metal, if the coating is more noble, or undue wastage of the coating metal, if that is less noble than the base metal. For this reason, drilling and shearing should precede electroplating or other coating processes rather than follow them and precautions should be taken against wear or other mechanical damage to the coating. Corrosion at or around gaps in a coating is not discussed further in this chapter.

Contact of a metallic coating, e.g. zinc on steel, with a more noble metal, e.g. aluminium, may accelerate wastage of the coating so as to cause exposure of the basic metal, e.g. steel, which may then accelerate corrosion of the second metal (aluminium in this example).

Area relationships

The danger of corrosion at a bimetallic junction is greatest if the area of the nobler metal or cathode is large compared with the area of the less noble metal or anode. The danger of very intense attack, when the ratio of the cathodic to anodic areas is high, is greatest under immersed as opposed to atmospheric conditions. Aluminium alloy bolts used to join copper alloy plates would be likely to suffer rapid corrosion; aluminium alloy plates joined with copper bolts would probably suffer less intense attack. Neither arrangement, however, represents good practice. It must also be borne in mind that the life of a structure or the functioning of a mechanism may be more seriously affected by the corrosion of one component than by that of another; in such cases, the more vulnerable component should preferably be cathodic to any less vulnerable part with which it makes contact.

Microgalvanic effects

In certain metallic pairs, accelerated corrosion of the bases member can occur even when there is no metallic contact between the dissimilar metals. For example, in a system conveying water small amounts of copper, dissolved by the water from copper or copper alloy components, may be deposited on less noble metals in the system such as zinc, aluminium or ferrous alloys, with which the water may subsequently come into contact. The microscopic bimetallic couples so formed are liable to result in severe pitting of the base metal.

Effects of graphite, coke or a carbonaceous film

Left on metal after fabrication these may act as a noble electrode in a couple and accelerate corrosion. Graphite in packing has also caused trouble and has been replaced, in some cases, by mica.

Other effects

In some circumstances the corrosion of a metal can be accelerated by contact with an identical metal, a metal of the same group of the table, or a non-metallic material. When a metal is joined to another material in such a way as to form a crevice, corrasion may be accelerated in the crevice, e.g. by differential areation or by trapping corrosive liquids.

Corrosion at crevices where there is a poor oxygen replenishment is particularly dangerous for stainless steels and certain aluminium alloys. A deposit, e.g. of sand, cotton waste, filings, etc., on a metal may cause attack under the deposit due to differential areation or interference with the formation of a protective film on the metal. Corrosion at two metal surfaces in

P

contact may be accelerated when there is slight repeated relative movement between them, giving rise to 'fretting' corrosion.

Certain timbers and phenolic materials, including some varnishes and organic insulating materials, etc., may give off corrosive agents which are frequently volatile; such materials may cause corrosion not only of metal components with which they are in contact, but of other metals in the same enclosed space. Cadmium, zinc, lead and magnesium are especially susceptible to some vapours.

Prevention of bimetallic corrosion

Features of design

The most effective methods of prevention lie in careful design and assembly practices. Full realisation of bimetallic contact corrosion risks from the outset will enable the designer to avoid making a decision which immediately places the design in jeopardy; to provide adequate ventilation and drainage; and to make due provision in the initial design for the use of the protective measures suggested below and to ensure their proper application during assembly. When protection is achieved by painting and it is impracticable to paint both metals, then it is preferable to paint the more noble metal.

Insulation

The fundamental method of preventing corrosion arising from dissimilar metals is to prevent the flow of the corrosion currents. This may be achieved by:

1 Effectively insulating the dissimilar metals from each other, e.g. breaking the metallic circuit.
2 By preventing a continuous bridge of water between the two metallic metals, by breaking the electrolytic path. In immersed conditions it is not possible, but may be used if insulation is not precluded by the need for electrical bonding for functional purposes.

For example, a base metal nut and bolt should be fitted with an insulation bush and washers where it passes through a nobler metal plate. Complete insulation is here essential and hence the bush is vital. In atmospheric conditions, 2 may be used. For instance, a large insulating washer may suffice to break the water film or to reduce the total corrosive action or to render it insignificant by distributing it over a wide area, and in this case bushing may not be necessary. All insulating materials should be impervious to water, e.g. felt should not be used.

Metallic coatings

Dissimilarity may be avoided by coating one metal (preferably the cathode) with the other metal or one compatible with it. Coating may be achieved, for example, by electro-deposition, hot dipping or metal spraying. Thus, in airframe assemblies for service aircraft it is recommended that all bolts, screws, pins, bushes, etc. made from stainless steels should be cadmium plated whenever such parts are used in contact with light alloy (magnesium — rich aluminium).

Notes

1 The exposure of iron, steel, magnesium alloys and unclad aluminium–copper alloys in an unprotected condition in corrosive environments should be avoided wherever possible even in the absence of bimetallic contact.

2 The behaviour of magnesium alloys in bimetallic contacts is particularly influenced by the environment, depending especially on whether an electrolyte can collect and remain as a bridge across the contact. The behaviour indicated in Table 3.4.1 refers to fairly severe conditions. Under conditions of total immersion or the equivalent, magnesium alloys should be electrically insulated from other metals. In less severe conditions complete insulation is not necessary, but steel, brass and copper parts should be galvanised or cadmium plated, and jointing compound used during assembly. Under conditions of good ventilation and drainage, contacts classified as D have given satisfactory service, e.g. brass and steel push-fit and cast-in inserts in magnesium castings.

3 Where contact between magnesium alloys and aluminium alloys is necessary, adverse galvanic effects will be minimised by using aluminium alloys containing little or no copper, 0·1 per cent max.

4 When contacts between copper or copper-rich materials and aluminium alloys cannot be avoided, a much higher degree of protection against corrosion is obtained by first plating the copper-rich material with tin or nickel and then with cadmium, than by applying a coating of cadmium of similar thickness.

5 The corrosion of mild steel may sometimes be increased by coupling with cast iron, especially when the exposed area of the mild steel is small.

6 Instances may arise in which corrosion of copper or brasses may be accelerated by contact with bronzes or gunmetals, e.g. the corrosion of copper, sea water carrying pipe-lines may be accelerated by contact with gunmetal valves, etc.

7 When magnesium corrodes in sea-water or certain other electrolytes, alkali formed at the aluminium cathode may attack the aluminium.

8 When it is not practicable to use more suitable methods of protection, e.g. spraying with aluminium, zinc may be useful for the protection of steel in contact with aluminium, despite the accelerated attack upon the coating.

9 It is not necessarily to discourage the use of the 'contact metal' as a coating for the 'metal considered', provided that continuity is good; under abrasive conditions, however, even a good coating may become discontinuous. In most supply waters at temperatures above 60 deg C, zinc may accelerate the corrosion of steel.

10 In these cases the 'contact metal' may provide an excellent protective

coating for the 'metal considered', the latter usually being electro-chemically protected at gaps in the coating.

11 When aluminium is alloyed with appreciable amounts of copper it becomes more noble and when alloyed with appreciable amounts of zinc or magnesium it becomes less noble. These remarks apply to bimetallic contacts and not to inherent corrosion resistance. Such effects are mainly of interest when the aluminium alloys are connected with each other.

12 In some immersed conditions, the corrosion of copper or brass may be seriously accelerated at pores or defects in tin coatings.

13 In some conditions there may be serious acceleration of the corrosion of soldered seams in copper or copper alloys.

14 When exposed to the atmosphere in contact with steel or galvanised steel, lead can be readily corroded with formation of PbO at narrow crevices where the access of air is restricted.

15 Serious acceleration of corrosion of 18/2 stainless steel in contact with copper or nickel alloys may occur at crevices where the oxygen supply is low.

16 Normally the corrosion of lead/tin soldered seams is not significantly increased by their contact with the nickel-base alloys but under a few immersed conditions the seams may suffer enhanced corrosion.

17 The corrosion product on zinc is, in certain circumstances, more voluminous and less adherent than that on cadmium. Where this is known to be the case, it should be borne in mind in making a choice between these two metals.

18 These joints are liable to corrosion in crevices where these are not filled with jointing compound.

19 Corrosion products from iron or steel reaching aluminium, or corrosion products from aluminium reach iron or steel may sometimes cause serious local corrosion through oxygen-screening or in other ways, even when the total destruction of metal is finished.

Chapter 3.5 **Multilayer alumina circuit boards**

The multilayer alumina circuit board is a new type of monolithic aluminium oxide substrate containing internally metallised wiring planes. It differs in that noble rather than refractory metals are used as the conductors. The buried metallisations are hermetic and highly conductive and the boards contain risers or vias of similar high-conductivity metallising to interconnect the buried planes with the top and bottom surfaces. Any of a large selection of air-fireable thick-film components can be applied to the surfaces of the board. Active metal and evaporated thin-film components are also compatible. The primary application is in the chip level interconnection of silicon integrated circuits.

A multilayer alumina circuit board differs from a 'print and fire' or crossover multilevel board in that the layers of dielectric isolation can be many times thicker than that obtainable with printing techniques. The dielectric layers are of the same high alumina composition as is the base substrate. The advantages of this approach include precise layer registration, low capacitance between wiring planes because thick dielectric layers can be used, high voltage levels, good physical strength and the ability to use the surface in the same manner as standard aluminas (see Table 3.5.1).

Table 3.5.1 **Properties of multilayer alumina circuit boards**

Physical

Composition	96% Al_2O_3
Water absorption	0·0%
Permeability (dye)	None
Permeability (He)	$<10^{-8}$ std. cc He/sec as-fired
Specific gravity	3·7
Layer adhesion (100% metallised)	>48 MN/m²

Electrical — *Alumina*

Bulk resistivity (25 °C)	$2 \times 10^{14} \, \Omega$ cm
Permittivity (1 MHz, 25 °C)	9·2
Dissipation factor (1 MHz, 25 °C)	$<0·0005$
Permittivity vs. frequency	Less than 1% change to 100 MHz
Permittivity vs. temperature (– 20 to +140 °C)	$< +2\%, -1\%$
Dielectric strength (DC) (6·35 mm specimen)	$>24,000$ V/mm

Metallisation

Buried conductor resistance	0·007–0·010 Ω/cm²
Top conductor resistance	$\geqslant 0·003 \, \Omega$/cm²
Risers	$<0·002 \, \Omega$

A simple three-layer circuit board has one layer of buried metallisation which appears in Fig. 3.5.1 as the top surface of the lower piece. Risers connect the buried metallisation to the top and bottom of the assembled board. The hybrid circuit comprises two phase-shift oscillators. The passive circuitry is air-fired thick film with five transistors eutectic back-bonded to gold pads. The transistor contacts are wire-bonded to the conductors.

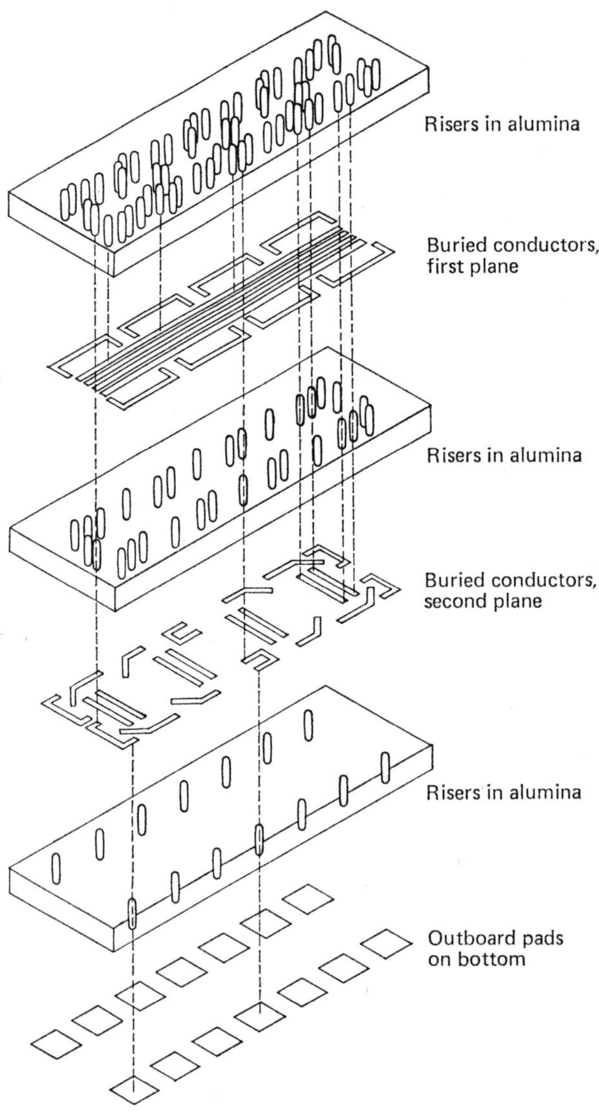

Fig. 3.5.1   Multilayer ceramic board, general purpose for four digital integrated circuits

Uses

This technology is used to greatest advantage for applications requiring high-density *packaging* of active devices. It is also compatible with all the present techniques for manufacturing hybrid integrated circuits. Pins, lead frames, soldered-on leads and various interconnection techniques can be used with these substrates along with all of the standard sealing methods for single- and multichip alumina packages.

The contributions of the multilayer substrate to advanced packaging technology are improved performance and economy. Functional groups of ICs can be mounted and interconnected in a single package. This hybrid, or multichip technique is an approach to MSI and LSI that circumvents the penalty of the exponential decrease of die processing yield with increasing chip area. Interconnecting lines are short so chip density on the substrate is high. Line resistance is low, typically 1 Ω.cm; for buried conductors of 0.25 mm width it is 0.01 Ω/cm^2. These factors make for minimum signal delay in high-speed circuits. Tremendous versatility is available by designing the buried conductors for general-purpose use in a whole class of circuits, then varying only the top surface discretionary conductor pattern to produce a given circuit.

The general-purpose multilayer substrate shown in Fig. 3.5.1 is designed to accommodate four digital ICs in a single 14-lead dual in-line package. The substrate bears conductive riser outlets but no surface conductors. The first buried level carries conductors running the length of the package. Risers connect these conductors to the top surface near each IC position. These can be used for service functions such as ground, voltage and clock. The second buried level provides crosslinks under each IC position and also makes connections to the risers from the outboard connection pads on the bottom of the substrate.

Properties and characteristics

A unique feature of this alumina multilayer board is its ability to tolerate subsequent firing in either oxidising or reducing atmospheres. This means that it can be used in conventional thick-film circuitry or with metallisations of the molymanganese–refractory metal type. With proper surface preparation, thin-film components can also be applied. Another important property is the resistance of the buried metallisation which is typically 0.007 to 0.010 Ω/cm^2. Riser resistance is below 0.002 Ω for any diameter–thickness combination within the limits listed in Table 3.5.2.

Table 3.5.1 lists some pertinent properties of the alumina, available metallisations and the multilayer composite. The alumina itself is typical of the many 96 per cent aluminas available for electronic applications today. The internal metallisation is contiguous with the ceramic and is bonded directly to the ceramic. Adhesion of a buried metallisation covering 100 per cent of the area of a substrate is in excess of 48.26 MN/m². The top conductor can be any of a number of thick film metallisations or combinations and sheet resistances as low as 0.003 Ω/cm^2, are available using gold conductor systems.

Table 3.5.2 lists dimensional parameters on multilayer boards of numerous

Table 3.5.2 **Dimensional parameters**

Board	
Dielectric layer thickness	0·13–0·76 mm
Total thickness	0·61–3·17 mm
Internal Metallising	
Line width	$\geqslant$ 0·13 mm
Line spacing	$\geqslant$ 0·13 mm
Fraction metallisable	100%
Conductive risers	
Riser diameter	>0·25 mm
Riser length (layer thickness)	$\leqslant$ 2 diameters
Riser spacing (centre to centre)	$\leqslant$ 2 diameters
Surface finish	
As-fired	20 μm (CLA)
Polished	4 μm (CLA)

designs. Units with six layers of ceramic and areal dimensions of 12·700 × 12·700 mm to 25·4 × 50·80 mm have been produced and no new problems are foreseen for larger units. Internal metallising is typically 0·254 mm lines and 0·254 mm spaces. Routine tolerances on finished pieces are shown in Table 3.5.3. In most cases closer tolerances can be obtained by using additional processing techniques. The as-fired surface is 20 μm (CLA) which can be polished down to 4 μm (CLA) for use with thin films.

Table 3.5.3 **Standard tolerances**

Length and width	±0·13 mm
Thickness	±0·076 mm
Camber	±0·0102 mm

Accessory developments

The multilayer alumina substrate with buried noble metal conductors can accommodate a wide variety of accessory components and attachments. Thick-film conductors, capacitors, resistors and crossovers can be screen-printed and fired on the substrate exterior by conventional thick-film techniques. Active devices such as integrated circuits can be attached by conventional thick-film techniques. Active devices such as integrated circuits can be attached by conventional eutectic die bonding, low-temperature brazing or soldering. Joining alloys such as gold–tin (melting-point, 280 °C) and gold–germanium (melting-point 356 °C), which melt at temperatures above tin–lead solder but below conventional silver–copper braze, are referred to here as low-temperature brazes in keeping with current electronics industry usage.

Lid seals can be made by gold–silver braze or by 'solder glass'. Outboard lead connections can be made by means of wires, pins, or lead frames secured by soldering, low-temperature brazing, e.g. gold–silver, gold–germanium, Au–Sn, Au–Ge, or high-temperature brazing (gold–copper). It is essential that the accessories and processing steps be planned so that they are compatible among themselves in temperature sequence and atmosphere requirements.

Table 3.5.4   **Connection methods for multilayer ceramic boards**

External	
Solder	Compatible with thick film conductors, resistors and capacitors
Low-temperature braze (Au-Sn, Au-Ge)	Compatible with thick film conductors* and resistors†
High-temperature braze	Compatible with thick film conductors
Internal	
Reflow solder	Selectively compatible with the listed external connection methods, thick film conductors, resistors and capacitors
Die bond	
Thermocompression bond	
Ultrasonic bond	

* Many commercially available thick film conductors are incompatible with exposure to brazing cycles. Special formulations are required
† Inert brazing atmosphere is required

A summary of connection methods and compatibility considerations is shown in Table 3.5.4. Emphasis is given to interconnection techniques employing thick film hybrid components at the multichip package level. The method for securing the external leads to the substrate or package influences all other processing steps. If tin–lead solder attachment of pins or wire leads is used, they are attached at the last step, in air, and are compatible with any of the conventional thick-film components. If gold–germanium or gold–tin braze attachment of pins or lead frames is used, all conductors and passive components already present must be resistant to forming gas (a nitrogen–hydrogen mixture) at a brazing temperature of 300 to 600 °C. Several new metallisations have been developed to meet this requirement. Palladium–silver based resistors cannot tolerate reducing conditions at this temperature; they can only be used if the brazing is done in inert gas, which is somewhat more difficult than in forming gas. If high-temperature brazing, e.g. silver–copper eutectic, of the external leads is employed, fired-on gold and platinum–gold conductors suitable for die wire bonding may be used on the alumina provided they are selected to be resistant to the conditions of the 800 °C brazing cycle.

Chapter 3.6 **Steatite and alumina insulators**

During the last ten or fifteen years there has been a considerable increase in the demand for ceramic insulating materials for use in the electrical and electronics industries. The range of shapes and sizes has increased, and there has been a demand for ever-improved quality.

Porcelain has found widespread use in the electrical industry for both high and low voltage application since it can be produced in a wide range of sizes having zero porosity. Although porcelain is an excellent insulator at low frequencies, its use at high frequencies is limited due to a high dielectric loss factor, particularly at elevated temperatures. The electronics industry relies to a large extent on low-loss steatite bodies which gives reduced losses at very high frequencies. For areas demanding even greater performance, alumina is used since it is superior electrically and possesses far greater mechanical strength.

Low-loss steatite ceramic materials

Manufacture and characteristics

Steatite bodies consist mainly of magnesium silicate ($MgO.SiO_2$). The main raw materials are usually talcs or soapstones which occur naturally in various parts of the world. Since the chemical and physical properties of different deposits vary, manufacturers often have a preference for a particular deposit or may obtain the desired properties by blending two or more talcs.

In order to control the firing temperature and to ensure a dense non-porous product, it is usual to add clay, feldspar or barium carbonate. These act as fluxes and help to ensure complete densification on firing.

In the manufacture of steatite, the ingredients are weighed and mixed together, either by wet ball milling or in a plastic state in a Z-blade mixer. The mixed material is dried and an organic binder is added, after which the powder is made free flowing by granulation so that it may be fry pressed. For extrusion, some 10–20 per cent moisture must be present to give the material plasticity.

The majority of steatite pieces are made by dry pressing freely flowing powder on mechanical presses at a large range of pressures. Hardened steel or tungsten carbide dies are used to reduce wear to a minimum. Several thousand pieces per hour can be produced on one machine.

Firing is carried out in electric, gas, or oil fired kilns, and tunnel kilns are extensively used for large scale production. Steatite bodies have a small firing temperature range of 20 to 30 °C so careful control of the kilns is important. The firing temperature naturally depends on the composition, but is in the range of 1,250 to 1,400 °C. Since the firing shrinkage may be as high as 20 per cent, it is obvious that the control of shrinkage plays a vital part in steatite manufacture. All the pressing tools have to be made based on a known shrinkage value, and hence it is difficult to hold very tight dimensional tolerances on fired pieces. In practice ± 1 per cent can be held, but tolerances on ground pieces can be economically produced to ± 0·025 mm.

Use

Steatite is the most widely used ceramic material for electronic applications. It is used for:

1 Aerial insulators.
2 Wave-band switches.
3 Tube sockets and supports.
4 Trimmer capacitors.
5 Coil formers.
6 Resistor shafts.
7 Relay insulators.
8 Valve-holders and many other applications.

Alumina ceramic materials

Manufacture and characteristics

Alumina is characterised by its extreme hardness and chemical inertness. It possesses excellent electrical properties and is used where maximum performance is required. Although naturally occurring alumina consisting of single crystals have been used for many years for watch and instrument bearings in the form of gems and sapphires, cost and size limitations prevent any widespread use.

For mechanical and electrical applications, alumina is almost invariably used in the form of a polycrystalline material consisting of tiny crystals 3–10 μm in size. A typical alumina body contains at least 85 per cent of aluminium oxide (AlO), but for special applications the purity may be as high as 99·9 per cent. Generally, the higher the purity the greater the electrical performance, but the cost increases considerably for 99 per cent purity or over. Another alumina body containing 90 per cent (AlO) which has approximately the same electrical properties as steatite at room temperature, but considerably improved performance at elevated temperatures, it is twice as strong mechanically and has greater thermal conductivity than steatite. For general electrical and electronic applications a 90 per cent body represents a good compromise between performance and cost.

In the manufacture of polycrystalline articles, the starting material is usually aluminium oxide powder of 98–99 per cent purity. This material is wet or dry ball milled to a very small particle size 1–3 μm, while additives such as silica, magnesia or clay are made to control the properties, and to reduce the firing temperature. The dried milled powder is mixed with an organic binder and made free flowing by granulation for dry pressing.

Alumina pieces may also be formed by extrusion or by isostatic pressing or slip casting in plaster moulds. The firing, during which sintering and shrinkage take place, is usually carried out in gas kilns at 1,500 to 1,650 °C. In the sintering process, the individual alumina crystals exhibits some growth at the expense of the smallest ones, and these crystals become bonded together to form a dense homogeneous structure which is non-porous. Once again, large shrinkage occurs on firing and the remarks made earlier on steatite dimensional tolerances also apply to alumina.

The maximum performance, both electrically and mechanically, is obtained by having the greatest purity level and the finest crystal structure. Whilst impurities and additives usually lower the firing temperature required to produce complete densification, they nevertheless tend to impair performance due to the formation of glass phases which surround the alumina crystals. Since the electrical and mechanical properties of these glass phases are inferior to those of the alumina crystals, excessive glass must be avoided particularly if it contains alkalies.

Applications of alumina bodies

Alumina can be used for all the applications listed for steatite since it has properties equal to or superior to such a material. It is however more costly due to material costs and higher firing temperatures used, so that in practice only the more severe applications warrant its use. For situations where strength and resistance to mechanical shock are needed, alumina is often successful where steatite would fail. Moreover, alumina can be used at fairly elevated temperatures without significant loss in performance. The high thermal conductivity of alumina has led to its widespread use as a substrate material and for envelopes and transmission windows in high-power microwave valves.

Glazing and metallising of insulating ceramic materials

Glazing

The surface condition of a fired ceramic article is of considerable importance to the user, and in general a smooth surface is desirable. Apart from surface finish, however, the porosity of the ceramic material can be vital in determining the usefulness of the product. Since steatite and alumina articles have a polycrystalline structure with grain boundaries, a glass smooth finish cannot be obtained, and finishes better than 50 μm are difficult to obtain on a production basis. Moreover, the existence of minute grain boundaries can facilitate the pick-up of surface moisture and contamination, even though the body is non-porous as a whole.

Where improved surface finishes are required, glazing is carried out imparting a smooth glossy surface onto the article. Such a surface is non-porous and moisture proof and may have a better finish than 2 μm. In the glazing operation a low melting ceramic composition is applied to the surface and this becomes vitreous or glassy when fired in a kiln. The glazing process coats the ceramic surface with a layer of glass having a smooth glossy finish.

It is necessary to match the glaze ingredients to give a similar expansion coefficient to that of the main body, otherwise cracking or crazing will occur.

The thickness of glaze varies according to the application from 0·08 to 0·25 mm. Tolerances on glaze thickness can be kept within 0·05 to 0·08 mm. Where glazing is not practical, silicon compounds can be used to make the ceramic surface moisture repellent. However, silicon treatment does not prevent contamination from dirt and is inferior to glazing.

Metallising

Ceramic materials can be metallised in order to facilitate subsequent attachment to metal components. Two methods can be used:

1. Fired-on silver paints followed by copper plating and tinning so that the surface can be soft soldered to metal.
2. Fired-on nickel paints which can be tinned without the need for copper plating.

The use of nickel is recommended since this form of metallising is not readily soluble in solder baths and can therefore remain in contact with hot solder for extended periods. Silver, on the other hand, will dissolve rapidly in hot solder.

Insulating ceramic materials made from steatite and alumina are widely used in the electrical and electronics field. They can be produced in large quantities by pressing and extrusion to relatively close tolerances and at competitive prices, while small numbers for special applications can be made by individual machining although at extra cost. The use of these ceramic materials has been further increased by the relative ease with which they can be metallised and soft soldered.

New applications often call for tighter specifications and improved performance, which must lead to the greater use of alumina in the future. In this connection, improved manufacturing methods are continually being developed, including the use of spray drying which is able to convert a wet ball milled slurry into a very uniform free flowing powder ready for pressing. Improved powders and forming methods together with closer temperature control of kilns will ensure that insulating ceramic materials can meet future demands over an ever-widening range of applications.

Section Four

Chapter 4.1   Sealing expansion alloys

Glass-to-metal hermetic packages for electronic devices are now used in several popular designs. Discussed here are the problems in the use of alloys in this regard and the development of new or modified alloys for packaging.

Engineering throughout its history has often solved design problems by utilising the complementary properties of dissimilar materials. Many examples could be given, but here we are interested in combining glass and metal: glass for its insulating characteristics and metal for its conductivity, and the combination must provide a hermetic environment within its capsule. We are thus faced with the problem of joining these two dissimilar materials and the methods available for joining glass and metal. Typically, these are the most difficult seals to form.

Glass-to-metal seal

The most permanent and reliable type of seal is that made by fusing the glass directly to the metal, usually referred to as 'glass-to-metal' seal. For optimum thermal and mechanical integrity of the seal it is necessary that expansion of the glass and metal be substantially the same over the temperature range in which both materials behave elastically. Table 4.1.1 describes the stress characteristics of actual seals as determined by the differential contraction below the strain point of the glass that temperature below which the glass is rigid. If differential contraction is less than 100 ppm, the integrity of an assembly can be assured.

Table 4.1.1   **Mechanical characteristics of glass-to-metal seals**

Differential contraction from strain point of glass, $\Delta l/l$	Stress characteristics of seal
$<10^{-4}$	Stress conditions excellent
$1{\cdot}0$ to $5{\cdot}0 \times 10^{-4}$	Stress conditions satisfactory for medium-sized seals
$5{\cdot}0$ to 10×10^{-4}	Stress conditions suitable only for small seals or for seals in which critical stresses in the glass are compressive

The particular application for a glass-metal composite specifies both the type of glass and the metal to which it can be sealed. Soft glasses have thermal expansion coefficients α of about 9×10^{-6} cm/cm.degC, whereas hard glasses

Q

have lower α of the order of $4\text{–}6 \times 10^{-6}$ cm/cm.degC. The latter are often required for thermal shock resistance: if this be the case, then a metal with similar values of α must be specified for sealing.

Such a case was historically represented by a rapidly expanding need for hermetically sealed gaseous or vacuum electron tubes. Although Fe–Ni and Fe–Ni–Cr alloys were known to have low values of α none of these had coefficients matching that of a hard glass up to the strain point of the glass. Molybdenum and tungsten were considered, but their utility was limited by attendant fabrication problems. An alloy of 54 per cent iron, 29 per cent nickel and 17 per cent cobalt (Kovar) was developed which displayed a match with hard glasses (see Fig. 4.1.1).

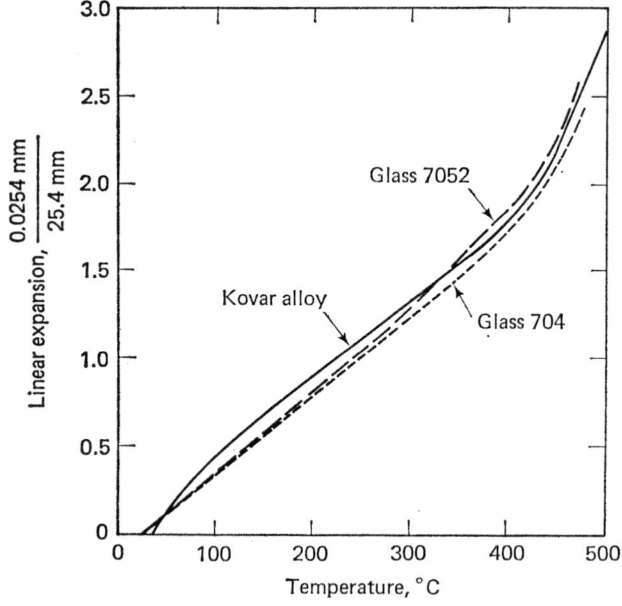

Fig. 4.1.1   Thermal expansion of Kovar compared with two grades of glass

Although thermal expansion is the most important requirement for making good glass-to-metal seals, other metallurgical parameters are of prime importance as well. Because of deleterious effects of non-metallic inclusions, residual gas content, transformation products or excessive carbon content on either sealability or processing, stringent controls are required during all stages of manufacture. This has been primarily accomplished by using pure raw materials and through eliminating the use of strong deoxidisers by a vacuum remelting process. The consumable-electrode remelt process has inherent zone-refining characteristics and also eliminates possible crucible contamination. By comparison with the original alloy, present-day alloys offer improved seal integrity (by virtue of significantly lower residual carbon), substantially lower gas evolution with time enhancing performance of 'long life' electronic devices and important improvements in machineability, weld characteristics, plating and etching properties by the elimination of non-metallic inclusions.

Another important metallurgical characteristic for glass-to-metal seals is oxidation; both bond strength and adherence are enhanced by accomplishing the seal through an intermediate oxide, which is primarily a mixture of iron oxides. The oxide is both adherent to the alloy and soluble in glass while its expansion coefficient does not represent a substantial mismatch to either. Figure 4.1.2 shows, schematically, the transition from metal-to-glass in a good seal.

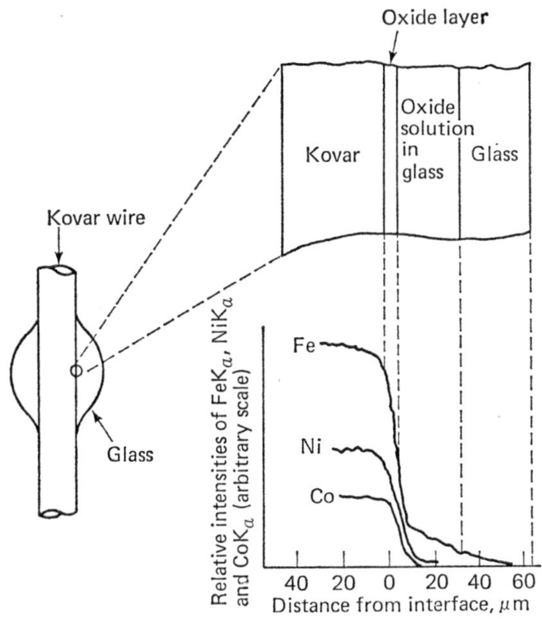

Fig. 4.1.2 Continuous transition from Kovar to glass in glass-to-metal seal

The impetus for much of the development of high-reliability alloy has come from its use in electron tube devices, but this application represents only a fraction of the entire market for glass-to-metal seals. Other large areas of usage are in speciality applications such as seals in vacuum systems, in the transistor industry, and most recently in molecular electronics.

With the advent of integrated circuitry and even more stringent requirements for hermeticity, this material has been a natural choice of a glass-to-metal sealing alloy for microelectronic packaging.

Utility in package design

In molecular devices, the package is not merely a safe storage container but an integral part of the device which can protect or destroy the device. The main parts of the package are the body, lead frame and sealant. To date, acceptable body materials for high-integrity military devices have been metallic or ceramic, although plastics are being seriously considered for application where hermeticity is not essential. Lead frame materials are metallic and must be specified for sealability with lesser emphasis on electrical conductivity and thermal conductivity.

Sealing materials are usually glasses for they provide the desired hermeticity. In turn, the choice of glass determines the lead frame material since the thermal expansion of the lead frame must be similar to that of the glass. Forming a functional package from the base, lead frame and sealant has led to three general designs which are currently in production.

The 'TO' pack has been adapted from discrete semiconductor technology and now houses the standard TTL or DTL digital logic circuit in one envelope. This package, which employs a Kovar-hard glass header enclosed by a projection welded can, is capable of providing environmental protection for full range specifications. For example, some specifications require operation at ambient temperatures over a range of -55 to $+55\,^{\circ}$C and specifies requisite resistance to moisture, salt spray, acceleration, shock and vibration. The typical variant on the TO-5 package, depicted in Fig. 4.1.3a, consists of 10 or 12 axial leads projecting from the cylindrical area. Over-all dimensions are of the order of 19·0 mm in length by 9·5 mm diameter.

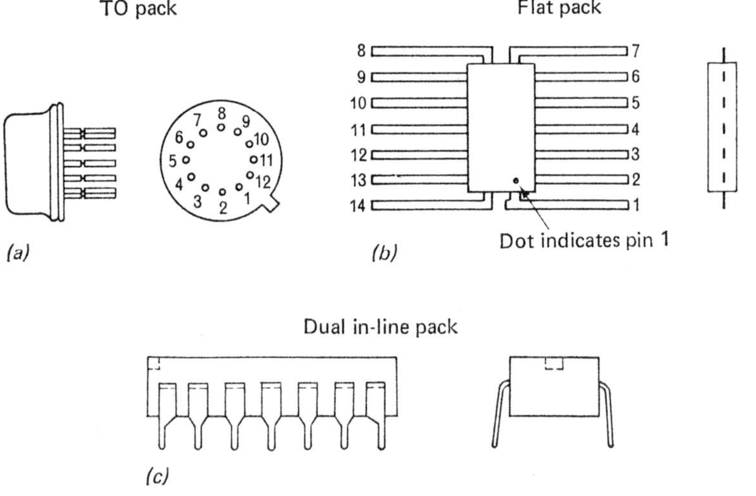

Fig. 4.1.3   Popular microelectronic package designs

The flat pack is presently the most popular molecular electronic package. It employs Kovar lead frames combined with glass or ceramic-sealed alumina (nominally 96 per cent Al_2O_3) version and decreases through the hard glass version to the so-called solder glass seal with its probability of reduced hermeticity by virtue of increased gaseous and ion diffusion. In its usual configuration (Fig. 4.1.3b) the flat pack provides 10 or 14 terminal leads to the various logic and differential amplifier circuits available. In general, it represents some space and weight saving over equivalent TO packs and some improvement in design accessability.

The third popular package type, the so-called dual in-line package (DIL), is an adaption of the flat pack design concept, but is capable of plug-in mounting (see Fig. 4.1.3c). Considerable development has been devoted to this package type and as a result we have all the hermeticity and reliability combinations denoted for other packages, plus some additional innovations. The prime change which has been investigated is the use of plastic encapsulants intended for reduced protection against humidity and operation at lower temperatures, typically o to 70 °C. Such encapsulation is presently being evaluated for a number of entertainment applications such as TV and high-fidelity components. Obviously, the use of plastic offers distinct cost advantages, but this appears necessarily coupled with somewhat reduced integrity and reliability.

Whether Kovar and hard glass can be replaced by other expansion alloys such as the iron–nickel alloys and solder glass (or plastic) without excessive loss in hermeticity and integrity has yet to be answered. The hermeticity provided by solder glasses is usually not as reliable as that of hard glasses, but whether it is sufficient to protect the material composing the integrated circuit and the thin wire bonds to the lead frame from oxidation and destruction is not known.

Lead-frame materials

Hermeticity requirements predictate the type of sealant material which in turn specifies the thermal expansion characteristics of the metal lead frames. If the thermal expansion is thereby fixed, secondary consideration of optimum thermal and electrical conductivity is made. As Table 4.1.2. shows, however, the common metals with desirable thermal expansion properties do not have very good thermal and electrical properties due to the high alloying.

Table 4.1.2   **Properties of some expansion materials**

Alloy	Thermal expansion coefficient (20–100 °C), $\times 10^6$ cm/cm.degC	Thermal conductivity, cal/cm.sec.degC (at 100 °C)	Electrical resistivity, MΩ/cm at 20 °C
Fe	12·2	0·163	9·7
Ni	13·3	0·145	9·0
Cu (OFHC)	16·4	0·92	1·72
Al	23·5	0·54	2·87
W	4·5	0·38	5·5
Fe–36Ni	1·0*	0·026	66
Fe–42Ni		0·026	78
Kovar	6·0	0·04	49
1008 steel	12·2	0·138	13·2
Si	7·6	0·20	10^5

* To inflection temperature

It is the consequence, however, of such alloying that leads to the desirable thermal expansion properties. With the original discovery of alloy it was realised that no rule of mixtures for thermal expansion was obeyed in alloying. Early explanations postulated the existence of Fe_2Ni (which has no basis in fact) or the coexistence of two forms of ferromagnetic material, one with normal dilatation while the other would increase in volume upon cooling. Indeed, the present understanding of low thermal expansion does involve ferromagnetism and the superposition of two phenomena. When ferromagnetic materials cool below the Curie temperature, there is a spontaneous lattice deformation caused by magnetisation. In alloys such as described, the magnitude and sign of the magnetostriction cause an expansion of the lattice. When these materials are heated, magnetostriction gradually disappears (which means the lattice will contract), thus counteracting the normal thermal expansion. The measured spontaneous lattice deformation of an alloy of composition 31Ni–5Co–64Fe superposed with its thermal expansion coefficient (see Fig. 4.1.4) illustrates this fact that the value of α is truly a combination of the normal (paramagnetic) value of α and the temperature dependence of the spontaneous lattice deformation.

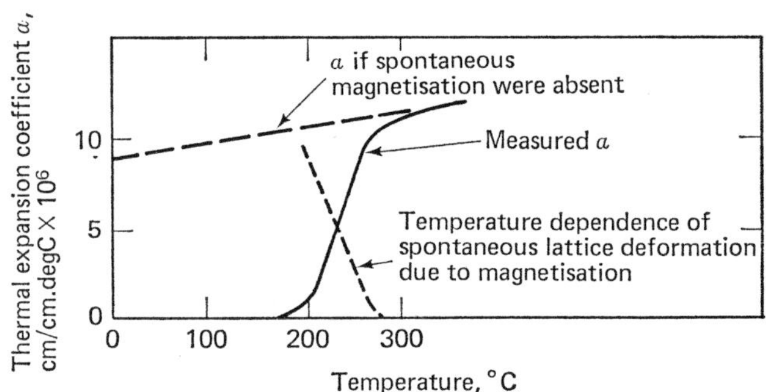

Fig. 4.1.4   Effect of spontaneous magnetisation on thermal expansion of Fe-Co-N alloy

Since low thermal expansion is intimately related to ferromagnetism, the temperature range for matching of glass and metal in sealing is quite significant.

Future for low-expansion alloys

It is unlikely that any new expansion alloys will be introduced commercially, although the knowledge necessary to design a specific expansion alloy is being researched at present. However, if new hard glasses were to become available or if package designers were to agree on the use of solder glass seals, it would then be possible, in fact necessary, to substitute a new or different alloy.

It is reasonable that package producers can expect some modifications of present lead frame or base materials which will reduce in-house processing problems and lead to fewer rejects because of poor seals. For example, it is at present common to employ a carbon fixture material during the sealing of molecular devices so that molten glass does not adhere to the fixture.

Stabilisation of the oxide in the presence of carbon is of current interest in expansion alloy research. Other modifications whether in composition or in processing, will most likely be directed towards improved thermal and electrical conductivity as well as seal quality.

Alloy-clad copper

Perhaps a sign of new expansion materials to come is the recent availability of alloy-clad copper. Such composite material improves the electrical and thermal conductivity of lead-frame material.

This does not justify the additional cost at present, however, for calculations show that heat dissipation through the leads is negligible in common packages, and that most heat is dissipated from the silicon chips through the base. Perhaps the package designer and material supplier will combine their efforts for improved heat dissipation by using a sealant alloy base which is clad to a good heat dissipator. This has been accomplished in power-device packages where large base areas, convection coolers and similar techniques are used.

Compatibility of glassy materials

Glass-to-metal seals can be fabricated between vitreous or crystallising glassy materials and elemental metals or alloys. Glass-to-metal seals can be logically segregated in terms of seal interface reaction or bonding. The packaging of monolithic devices utilises chemical bonding of vitreous or crystallising glassy materials.

Matched and unmatched seals

A matched seal is one in which the metal and glass thermal expansion and contraction coefficients are kept sufficiently close so as to prevent the

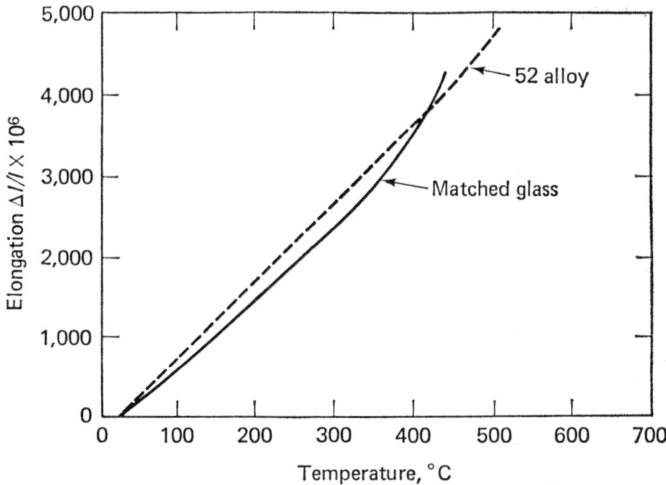

Fig. 4.1.5   Typical contraction diagram — matched seal

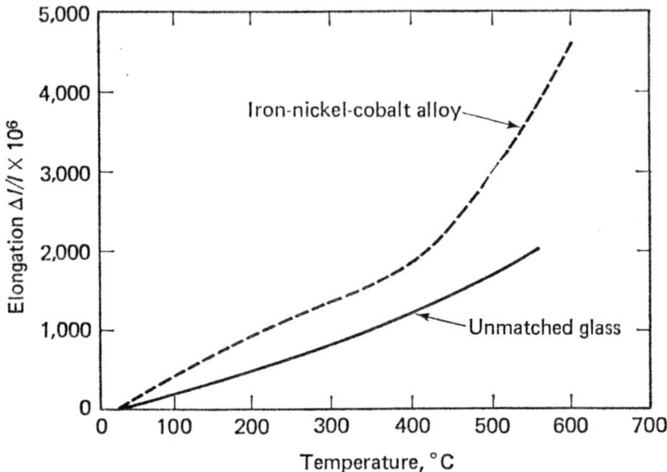

Fig. 4.1.6   Typical contraction diagram — unmatched seal

development of dangerously high stresses. Contraction curves of a typical 'matched' seal are shown in Fig. 4.1.5. An unmatched seal (Fig. 4.1.6) is one in which the coefficients are sufficiently different so that dangerous stresses are developed unless:

1 Seal geometry is controlled to maintain only high compression in the glass.
2 Metal is sufficiently ductile so as to yield and prevent high stress levels from developing.
3 Metal size or thickness is kept small so that the seal dimensions do not allow the development of high stresses.
4 A series of grading glasses is placed between the mismatched glass and metal. Thus, coefficients are controllably mismatched in steps so that no part of the seal experiences excessive stress.

Metal requirements

A metal used for chemically bonded seals must meet certain basic requirements:

1 The metal must have no adverse structural or phase transformations which would produce radical changes in thermal expansion/contraction characteristics below the annealing point or sealing temperature of the glassy material.
2 The melting-point of the metal must obviously be above the glass sealing temperature.
3 The metal should form an oxide that adheres to the base metal, and is sufficiently soluble in the glassy material so as to form a strong hermetic bond.

4 The metal should be free of uncombined, adverse reacting constituents such as carbon or organics which will produce reboil (tiny air voids in the glass at or near the interface). The reboil lowers seal strength by reducing the glass cross-sectional area near the seal interface.

5 The metal must possess the electrical and thermal conductivity required for the specific application. This minimises the formation of high operational stress.

6 The metal must have sufficient ductility to allow any required bending or reforming after sealing.

7 The metal should be amenable to plating and/or soldering, if these operations are required.

Glass requirements

The glassy material must also meet basic requirements:

1 The glass must wet and adhere to the oxide layer. This is the basic mechanism of bonding. The normal criterion for wetting is that the seal is not re-entrant. The seal is termed 'not re-entrant' when the contact angle is between 0 and 90°. The wetting is generally influenced by surface conditions, sealing atmosphere, composition of both the glassy material and the metal, sealing or bonding temperature and the method of heating.

2 The glassy material must not cause reboil. A chemical bond is formed by heating the glassy material so that it softens, flows, and wets the oxide layer which is intimately attached to the base metal. If the glassy material is a crystallising type, as the temperature increases crystals form and grow. The oxide, typically, is at least partially dissolved by the glassy material and taken into solution. The driving force for this reaction is the chemical potential gradient established across the oxide–glass interface.

Solder glasses

There are two basic types of solder glasses: vitreous and crystallising. A vitreous solder glass is simply a low-melting-point glass, which retains the characteristic 'glassy' appearance after curing. A crystallising solder glass is a low-melting-point glass in which crystals grow at the expense of the glassy phase during a thermal 'soak' or cure. The thermal contraction of a commercial crystallising solder glass is influenced by the time and temperature of processing as shown in Fig. 4.1.7. A longer curing time at a fixed temperature typically produces a lower thermal contraction than a shorter cycle at the same temperature.

Figure 4.1.8 illustrates the change in thermal contraction values of the crystallised solder glasses. The displacement between the vitreous and crystallised solder glasses is the result of thermal processing which produces crystallised material of lower thermal contraction.

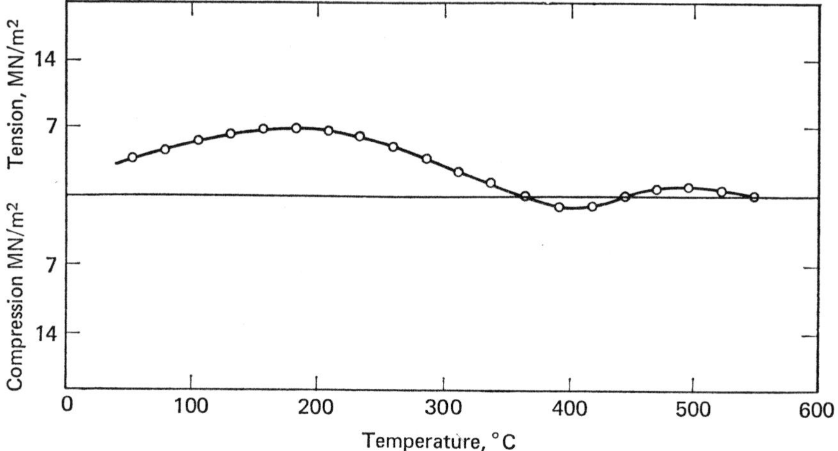

Fig. 4.1.7   Stress versus temperature diagram of B-Si glass to Fe-Ni-Co alloy

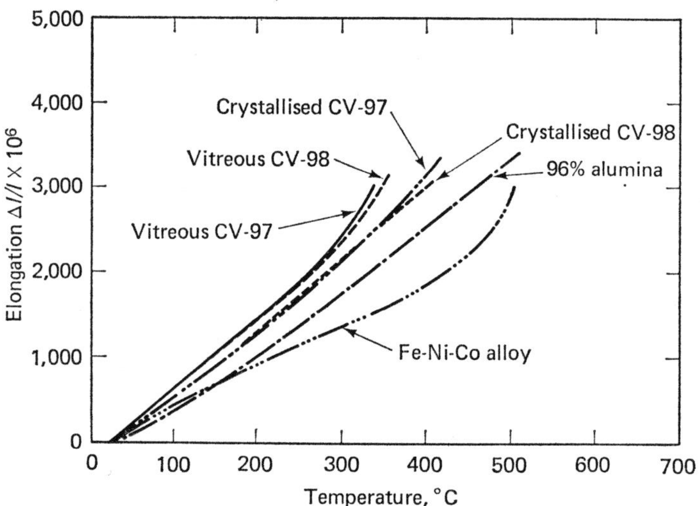

Fig. 4.1.8   Contraction data for Fe-Ni-Co alloy and CV-97 and CV-98 glasses

The basic properties and application characteristics are shown in Table 4.1.3 for both types of solder glasses. Solder glass seal properties are measured by DTA (differential thermal analysis), gradient boat tests, dilatometer measurements of the viscous flow which occurs during the thermal curing.

Crystallised solder glass seals

The properties of crystallising solder glasses are greatly influenced by processing. The principal sealing criteria are thermal contraction compatibility and uniformity of state throughout the seal volume. Thermal contraction

Table 4.1.3 Comparison of vitreous and crystallising solder glasses

	Vitreous solder glasses		Crystallising solder glasses
Method of application	Hot dip	Paste or slurry	Paste or slurry
Curing cycle, min, at recommended temperature	~5	60	60
Electrical resistivity, log ρ (Ω.cm), at 350 °C	10–11	10–11	7–8
Permittivity, K	~12	~12	~20
Separation of sealed parts	Easy-simple remelting		Difficult-chemically or mechanically

compatibility helps insure seal stability, long-term hermeticity and maximum resistance to thermal shock. The uniformity of state is necessary to insure that wetting is complete and uniform with no highly localised stress concentrations occurring in the seal material.

Glass-to-ceramic seals

Previous glass-to-ceramic seals used a glass with a contraction coefficient similar to that of the ceramic. The glass, therefore, is in compression at and near room temperature. This provides the greatest over-all vitreous seal strength.

The glass wetting occurs because of the solubility of the ceramic in the sealing glass. This solubility produces an interdiffusion between the sealing glass and the ceramic which promotes long term hermeticity and high bond strength.

Glass-to-ceramic seals may be treated as an extension of glass-to-metal seals. The increased brittleness and rigidity of ceramics requires more care on sealing, and the interdiffusion affect cited above promotes high interface compatibility.

Sealing the alumina package

In addition to the criteria discussed previously, package sealants must also meet the following requirements:

1 LOW CURING TEMPERATURE From the present generation of crystallising package sealants, those with thermal contraction coefficients higher than the alumina are selected to preserve the monolithic devices. Since the devices have definite time/temperature limitations, it is necessary to select those sealants which cure at the lowest temperatures and which still provide a 'matched' seal. The thermal contraction properties of two current sealants are illustrated in Fig. 4.1.9.

2 CHEMICAL DURABILITY The exposed leads of current packages are usually processed after the sealant is cured. Typical processes are electrochemical cleaning of the lead frame and metal plating or soldering. The conventional sealants meet these requirements as long as the reasonable pH

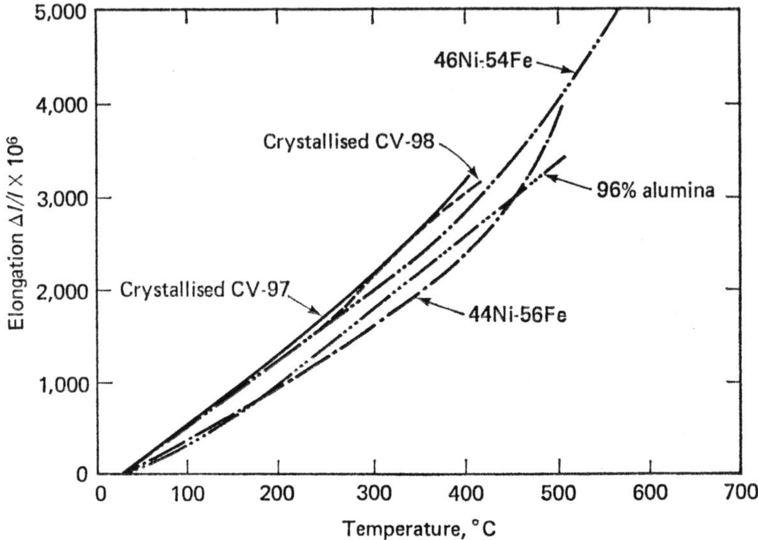

Fig. 4.1.9   Contraction data for 44Ni-56Fe and 46Ni-54Fe alloys and CV-97 and CV-98 glasses

levels of the solutions are maintained and etching times are not excessive. The electrochemical techniques utilise dilute solutions, short times and produce repeatable results.

3 STRENGTH The sealant must have sufficient strength to effectively resist stresses from torques and tensile stresses required in critical applications. The crystallising solder glasses are significantly stronger than vitreous systems under similar processing conditions and meet the stress requirements when properly cured.

The glass-to-metal seal will be considered now. It is apparent from Fig. 4.1.9 that the greater thermal contraction mismatch occurs between the crystallising solder glass and iron–nickel–cobalt alloy. This mismatch does not produce a strong, durable glass-to-metal seal. There are two obvious alternatives for improving thermal contraction compatibility:

1 Lower the thermal contraction characteristics of the solder glass.
2 Raise the thermal contraction characteristics of the metal.

There are commercially available alloys which closely approximate the thermal expansion/contraction characteristics of 96 per cent alumina. A very compatible alloy is 46 per cent nickel, balance iron plus trace elements. From Fig. 4.1.9 and Table 4.1.4 is is seen that 44Ni–56Fe and 46Ni–54Fe alloys are thermally compatible and the electrical resistivity of the 46Ni–54Fe alloy is lower than the iron–nickel–cobalt alloy.

The more-closely-matched thermal expansion/contraction coefficients provide less temporary and residual stress in the glass-to-metal seal. The improved electrical conductivity and thermal compatibility (Fig. 4.1.8 and

Table 4.1.4 Properties of various nickel alloys

Alloy	Electrical resistivity, $\mu\Omega$/cm	Thermal conductivity, units	Recommended operating temperature in air	Curie temperature, °C
Ni–Fe–Co	48·4	0·04	210	435
42Ni–58Fe	65·0	0·025	180	380
44Ni–56Fe	64·0	—	200	420
46Ni–54Fe	45·7	—	210	460
52Ni–48Fe	43·2	0·032	250	530

Table 4.1.3) lower environmental stress due to operation resistance heating of the lead frame through the sealant. This means that packaging yields will increase and sealant failures in use should decrease.

The iron–nickel alloys containing 44 per cent or more nickel are recognised as being good glass sealing alloys. Excellent glass sealing grades are available with a controlled carbon content of 0·10 per cent or less. The cost of these alloys is approximately two-thirds that of the iron–nickel–cobalt alloy.

Future compatibility

The next generation of package sealant should cure at lower temperatures. As was seen in Fig. 4.1.8 the 44Ni–56Fe and 46Ni–54Fe alloys match the thermal expansion characteristics of alumina more closely than does the iron–nickel–cobalt alloy now in vogue. They form a more compatible package.

Since losses in the 'packaging step' are the most expensive losses, it is highly desirable to use a fully compatible package. The most compatible package discussed would be 96 per cent alumina body, and 46Ni–54Fe alloy. This metal or a variant of it will also be acceptable for the next generation of sealants designed for the alumina package.

This utilisation of this metal will not only anticipate the generation of sealants, but will also improve the present package mechanically, electrically and thermally.

Chapter 4.2 **Insulation and resistance: potentiometer wire materials**

Insulation materials

The choice of enamel for use as an insulating coating on the wires of close-wound resistors and potentiometers can be as important as the choice of meterial for the winding itself. Different types of insulating enamel from which users of resistance and potentiometer wire may select are available, but it is necessary to employ coatings that are suitable for all conditions of manufacture and service.

These enamels have a range of specially formulated parameters each containing useful properties; apart from selecting compatible enamels, coating must have an even thickness, no matter how fine or delicate the wire.

To enable the designer or engineer using resistance and potentiometer wires to select the most suitable type of enamel for any particular application, a brief statement of the properties of insulating enamels are given:

Grade 1 This grade is a form of epoxy-based synthetic enamel that has a combination of excellent durability, high breakdown voltage, good resistance to solvents and heat.

Grade 2 This enamel is a high temperature synthetic polyimide ester. The British Standard does not, however, specifically refer to polyimide enamels but it provides a suitable basis for the determination of the enamel's electrical and other characteristics. The maximum service temperature is 200 °C but here again this may be exceeded for periods of short duration.

Grade 3 This is an epoxy-based synthetic enamel which has been developed to provide a combination of good durability, high breakdown voltage, resistant to solvents and heat; the coating is particularly free from pinholes, is flexible with good adhesive properties. This enamel can be applied as an extremely thin, even coating that ensures high resolution and the close tolerance required in the manufacture of high-precision wire-wound potentiometers; such potentiometers normally operate at low voltages and therefore requirements of high electrical insulation for resistance wires are not so necessary.

Because this enamel can be readily removed from slider tracks by means of a rotating rubber bonded abrasive or 'blasting' with a powder such as sodium bicarbonate, makes for easy production.

Where connections at the ends of the wires have to be made, immersion of the coated wire in methylene chloride for about 2 min enables the enamel to be removed with ease.

Grade 4   This type is a polyurethane enamel and complies fully with the requirements of BS3188 relating to a self-fluxing enamel with a polyurethane base. Wires coated with this enamel can be readily stripped by immersion in a bath of molten tin-lead solder at a temperature of 360 °C. Where the diameter of the base wire exceeds 0·4 mm, the temperature should be increased to 400 °C.

Grade 5   This enamel is the oleoresinous type: it is considerably softer than Grade 1 type also other epoxy-based enamels and in consequence less resistive to abrasion. Although, for the majority of applications it is a preferred coating because of the advantages to be gained by its use on certain types of close-wound precision potentiometers where mechanical and electrical properties are not critical. This enamel can be removed mechanically with less risk of damaging the wire; it also conforms with BS156 relating to oleoresinous insulating enamels.

Resistance and potentiometer base wire materials

Selection of bare metal-resistance wires must be made with respect to their characteristics and uses; firstly then let us look at the materials most suited to a particular application:

Nickel—chromium 80/20

This type of alloy combines the high specific resistance and moderately low temperature coefficient with resistance to oxidation at elevated temperatures; this combination is particularly suitable for high-value resistors and potentiometers used in electronic equipment. Nickel—chromium is available as hard drawn, bright annealed, or oxidised annealed. It is important to remember that the specific resistance can be increased and the temperature coefficient reduced by a stabilising heat treatment that minimises the tendency for the resistance to change with time.

Copper—nickel 56/44

This is a copper—nickel alloy and is well known under a number of trade names. It is characterised by moderately high resistivity, very low temperature coefficient and good resistance to corrosion. This material is suitable for moderately rated resistors also for potentiometers where the change of resistance with temperature must remain low, there is also a low temperature grade available having half the normal temperature coefficient.

Copper alloy

This wire is a copper-based alloy having a relatively low specific resistance together with a low temperature coefficient; the latter is only obtainable normally in high resistance alloys. The alloy is used for low value resistors to provide the same value of resistance as other resistance wires which may have inconveniently larger diameters. This alloy is normally annealed and may become work hardened during winding, but these effects can be overcome by heating the completed component at 140 °C for at least 10 hr in a neutral atmosphere. To minimise the amount of cold work during windings the diameter of the former should be as large as possible in relation to the wire, also the former material should have a low coefficient of expansion. The wire itself should have a minimum of stress during winding, stretching or sharp bending must be avoided.

Copper–nickel–manganese alloy

This is an ideal material for precision resistors, also resistance coils due to its low temperature coefficient characteristics and its low thermal e.m.f. against copper, but the outstanding electrical properties can only be obtained from an annealed material.

Nickel–chromium alloy 80/20 (modified)

The alloy is used in high value fixed also variable wire-wound resistors. It provides greater resistance stability over a wider operating temperature range also a higher resistance per unit volume with a lower temperature coefficient. Should it become work-hardened, a heat treatment will restore its properties.

Noble metal resistance wires

Requirements in servomechanisms and types of analogue equipment have led to stringent mechanical and electrical specifications for such devices; they include high accuracy, high resolution and low operating torque, as well as low contact resistance with freedom from electrical noise between the winding and sliding contacts.

The above requirements can be met in part by base-metal windings such as nickel–chromium, copper–nickel alloys, but in more critical applications the tarnish films that form on these alloys when exposed to normal atmospheres preclude their use and recourse must then be made to tarnish-free noble metal alloys.

Choice of materials

Resistance wires

Apart from a specific resistance of a required order, other main factors to be considered are the temperature coefficient of resistance, thermal e.m.f. of the alloy against a contact, and its mechanical strength. As a rule the specific resistance of materials varies inversely with the temperature coefficient. The lowest temperature coefficient is provided by a 40 per cent silver–palladium alloy.

It should be noted that although wire in the hard-drawn condition may be easier to wind, severely cold-worked alloys tend to relax with time, with a small change in their resistances.

Potentiometer wire materials

Potentiometer wires must be manufactured to an extremely stringent specification in order to meet the exacting demands of modern precision wire-wound components. Designers or engineers have to make selection of wire materials to meet their specific resistance requirements for their particular applications; it is necessary in these cases that tolerance should be maintained to the closest limits so that the component can achieve very high accuracy, linearily, and resolution in the finished product.

The roundness of the wires, the care with which they are spooled, also the steps taken to minimise curl or twist ensures trouble-free winding. It is very important that the wire should have clean, bright surfaces, free from inclusions and imperfections if maximum reliability and long service life is to be achieved.

If the wire is to be enamelled it should be applied in a thin, even coating also to close limits along the whole length of the wire. In d.c. applications

Table 4.2.1   Noble metal resistance wires—typical physical properties

Alloy	Specific resistance, $\mu\Omega.\text{cm}$	Temperature coefficient of resistance 0–100 °C, degC⁻¹	Thermal e.m.f. at 100 °C, mV		Tensile strength		Minimum diameter, mm	Specific gravity
			Against 625 alloy	Against copper	Annealed, kg/mm²	Hard drawn, kg/mm²		
% palladium-silver*	6	0·0008	− 0·45	− 0·62	23	55	0·05	10·6
% palladium-silver*	10·5	0·0005	− 0·71	− 0·88	28	60	0·02	10·7
JMM 625R alloy	12·5	0·0005	neg	− 0·17	47	94	0·018	14·4
% rhodium-platinum	19	0·0017	+ 0·07	− 0·10	47	117	0·013	20·0
% iridium-platinum	24·5	0·0013	+ 0·72	+ 0·55	55	125	0·018	21·6
% ruthenium-15% rhodium-platinum	31	0·0007	+ 0·20	+ 0·03	101	173	0·02	18·6
% iridium-platinum	32	0·00085	+ 0·78	+ 0·61	70	165	0·013	21·7
% silver-palladium	42	0·00003	− 4·03	− 4·20	38	110	0·02	11·0
% ruthenium-platinum	42	0·00047	+ 0·31	+ 0·14	79	140	0·02	19·9
% tungsten-platinum	62	0·00028	+ 0·88	+ 0·71	94	150	0·025	21·3
% molybdenum-platinum	64	0·00024	+ 0·94	+ 0·77	94	140	0·02	20·3
% copper-platinum	82·5	0·000098	− 0·50	− 0·67	60	140	0·025	16·3
% molybdenum-40% palladium-gold	100	0·00012	− 0·02	− 0·19	69	110	0·025	14·4

* Not completely free from tarnish

R

the thermal e.m.f. against the wire is important, because the friction between the winding and the wiper can generate spurious d.c. signals.

Noble metal and two base metal alloys constitute normal metals used in potentiometers and are listed in Table 4.2.1.

Tolerances

Tolerances can only be maintained on wire specified in terms of its resistance per unit length rather than in terms of its nominal diameter.

Wiping contacts

In precision potentiometers, the wiping contacts recommended are in the form of wire, strip or an alloy that can be formed into a complex shape depending on the application. The alloy usually employed comprises of copper–silver–gold alloy that has spring properties similar to phosphor bronze.

Chapter 4.3 **Carbonised polymer materials**

Engineers and designers have long sought a material that combines the properties of normal industrial carbons with the corrosion and abrasion resistance of glass and silicon. Now with the introduction of vitreous carbon such a material is available.

Vitreous carbon has become a generic term for several types of carbonised polymer, not all of which are of the high quality required for most electronic applications.

In this chapter we look at the physical and chemical characteristics and the way in which it differs from other carbons and graphites; then the manufacturing process and the applications for which the new material is well suited. Firstly, we will compare the common forms of carbon. The element carbon exists in many forms; but only two of them, diamond and graphite, can be precisely characterised. Both occur naturally and both can be produced synthetically. The normal industrial carbons are highly disordered materials which may be readily graphitised. Pyrolytic graphite, prepared at high temperature, has a high degree of conversion. Vitreous carbon is a recently discovered form of the element that is similar to glass both in appearance and fracture, the name vitreous originated from this similarity.

Structure of carbons

Graphite is a stable allotrope and consists of carbon atoms arranged in hexagons in parallel flat sheets, held together by Van der Waals forces at a distance of 3·55 Å. The distance between atoms in a hexagon is 1·42 Å. Two possible arrangements. of the layer planes are known; the more usual has alternate layers displaced by a distance equal to half a hexagon. Graphite has a specific gravity of 2·26 and is very soft — its hardness on the Mohs scale is 1.

In the diamond lattice each atom is linked to four others surrounding it at the corners of a regular hexagon, with an interatomic distance of 1·54 Å. Diamond, with a specific gravity of 3·52 and a hardness of 10 on the Mohs scale, is the hardest substance known: but it can be converted to graphite on heating to a temperature of 1,800 °C.

Graphite and the industrial carbons are generally prepared by heating a grist of ground coke and pitch to a temperature of about 1,200 °C for baked carbons and 2,500 °C for graphite. Their structural properties are dependent upon those of the coke grains and the carbon residue from the carbonisation of the pitch, and also upon the heat-treatment temperature.

Crystalline growth occurs during this treatment; the size of the crystallites in the product is a function of the properties of the original coke. Differences in their parameters enable a wide variation in properties to be obtained. In addition, in the manufacture of graphites, the grist is extruded into the form of large bars or rods; the process causes the coke grains to align themselves, giving rise to a marked anisotropy in properties measured parallel and perpendicular to the grain or axis of extrusion.

Pyrolytic graphite is prepared by the thermal decomposition of methane and benzine vapour on a hot carbon surface. Such properties of the product as crystallite size are influenced by the temperature of the substrate, and exhibit marked anisotropy. Materials with a specific gravity approaching the theoretical value for graphite can be obtained commercially.

Vitreous carbon is prepared by the thermal degradation of certain cross-linked polymers under carefully controlled conditions. The properties of the chars are dependent on the maximum heat-treatment temperature. A product heated to a temperature of 1,800 °C has proved a useful material for many purposes. X-ray diffraction suggests that its structure consists of graphite-like hexagonal layers arranged in small crystallites but with no orientation between the layers as found in graphite. The distance between the layers is approximately 3·5 Å for graphite. However, the specific gravity of vitreous carbon is only 0·47 — about two-thirds that of graphite — hence, if the structures of the two materials were basically similar, vitreous carbon would appear to be very porous, but, in fact, the porosity is extremely low.

The hardness, high strength and non-lubricity suggests that cross-linking present in the original polymer still persists during the heat-treatment process. This view is partially supported by the fact that this material is isotropic in its properties, and emphasised by the material's very similar specific gravity in water and in helium, showing that any pores present must be inaccessible to helium molecules.

Physical properties

It has been shown that vitreous carbon has an extremely low density. It has a permeability to gases lower by about 12–13 orders than that of graphite and about the same as that of quartz. Its porosity too is considerably lower than that of the industrial carbons and is comparable to that of the ceramics and quartz. The thermal conductivity is a little less than that of the ceramics and quartz; much less than that of pyrolytic graphite in the 'a' direction, and considerably less than those of the metals except titanium. The cross-breaking strength is considerably greater than those of baked carbon and graphite, a little higher than those of pyrolytic graphite and quartz, but less than ceramics. Its resistivity is comparable to that of baked carbon and it occupies a position between those of the ceramics and PTFE in one direction and the metals in the other. The maximum arc temperature is of interest; in air the threshold for oxidation is about 550 °C but in an inert atmosphere it can be used at 2,500 °C without phase change or deformation. Under these conditions it has advantages over metals and ceramics.

Chemical properties

Carbons are generally inert to a wide range of reagents compared to metals and ceramics, vitreous carbon particularly so, partly owing to its high purity and low porosity. Many reagents can be retained in it, up to a temperature of 150 to 200 °C, without appreciable reaction. It is even superior to platinum for use with fused alkali peroxides and other oxidising agents, aqua regia and oxidising solutions, and compounds including those of lead, tin, selenium, tellurium and phosphorous.

Vitreous carbon, as normally produced, is a very pure material and has an ash content of up to about 200 ppm. When made from pure polymers it has an ash content of less than 10 ppm. On examination for gaseous products, using proton resonance, hydrogen is barely detectable.

Table 4.3.1 shows a typical analysis of the ash.

Table 4.3.1

Impurity as oxide	Quantity, ppm
Silicon	80
Calcium	40
Iron	40
Aluminium	20
Barium	5
Sodium	5
Magnesium	5
Titanium	5
Strontium	5
Ash	200

The basic process is not intrinsically expensive — the cost of the raw material is relatively low, methods of shaping the articles are suitable for mass production and the carbonisation times are shorter than those used to produce graphite.

Applications

Pressure moulding is the most common method used for fabricating, but there is ocnsiderable scope for the designer with sufficient ingenuity to adapt or develop other fabrication methods. For example, careful thought at the design stage eliminates the problem of 'flash' in the production of magnetic disc heads or flying heads.

Vitreous carbon is similar in hardness to ferrite cores. Lapping and polishing of the heads with the ferrite cores cemented in place can be more easily accomplished when the heads and the cores are of a similar hardness.

While discussing magnetic heads, it is worth noting that whereas electrical noise can and does occur in heads made of alumina ceramic, vitreous carbon is a conductor of electricity and eliminates this problem. In addition to preventing static build-up, this material is a commercial proposition for this application because the manufacturing costs for the complicated shapes involved are considerably lower than for ceramic heads.

The important point to remember is that no mass-produced component can be better than the reproducibility of its manufacturing tolerances. Reproducibility is usually easier to predict in thermoplastic materials and vitreous carbon is unusual in that it is produced from a thermosetting resin with this same characteristic. Once the die has been made, the moulds will be identical so that the process is not limited by the accuracy of the operator; inspection, too, is less costly.

Electronics

An early development was the manufacture of jigs for making the glass-to-metal seals for transistors. Glass powder is poured into the jig, the metal strip inserted into the powder and the whole assembly then passed through a furnace to complete the seal. Because of the non-wetting, non-sticking properties of vitreous carbon, it is an excellent jig-material for this application.

An extension of this principle to the manufacture of flat-packs has eliminated many of the disadvantages of ordinary graphite or carbon fixtures. These jigs are used to furnace-seal cobalt components to glass. They consist of two parts which are located by spigots and sockets. Although vitreous carbon jigs cost more than traditional material in this application, they have several advantages:

1 Insertion wear on the fixture is negligable so that the spigot location is always accurate and the fixture lasts longer.
2 Glass does not stick to the non-porous surface of vitreous carbon, as happens with other jig materials. The package is cleaner and alignment is more positive.
3 There is no carbon contamination in the package, as there may be when using moulds of other forms of carbon.
4 The fixtures themselves pick up contaminants during the process but these can be removed chemically; this is not the case with graphitic carbon fixtures.
5 No out-gassing of vitreous carbon fixtures occurs in the furnace during sealing.
6 The fixtures are easily moulded; this is not possible with other forms of carbon graphite which requires costly and difficult machining operations.

This principle applies equally to furnace- and vacuum-brazing methods, and to vacuum-quenching processes where components tend to stick to stainless steel fixtures.

Deposition techniques features strongly in electronics. In transistor manufacture, for example, radio-frequency induction heating is used to deposit one type of silicon on top of another. Silicon disks are placed in a furnace on a flat electrode heating element, in an atmosphere containing a gaseous compound of silicon. A graphite element is normally used, but being porous by nature gives rise to out-gassing problems and the silicon deposited on the graphite during the process can only be removed by acid vapour. Potentially there is an outlet for this material in the manufacture of electrodes for protecting telephone circuits from lightning strikes. The electrodes carry high voltages which they discharge to earth, and must therefore be effective for long periods without attention.

Some other important uses of this material

Its ability to conduct electricity makes possible numerous other applications. In its simplest form, the material can be used to deposit metals in the form of thin films, because the plating electrode is non-porous, the thin metal film does not adhere and can be easily removed.

Spot-welding electrodes are another potential use, current practice is to take down the electrode and reshape it every hour. Engineers have found that electrodes made of this material lasts ten times longer than conventional electrodes. The problem is to compete financially with conventional electrodes.

Platinised titanium electrodes are currently used for the electrochemical preparation of organic compounds. Pin-holes inevitably appear in the platinum; then, as electrolytic oxidation is taking place at the anode, the titanium oxidises through the pin-holes to form titanium dioxide, which is an insulator. This problem can now be overcome by using vitreous carbon electrodes.

Potentiometer brushes, like those used in radio volume controls, are normally made of carbon black with a suitable binder. These brushes move across a carbon track, reinforced by glass and resin binder, but this type of track is abrasive and can wear away the brushes; carbon dust then changes the resistance value of the track. Vitreous carbon brushes tested over a period equivalent to several times the normal life of these components proved satisfactory; they did not wear and the resistance value of the track was unchanged.

The material's low coefficient of friction has also suggested its use for rotating contacts on slip rings. Work done so far suggests that the use of vitreous carbon could lead to lower starting torques and lower friction when running on stainless steel than are possible with any comparable material.

On the domestic side, it can be used for the heating elements in coffee pots and to replace the ceramic plates which are used to improve heat transfer between cooker heating and saucepans.

Chapter 4.4 **Electrically conducting composites**

Electrically conducting composites may be made from insulating plastics by blending with metals or other good conductors, when the performance of the material depends critically on inter-particle contact of the conductive filler. Fillers in the form of flakes and fibres give improved results over spherical particles. In particular, metal-coated glass fibres provide a very efficient way of using a metal to confer conductivity on a composite, and thermoplastics and thermosetting resins containing silver coated glass fibres have been examined. Resistivities in the range of 0·01 Ω.cm have been achieved with only 1 per cent added silver. Cheaper products based on copper and nickel coated fibres have also been made.

For a long time much effort has been spent in trying to combine the property of electrical conductivity with the properties characteristic of plastics. It would be very useful to have a readily-mouldable, lightweight material with high electrical conductivity. Such materials almost constitute a contradiction in terms because plastics based on organic polymers are usually extremely good insulators. The most common exceptions are cases where ionic conduction is possible, e.g. in nylon 6 which has a resistivity below 10^6 Ω.cm at a temperature above $150\,°C$. This type of conduction is not desirable for most applications since ion transport cannot proceed indefinitely in a specimen, and we concern ourselves only with cases of electronic conduction.

There are two approaches to the problem of making a conducting plastic. The first is an attempt to synthesise a polymer which is intrinsically conducting. Some success has been achieved in making wholly organic polymers with resistivities as low as 100 Ω.cm but so far these polymers no longer have useful mechanical properties. The alternative approach uses composite materials in which a good electrical conductor, a metal for example, is mixed into an insulating plastic.

Composites have a definite limitation; since the conductor is 'diluted' by the plastic the volume resistivity of the product is necessarily below that of the original conductor. However, electrical resistivity, p, spans about 25 decades so there is plenty of room for manoeuvre in composite formulations aimed at conductivity:

$$p\ \text{(polythene)} = 1 \times 10^{19}\ \Omega\text{.cm}$$
$$p\ \text{(silver)}\qquad = 1\cdot6 \times 10^{-6}\ \Omega\text{.cm}$$

Different applications require materials of different resistivities. For example, heating applications often require values in the region of 1 Ω.cm.

Probably the simplest conducting composite consists of a fine metal powder dispersed in a plastic. The main disadvantage of this system is that the approximately spherical particles of metal remain isolated from each other, unless they are present in very high concentration; high concentrations generally destroy desirable mechanical properties of the plastic.

It is apparent that inter-particle contact is the most important factor in determining the success of a conducting composite and inter-particle contact is governed largely by the configuration of the particles. Spheres may be expected to make a minimal number of contacts, because surface to volume ratio for a given particle size is at a minimum. It will consequently be relatively difficult to form a three-dimensional, interconnected array with spherical particles at low concentrations. Indeed, with metal powders having a particle diameter of about 10 μm conductivity is lost catastrophically below a concentration of about 20 per cent by volume.

In this chapter the effect on electrical conductivity of shape and disposition of conductive fillers is considered, platelets possess a high surface to volume ratio for a given particle size and the opportunities for contact are therefore enhanced. Fibres, which may possess a similar high surface to volume ratio, are even more efficient, since less material is wasted between contacts.

A system based on metallised glass fibres, where the metal is effectively distributed as thin-walled tubes, affords a particular efficient way of using a metal to make an electrically conducting composite.

Spherical particles

As mentioned above, conductive fillers in the form of spherical particles are, in general, not very satisfactory because a high loading is required before any appreciable conductivity is obtained, and this high loading is expensive and is detrimental to other properties of the plastic. Nevertheless, silver powder is used in the manufacture of electrically conducting resins, e.g. silver araldite. At a concentration of 85 per cent by weight a resistivity of 10^4 Ω.cm is obtained. A new composite of this type has been reported where silver-coated copper powder is used. This combines the excellent electrical contact properties of silver and the cheapness of copper.

The resistivity of a compaction of spheres has been treated theoretically. When the radius a of the circular area of contact between two spheres is small with respect to the radii of the spheres, the resistance of the constriction is given by Holm's expression, p/wa, where p is the resistivity of the material of the spheres. The contact area is dependent on the compressive forces which cause elastic and yielding deformation of the spheres. That the resistivity of a compaction of spheres is dominated by the resistance of the constrictions at inter-sphere contacts has been demonstrated by examining the pressure dependence of the resistivity of carbon powders. The behaviour is adequately explained by Holm's contact theory. The same theory may be expected to apply also to composites containing high loadings of metal powders, where a certain proportion of particles in contact form the conducting paths through the material.

Carbon blacks

Carbon black has been used extensively in the manufacture of conducting rubbers. Particles of carbon black which are very small (typically 200 Å in diameter) have a disordered graphite structure with the molecular planes slightly further apart than in graphite itself. Some carbon blacks are especially good in conferring conductivity, because their particles collect into chains or 'necklaces' which give a pseudo-fibre structure to the composite. The effect of chain formation is most pronounced at low concentrations where particles would normally be too well separated for conduction to be possible.

The mechanism of conduction in carbon black systems is essentially different from powder composites. The pressure dependence of the resistivity of carbon black composites shows that conduction depends exponentially on interparticle separation, indicating that conduction occurs via tunnelling across gaps between particles. This tunnelling mechanism is a function of the

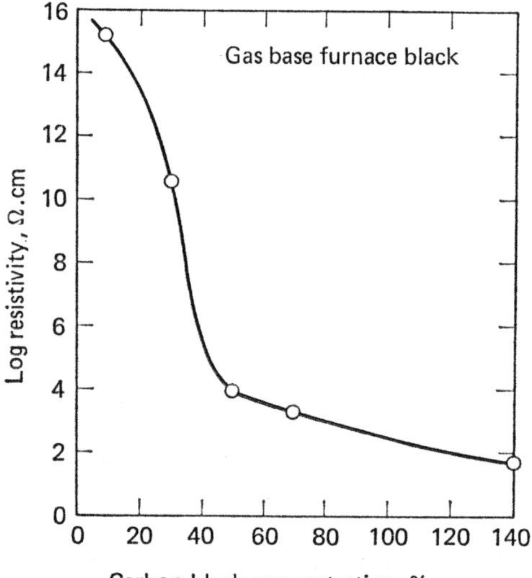

Fig. 4.4.1   Dependence of resistivity on the concentration of carbon black for a typical conducting rubber

small size of the carbon black particles; as particle size is reduced the average distance between particles is also reduced and eventually, when the particle separation is very small tunnelling can occur. It should also be noted that Brownian motion of such small particles as are present in carbon black means that particles do not remain in contact with each other all the time anyway, but rather make frequent collisions with each other.

A typical curve of resistivity against concentration is shown in Fig. 4.4.1.

The minimum resistivity achieved with any carbon black is above 1 Ω.cm (at about 50 per cent by weight). The temperature dependence is variable, although the resistivity of polycrystalline graphite decreases with increase in temperature in the range 5–500 K, the resistivity of a composite may increase with increase in temperature as a result of thermal expansion of the matrix. Composites based on a silicone rubber plus carbon black, have this useful property, the temperature coefficient being $+0\cdot003$ deg^{-1}. Quantitative studies of carbon black systems are not very easy because the 'particle structure' is a variable quantity which is difficult to determine. Very fine metal dispersions obtained by reduction of metal salts seem to behave in a similar way to carbon blacks.

Flakes

A case intermediate between particles and fibres, which can be more easily controlled than structural blacks is a system based on flake fillers. Shorokhova has reported enhanced conductivity using copper flake ($2 \times 2 \times 0\cdot001$ mm) over copper powder ($0\cdot1$ mm). At a concentration of $56\cdot4$ per cent by weight the conductivity was improved by more than a factor of 10. Use of flake may be regarded as an introduction of ready-made portions of the three-dimensional conducting network.

Flake may be oriented in composites to some extent by compressing in one direction, and this may impart anistrophy to the system. The development of composite materials incorporating graphite with the deliberate aim of achieving marked anisotropy of physical and electrical properties. An electrically anisotropic composite material from synthetic graphite flake and polyethylene has been obtained. The best results were found with fine graphite (particle size about 8 μm) which was milled into polythene and the crepe subsequently biaxially calendered. The resistivities in the plane of the sheet and in the perpendicular direction were $5\cdot5 \times 10^3$ and $1\cdot26 \times 10^5$ Ω.cm, respectively, for 40 per cent concentration. The alignment of graphite flake in the plane of the sheets means that through-conducting paths are probably more common in the plane of the sheets than across them. The anistropy will also be enhanced by the inherent anisotropy of the graphite itself.

Fibre composites

There are several different types of fibres which may be used to make an electrically conducting fibre composite. Fine metal wire is the most obvious example, but fine metal wire is difficult to produce, even wires with a diameter of the order of $0\cdot01$ cm, which would give a rather coarse composite, are very expensive. However, the recent advent of new processes for making metal fibre may mean that the use of metal fibres will eventually become economically feasible. A sample of steel fibre (10 μm diameter and 1 cm long) was milled into polythene in order to meet the effectiveness of such fibre in conferring conductivity.

Carbon fibres which have been developed recently are also an obvious alternative to metal fibres. They have a graphite structure and consequently have high conductivity, at present these fibres too are expensive, and possess corrosion risks between the structure and metals.

One way of avoiding the difficulty of having to make fine metal fibres is to coat glass fibres with metal. This is the system which has been developed.

Other metal-coated fibres

Coatings other than silver may be used on the glass fibre for use in conducting composites. Fibres have been coated with copper and nickel by established electroless plating methods. Reduction of Fehling's solution with formalin and reduction of complexed nickel salts with hypophosphite were used for copper and nickel, respectively. Copper coating greatly reduces the cost of the composites without reducing the performance in conductivity. Nickel gives a corrosion resistant product with a somewhat higher resistivity.

Applications and cost

Although the compositions described above have higher resistivities than metals they offer the advantages of low density, corrosion resistance and ease of fabrication. The physical properties can be controlled by choosing a suitable polymer while the metal can be varied independently to suit a given chemical environmental.

Chapter 4.5 **Metal platings**

Platinum metals

In the platinum metal context the term 'engineering' is taken to refer to all applications other than the purely decorative, and hence to coating thicknesses that may vary from 0·5 to 25 μm.

The platinum metals comprise platinum, palladium, rhodium, ruthenium, iridium and osmium, and until comparatively recent years it was probably true to say that rhodium alone was of significant interest in the industrial electrodeposition field, where it found very well-defined applications primarily as an electrical contact surface. During the past five years, considerable diversification has occurred in plating materials and techniques, not only in respect of other members of the group which are finding important engineering uses, but equally in the variety of processes available, including a number of the electroless type.

Developments along these lines have been stimulated by an increasing sophistication on the part of users. Ten years ago the problems of platinum metal deposition were in the main process, very often in the form of poor adhesion or excessive cracking of deposite. Improvements in plating technique, increasing industrial experience, and the availability of a useful volume of technical literature have gone far towards meeting practical difficulties of this nature. It now appears that precious metal plating in general is entering a phase of development in which the mechanical and electrical properties of the coatings are coming under close scrutiny by engineers in relation to specific applicational requirements and against the background of an ever-increasing emphasis on the reliability of components and performance.

This situation imposes stringent demands on the precious metal plater, which must be successfully met if the deposits are to function satisfactorily in their critical roles.

Rhodium plating

The general applications of rhodium plating, in the industrial field, are so well-known at the present day as to call for little further comment. All depends on the essential properties of the electrodeposited metal of extremely high hardness hence, good wear-resistance, corrosion and tarnish-resistance, and good reflectivity.

Use

The main use of the metal is as a surface for electrical contacts, especially sliding contacts or contacts required to operate reliably after periods of idleness. In the former category are numerous types of slip-rings and switches used in radio-communication and radar control gear, where electrical signals are small and voltages insufficient to break down tarnish films on silver or base-metal contacts.

High-speed computer switching represents another important field of application. In many cases this involves printed copper switch patterns on laminates, which pose special plating problems. The high internal tensile stresses that characterise rhodium deposits from the conventional sulphate electrolyte, and lead to the well known cracking of coatings at critical thicknesses, have provided a long standing problem of rhodium plating.

The exfoliation of deposits, which is frequently attributed to internal stress, is more often due to faulty pre-treatment of the basis metal. Where solid metal substrates are concerned, it in fact seems a reasonable view that for coating thicknesses up to 5 μm, which meet the majority of contact requirements: the conventional sulphate type of electrolyte is a satisfactory basis for operations. It is also being used in the plating of copper-clad laminate with deposits up to at least 1 μm thick, but where thicker coatings are required it is generally preferable to use one of the low-stress electrolytes, since otherwise the stress in the rhodium, coupled with slight edge attack of the adhesive, may result in lifting the copper.

The electrolyte is based essentially on modification of a sulphate solution by the addition of selenic acid. The solution furnishes, under the preferred operating conditions, bright deposits of very low residual stress, and is becoming increasingly adopted for the plating of printed switch patterns and similar components. Stress-relief of the deposits probably occurs as a result of extremely fine cracking, which, however, is quite different in character from that shown by conventional deposits. It is not revealed, for instance, by electrographic testing, and at thicknesses of 2·5 μm and upwards the coatings show a much greater protective value than conventional deposits. For example, a low-stress deposit of 2·5 μm will function successfully as a resist against ferric chloride etching in the production of printed switches by the 'reversal' technique, while a conventional coating of similar thickness may fail due to the presence of cracking. Care should be taken in using this particular type of low-stress solution to ensure extremely good cathode connections, particularly in the case of a number of items wired on a jig, otherwise breaks in the plating current will result in severe exfoliation.

Another problem encountered in printed-circuit plating is bath contamination by organic materials from the laminate. This is likely to be particularly serious when laminates are immersed in the electrolytes in a partly cured condition. This is likely to be particularly serious when laminates are immersed in the electrolytes in a partly-cured condition. This occurs in the production of flush-bonded switch patterns, where 'flushing' is carried out by a pressing operation at the same time as final curing. Laminates impregnated with melamine resin are particularly troublesome in this respect, but much depends on the extent of the first curing treatment. If this is such as to produce a uniform brown shade of the (originally white) laminate, rhodium plating can be carried out fairly satisfactorily, but if white areas remain, bath contamination will be severe. Contamination can be reduced to some extent by lowering the plating temperature, but even so it may be necessary to employ frequent treatments with activated charcoal to keep the electrolyte operative.

Rhodium has sometime been criticised as a slip-ring finish on the grounds of variable performance. Clearly, good adhesion is an essential for successful operation, since if flakes of the extremely hard coating become dislodged in service, failure is likely to be catastrophic. Other important and critical requirements are adequate thickness of the deposit in relation to the basis metal, and surface smoothness. With an intermediate hard supporting coating, e.g. nickel, even lower thicknesses may prove satisfactory. The limiting factor on the life of the contact combination then becomes the rate of wear of the wiper material, which will depend critically on the degree of surface finish. Since it is neither easy nor desirable to impart a high degree of surface finish, to a rough rhodium plate by buffing, it is essential that the as-plated finish should reach the necessary standard, which can be ensured by careful control of electrolyte and plating conditions.

Notwithstanding the wide usage of rhodium plating in the contact field, there is little published information on details of service performance, particularly in comparison with other materials. The more interesting, therefore, are the results for relay contacts and computer switches. In the former case, under conditions of 10 to 100 mA, 26 V, the life of solid palladium–gold alloy contacts was 10×10^6 cycles, which was increased to 15×10^6 cycles for a 2·5 μm rhodium plate on a nickel undercoat. Computer switches operating at 1,500 rev/min gave a life of 3 to 4 hr for silver, 20 hr for a bright gold plate of unspecified thickness and 100 hr for a rhodium plate of 5 μm, again on a nickel undercoat. A palladium alloy wiper was employed in each case. A 2·5 μm layer of rhodium on nickel-plated beryllium–copper has been used for vibrator contacts breaking a 48 V d.c. current of 10 mA at 200 Hz.

A development of considerable interest is that of the sealed reed switch. Designed primarily to avoid the contamination of contact surfaces, it comprises two nickel–iron alloy strips sealed into opposite ends of an evacuated glass envelope, and actuated magnetically. A precious metal finish is required on the contacting areas and rhodium is used, among other finishes, for this purpose.

From the process viewpoint the method for the direct rhodium plating of nickel–iron–cobalt alloys with rhodium, is based on the deposition of a thin (0·1 to 0·5 μm) coating from a sulphate or phosphate bath, heat-treatment at 700 to 1,100°C in hydrogen, followed by a further rhodium plate.

Rhodium has also been proposed as a diffusion barrier layer between gold and copper, and between silver plate and nickel–iron alloy. In the former case the application is to the copper anodes of thermionic valves, which are gold-plated to protect them against oxidation during processing and sealing into their envelopes. A thin layer of rhodium or nickel is used to prevent diffusion of gold into the copper during the baking-out process. The other application mentioned refers to sealed reed switches.

With few exceptions, the applications so far mentioned involve use of coatings at ambient temperature or only slightly above. The use of thin rhodium deposits on nickel as high-temperature reflectors is well established, with the upper limit of service temperature at 450 to 500 °C, above which the reflectivity is impaired, short-time protection of tungsten wire at temperatures up to 1,650 °C by the use of a complex chromium–silicon–chromium–rhodium coating, has been reported but it is clear that in this highly specialised field much more work remains to be done before rhodium, or indeed any other single electrodeposited coating, can claim successful application.

Palladium plating

Palladium is readily electrodeposited from a variety of aqueous electrolytes with good cathode efficiency and it is rather surprising at first sight that the metal has had to wait until comparatively recent years to acquire a position of importance in the industrial field. With the development of satisfactory plating solutions, rhodium won the day on the grounds of its whiter colour, and, on the strength of this earlier advantage, was more favourably placed to meet later developing industrial needs, which, of course, received strong impetus from technical requirements. Palladium was suggested, and used to some extent, as an alternative to rhodium when the latter metal was in short supply as a strategic material, but rhodium continued to be of major importance in post-war developments.

Renaissance of interest in palladium came with the development of printed circuits, in relation to the need for a precious-metal finish on board end-connectors. In early developments, the established finishes of gold or rhodium posed process problems due first to attack of copper-to-laminate adhesives by conventional cyanide electrolytes and secondly to the lifting of the copper foil as a consequence of the high internal tensile stress in the rhodium coating. Palladium offered solutions to each type of difficulty, in the form of neutral or mildly alkaline electrolytes, and deposits with a relatively low stress, yet with a hardness in the order of 300 DPN. Despite subsequent improvements in the shape of cyanide-resistant adhesives, acid gold processes, and low-stress rhodium electrolytes, interest in palladium has persisted, a potent spur, in addition to its technical attributes, being its favourable cost factor in relation to the alternative finishes.

At present palladium is finding considerable use in the end-connector application, and in related cases, such as the plating of printed switched patterns.

There seems, in principle, no reason why palladium should not replace the more expensive gold more widely in the electronics field, but developments in this direction have been hindered by two factors, firstly, the lack of systematic comparative data on deposit properties, and secondly, by doubts concerning the reliability of existing deposition processes. In the former context, some concern was caused by the formation of insulating films on rubbing palladium contacts due to catalytic polymerisation of organic vapours, but the general view now appears to be that this effect is likely to be of significance only under highly-specialised conditions. Tests for instance, found only a slight increase in average resistance between the palladium plated contacts of a 32-contact two-motion selector plug and socket assembly, from 4·9 to 5·1 mΩ, after 500 insertions and withdrawals in benzene vapour in air followed by 100,000 vibrations in the same atmosphere. While the technical merits of palladium are now fairly generally accepted, doubt is still sometimes expressed about the state of the technique with regard to plating processes. In the end-connector application two electrolytes in established use are the so-called 'P' salt solution, based on diaminodinitrito-palladium, and the electrolyte based on tetramino-palladous nitrate, and experience indicates that, although sometimes criticised on grounds of variable performance, both solutions are capable of furnishing satisfactory coatings. The normal thickness requirement is 0·005 to 0·007 mm, but in at least one case the 'P' salt electrolyte is used to deposit a relatively short time.

Platinum plating

Platinum has a well established use in the electrical contact field, but mainly in the solid form, and in general, there would seem to be few applications, excluding, perhaps, high-temperature contacts, which are not equally effectively and more economically met by palladium, rhodium, or gold.

The considerable interest in platinum plating processes has been stimulated by the development of platinised titanium as an insert anode material. If bare titanium is made anodic in, for example, a chloride solution, a stable resistive film is formed. Increase in applied potential merely serves to thicken the film, until the value in the order of 15 V is reached, when complete film break-down occurs and the metal dissolves rapidly with the passage of heavy corrosion currents. If a thin coating of platinum is applied to the titanium, heavy anodic currents can be passed at potentials well below the break-down voltage of the titanium oxide film and development of the latter at possible discontinuities in the platinum deposit acts as a seal at these points, rendering the composite material comparable in corrosion-resistance to massive platinum.

Although this type of application generally makes no great demand on the continuity of the platinum deposit, more sophisticated requirements are imposed in terms of the corrosion rate and over-voltage characteristics of the coating, especially in connection with the very large potential application as an anode in the electrolysis of brine for chlorine production where a low-over-voltage for chlorine evolution is essential to restrict the electrical power requirements. Adhesion is likewise more difficult to achieve on titanium than on the commoner basis metals, and a variety of techniques have been proposed, including pre-treatment in hydrochloric acid solution or vapour, etching in ammonium fluoride-phosphoric acid solution, and heating in hydrogen at 600 °C. Bonding by post-heat-treatment in hydrogen at 400 to 800 °C for 15 sec to 1hr has also been suggested.

It seems doubtful whether there are any significant difference in the over-voltage characteristics in the as-plated condition of platinum deposits from either of the two best established plating solutions, namely, the alkaline 'P' salt and sodium hexahydroxyplatinate electrolytes, or from more recently developed solutions such as the DNS electrolyte or acid 'P' salt formulations. Modification of over-voltage has, however, been reported as a result of

s

various post-treatments of coatings, either in the massive form, or as deposited from platinic chloride solution in the form of 'black'. In the former case, this comprises charging the deposit with hydrogen, followed by heating to cause partial recrystallisation of the coating, with description of part of the hydrogen: in the second case 'activation' of the coating is effected by heating in a stream of air containing hydrocarbon vapour, to bring about localised catalytic combustion on the platinum surface.

Platinum is the only member of the group which is free from oxide formation at high temperatures, but application of the electrodeposited coating to the protection of base metals is limited at very high temperatures by the possibility of diffusion of oxygen through the coating, or of interdiffusion between the deposit and the substrate. Electrodeposits seem to be more prone to failure from these causes than clad coatings, due to greater rates of diffusion associated with the deposit structure or with the presence of impurities. Cracking of the relatively thick coatings required is also likely, and in this respect it is of interest to note that platinum coatings of greatly increased protective value can be produced by the application of a periodic reverse current cycle (5 sec cathodic, 2 sec anodic) to an alkaline 'P' salt electrolyte of conventional type. A 5 μm deposit produced in this way on zirconium effectively protected the basis metal in boiling 20 per cent hydrochloric acid, while a d.c. plate of similar thickness failed in a few minutes. No similar improvement was noted for deposits from the hexahydroxyplatinate electrolyte. Uncracked and ductile platinum coatings have been produced from a strongly acid chloride electrolyte and, despite the aggressive nature of the solution.

An established application of platinum plating is in the coating of tungsten or molybdenum valve grids to prevent secondary emission of electrons. Such coatings are usually applied from platinic chloride solutions containing a trace of lead acetate, to produce a smooth black deposit which is subsequently consolidated by sintering. A recent improvement in technique is the replacement of lead by mercuric chloride, mercury being more readily eliminated from the coating during the sintering treatment.

Ruthenium and iridium plating

Ruthenium plating is at the applicational stage at the present time. Electrodeposits from a nitrosyl sulphamate electrolyte, show physical properties closely comparable to those of electro-deposited rhodium, and since the cost factor for ruthenium is only about one-half that of rhodium, initial interest is in the type of application currently met by that metal. Tests have confirmed that ruthenium deposits of 1 to 2·5 μm are in fact comparable to rhodium in wear resistance, and this applies also to corrosion resistance at normal temperatures. Like rhodium, ruthenium suffers superficial oxidation at elevated temperatures but the oxide film shows a conductivity in the same order as that of the metal, which suggests that the coating may be particularly suitable for electrical contacts required to function at higher than ambient temperatures.

Iridium

Iridium is deposited from aqueous electrolytes only at extremely low cathode efficiencies (1 per cent) and although there are potential applications for the coating on valve grids and for high temperature protection, these can be met by electro-deposition only by having recourse to a fused cyanide electrolyte. The process has been used for the protection of molybdenum at high temperatures, with promising results, but, since it involves deposition from a melt at 600 °C, with an inert gas blanket, is clearly not attractive from the viewpoint of general utility.

Electroless plating

In the electroless plating field the chief interest has been in copper, gold, nickel, etc. Palladium in view of its relative cheapness has therefore a wider applicational potential. A variety of processes are currently available, including both chemical replacement and autocatalytic types. In the first category is the process based on the use of strongly acid chloride solutions, and applicable also to the deposition of rhodium, platinum and ruthenium. In this case the coatings are not of the self-sealing type, and hence require post-treatments in ammonia, boiling water, or an immersion gold solution to develop maximum protective effect. By the nature of the process, the deposits produced, although considerably thicker than those obtained by self-sealing processes, have a rather porous structure and do not therefore appear to be suitable for applications calling for a significant level of wear-resistance. Excellent results have however, been reported for palladium deposits of this type in improving the solderability of copper printed circuits, and tests on nickel–iron reed switch tags in a similar context are also satisfactory.

Apart from the solderability aspect it would seem that the likely sphere for this and other replacement palladium processes may well be as activating treatments a prior to electroplating or immersion plating of other metals, in line with the well-known use of a palladium 'dip' in electroless nickel plating.

Of more interest from the functional aspect are processes of the auto-catalytic type. A process of this nature, based on the use of hydrazine as a reducing agent in an ammoniacal solution of palladous chloride. Another process utilises a solution prepared by reacting palladium 'P' salt with sulphamic acid. The first process is particularly adaptable to barrel plating, the second process is suitable for the provision of thin palladium deposits on a wide variety of base metals, the main interest is in the possibility of producing relatively thick deposits which may be suitable as alternatives to electro-deposited coatings.

One potential problem, however, in the important field of printed circuits is the tendency with increasing metal thickness, for the coating to grow laterally from conductor edges over the insulating surface, which in extreme cases, may lead to short circuiting of adjacent connectors.

Gold-plating materials

Considerable progress has been made over the last two decades in gold and alloy gold electroplating, not only in the formulation of processes capable of depositing binary alloys but also ternary and quarternary alloys. Characteristics of such alloys include uniformity of coating, minimum porosity, controlled hardness, reduced corrosion and stable electrical properties.

With the increasing use of precious metal electroplating processes has come the necessity to use more sophisticated plants in modular form, which consist of an assembly of combinations of rectifiers and controls together with pumps, filters, heat exchangers and thermostats. These plants can be manually operated although the current tendency is towards automatic plants based on a fixed programme, or on plants wherein the control has been replaced by a programmer of the sequential type.

In parallel with this growth has come the stringent use of specifications. Not only has the electronic industry introduced these documents but it also vigorously checks the results obtained from the processes. To this end, test equipment is being developed which, although it will not correct a poor solution or electroplated article, will detect and indicate good or bad functioning of a process and, in doing this will aid process control and evaluation.

The ability to co-deposit base metals such as cobalt, nickel and indium with gold has led to considerable developments in alloy gold solution. Cobalt will not be co-deposited with gold from cyanide solutions but does co-deposit from acid electrolytes. However, the stability of the cobalt complex used is of major importance in obtaining consistent results. Variations of a very small order cause considerable modifications to the deposit's properties. It is possible, under certain conditions, for gold–cobalt alloys to be solid solutions. In the case of gold–indium alloys, provided the indium complex is correct, co-deposition takes place. However, the hardening properties of indium are not so good as those of cobalt or nickel.

There are some fifteen to twenty formulations which permit the use of a specific process to suit a particular application. In the main the processes produce mirror bright deposits with excellent resistance to corrosion, oxidation and galling together with constant electrical resistivity and good solderability.

Fields of application are the plating of connectors, circuit boards, contacts and semiconductor devices. Table 4.5.1 indicates the general characteristics of non-alloy and alloy gold deposits.

Processes

In order to assess the choice of process, the requirements of the component to be plated must be examined. Gold and alloy gold depositions are applied to the whole article or to a 'significant' or 'contact' area, to provide qualities as required under given operating conditions, e.g. maximum corrosion resistance, minimum contact resistance, maximum wear resistance, ductility, low coefficient of friction and good appearance. Where wear resistance is of prime importance, as in the case of most plug and socket-type connectors, gold–cobalt is the obvious choice. If wear resistance is coupled with temperature resistance, gold–nickel would be chosen or a combination of gold–nickel and pure gold. For a very heavy electroform deposit or a high-purity thin film, pure gold would be used.

Table 4.5.1   **General characteristics of non-alloy and alloy gold deposits**

Alloy	Au/Co	Au/Co	Au/Ni	Au/Ni	Au/Co/Ni	Au/In	Au/Ni	24 K
Gold	99·0	99·9	99·0	99·9	96–98	98	75–85	Pure
Hardness (Knoop)	235–245	150–160	190–200	125–135	265–285	180–200	400–450	80–85
Resistivity at 0 °C, Ω.cm	16·75	10·25	—	7·0	—	11·75	—	2·2
Contact resistance, MΩ	0·6	0·6	0·3	0·3	0·8	0·8	2·0	0·3
Crystal structure lattice constant, A	fcc 4·066	fcc 4·069	fcc 4·068	fcc 4·070	fcc 4·056	fcc 4·080	—	fcc 4·070
, A min	30	45	30	45	30	30	8	123
Temperature resistance	—	−80 deg C/ +160 deg C	—	−80 deg C/ +200 deg C	—	—	—	−80 °C/ +300 °C
Application (see notes below)	1	2	3	4	5	6	7	8

1. Contacts, connectors, printed circuits for extra wear resistance
2. Printed circuits, connectors, normal transistor headers
3. Transistor headers, where resistance to tarnishing is required at high temperatures, contacts, connectors
4. Excellent machinability
5. Decorative
6. Bonding semiconductor chips to headers
7. Used for circuit boards, connectors, switches requiring hard wearing surfaces. Excellent solderability, etch resistance, anti-galling characteristics
8. Used for high reliability transistor headers etc. High-temperature durability, high ductility, fine general cohesive structure

Equally important to the choice of process is the type of pre-treatment given to the component. This is best illustrated by a typical, abridged process specification for a fully heat-treated beryllium copper spring connector.

Uniformity of plating quality is a basic requirement, and on a production basis the only satisfactory method of achieving this is by maintaining fully automatic control. Obvious advantages include the precise execution of specification requirements, batch to batch reproducibility, decrease of rejects owing to wrong processing, labour saving, controlled 'significant' area plating. There is also the possibility of processing different base metals, to plate different processes and carry out both vat and/or barrel plating, and versatility, whereby cycles can be processed by means of manual control or the use of a universal automatic equipment.

Chapter 4.6 **Beryllium and its alloys**

Beryllium and beryllium alloys have gained increasing fields of use in recent years, especially in aerospace and electronics. Some forms of the basic material include copper, nickel and aluminium alloys: in this chapter, the many unique properties of these materials are discussed, together with a review of some of the applications. Beryllium is rapidly evolving from a metal used in highly specialised applications to a structural material of broad utility. The high strength and stiffness which characterises today's wrought products, together with the metal's density, are attractive properties to the designer and engineer — as are beryllium's thermal properties and excellent dimensional stability. With increasing experience and continuing accumulation of engineering data, the design of aerospace structures have advanced to the point where beryllium mill products are being used in routine production fabrication of structural components.

Properties

The new wrought forms of beryllium-sheet, extrusions, wire and forgeings represent significant advances over the powder metallurgy blocks developed for nuclear applications. Mechanical properties are substantially improved, there is a wider choice of fabrication methods, and scrap loss is reduced. While broader product availability is contributing to the increasing number of beryllium applications, the engineer's interest in the metal's combination of unique properties remains the impetus to further developments in metallurgy and mill processing. In this one material the design engineer finds the lightness of magnesium, a stiffness substantially greater than steel, a coefficient of expansion matching ferrous and nickel-base alloys, a thermal conductivity two-thirds that of aluminium and useable strength properties beyond the melting points of magnesium and aluminium. Beryllium is a true aerospace material.

Density and modulus of elasticity

In comparison with other aerospace materials, only alloyed magnesium and aluminium approach the low density of beryllium. Titanium, structural and high alloy steels, and high temperature alloys range from two and one-half to five times the density of beryllium. In addition, beryllium is the lightest available metal that will not vaporise appreciably in the vacuum of space.

While alloyed magnesium and aluminium compare favourably with beryllium on a density basis, these materials have tension moduli significantly below that of beryllium. At room temperature, the modulus of beryllium is higher than that of the more widely used structural metals — and this relationship persists up to the temperatures at which beryllium loses useful strength. This high elastic modulus makes practical the design of lightweight rigid components.

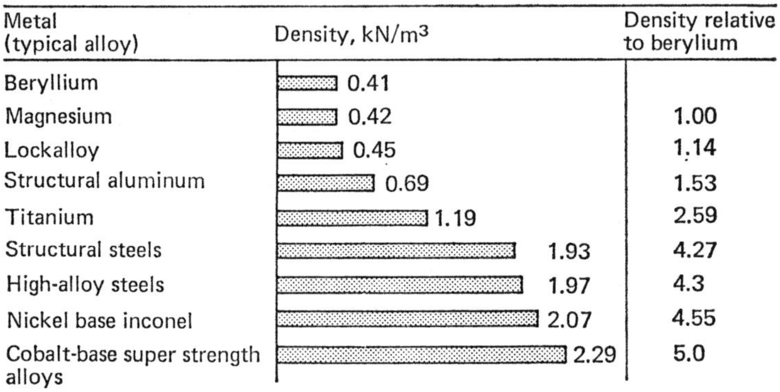

Metal (typical alloy)	Density, kN/m³	Density relative to berylium
Beryllium	0.41	
Magnesium	0.42	1.00
Lockalloy	0.45	1.14
Structural aluminum	0.69	1.53
Titanium	1.19	2.59
Structural steels	1.93	4.27
High-alloy steels	1.97	4.3
Nickel base inconel	2.07	4.55
Cobalt-base super strength alloys	2.29	5.0

Fig. 4.6.1   Density of typical materials used in aerospace applications

Figure 4.6.1 shows a comparison of the densities of typical metals used in aerospace application, and Fig. 4.6.2 shows the elastic modulus for similar materials.

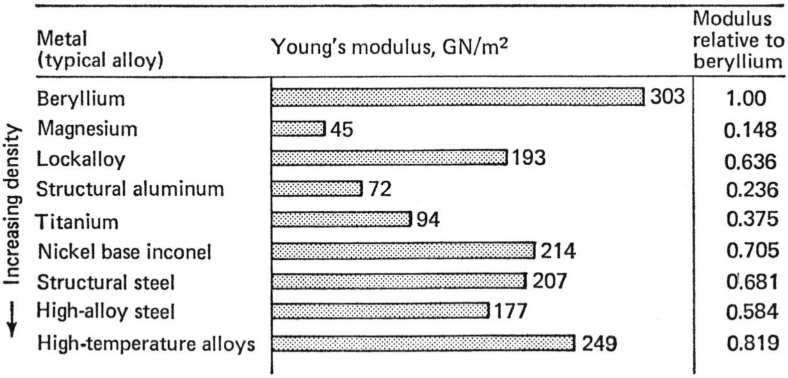

Metal (typical alloy)	Young's modulus, GN/m²	Modulus relative to beryllium
Beryllium	303	1.00
Magnesium	45	0.148
Lockalloy	193	0.636
Structural aluminum	72	0.236
Titanium	94	0.375
Nickel base inconel	214	0.705
Structural steel	207	0.681
High-alloy steel	177	0.584
High-temperature alloys	249	0.819

Increasing density (↓, left axis label)

Fig. 4.6.2   Elastic modulus of typical aerospace materials

Strength-to-density ratio

On a strength-to-density basis, beryllium is competitive with the more commonly used materials — magnesium, titanium, aluminium, steels and high temperature alloys — and it is superior to the other light metals at elevated temperatures, retaining its strength up to about 593 °C. In the use of most high-strength materials, the need for rigidity usually results in an overweight section of higher than necessary strength.

The high elastic modulus and low density of beryllium thereby enable the engineer to design lightweight structural members to minimum thickness. In a purely buckling application, a beryllium column will be lower in weight than a column made of any other metal of equal length and load-carrying ability.

Notch sensitivity

Notches with concentration factors (KT) less than 3·0 have little influence on the strength of beryllium sheet. In fact, increased temperatures reduce the notch sensitivity so that the strength of specimens containing these low concentration factors remains almost constant to 260 °C. More severe notches also show lesser effect on fracture strength as the temperature increases. Beryllium's notch sensitivity, crack propagation and impact characteristics must be considered in applying beryllium sheet to dynamically loaded structures.

Under cyclic loading, failure occurs in almost all metals at stresses well below the values indicated by simple tensile testing. In such fatigue conditions, the ratios of steady-state and cyclic stresses have a major influence on the performance of the material.

Thermal properties

Many of the first applications for beryllium resulted from the inability of other materials to withstand the extreme temperatures and to absorb the heat developed during a limited duration. The high specific heat of beryllium allows its use as a structure in areas where stronger and more refractory materials would be heated to temperatures well in excess of their service ranges. Figure 4.6.3 illustrates the specific heat of beryllium compared with that of several structural alloys.

The coefficient of expansion of beryllium approximates to that of stainless

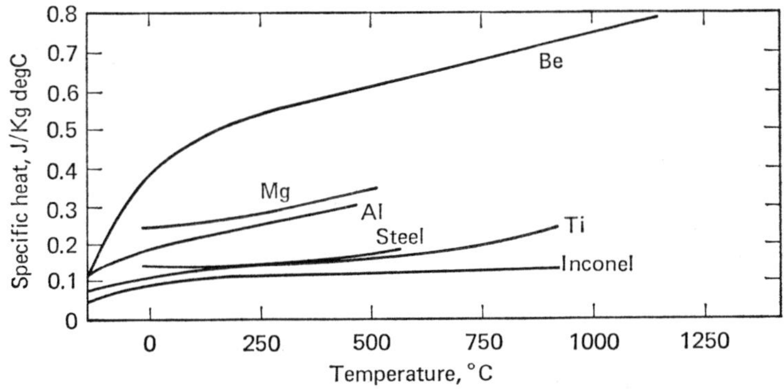

Fig. 4.6.3   Specific heat of beryllium compared with other metals

steel and nickel and cobalt alloys. This compatibility of expansion eliminates many of the design problems arising from the high expansion of other light metals.

Of the commonly used materials, only aluminium exceeds beryllium in thermal conductivity. Where heating of a structure is involved, high thermal conductivity helps to eliminate thermal gradients and protects the part against warpage and build-up of surface heat. Its advantage in heat sink applications is apparent.

The combination of beryllium's unique thermal properties contributes greatly to the metal's excellent dimensional stability — even under difficult service conditions.

Beryllium is available in an increasing number of forms. Representative product forms currently available are shown in Table 4.6.1.

Beryllium–copper alloys

The most common alloys of beryllium are the beryllium–copper alloys. These alloys have many unique properties and can be alloyed for high strength, high electrical conductivity or a choice compromise of each. The most important general characteristics of beryllium copper are itemised below.

Conductivity

The electrical conductivity of beryllium copper alloys is 22–60 per cent that of copper depending on the alloy and heat-treatment used. Electrical and thermal conductivity increase substantially during heat-treatment, making beryllium–copper alloys ideally suited for resistance welding electrodes and contact members.

Strength

The high strength of beryllium–copper alloys makes it possible to reduce the size of parts where ordinary copper alloys might be used. Strength is doubled by a simple low-temperature heat-treatment — an important factor in spring design and mechanical parts.

Corrosion resistance

Beryllium–copper alloys have excellent resistance to corrosion and are ideal for use in marine, industrial and rural atmospheres and many other corrosive conditions. Loss of endurance strength due to hydrogen embrittlement does not occur with these alloys.

Fatigue resistance

Another outstanding property of beryllium–copper alloys is an unusual resistant to fatigue. Excellent resistance to fatigue under vibration is also shown when other high-copper-base alloys show indications of early failure.

Table 4.6.1 **Typical beryllium product forms**

Product	Grade	Description	Availability
Block	Standard	The most commonly specified grade, produced by hot pressing beryllium powder. Offers a good combination of properties suitable for instrumentation, optical and nuclear uses	Beryllium block is available in a variety of sizes and configurations. Rounds, squares and rectangular cross-sections are ordinarily supplied and other shapes are available
	High purity I, high purity II	Higher purity materials produced by hot pressing beryllium powder. For specialised applications requiring higher chemical purity	
	High strength	A special material produced by hot pressing subsieve high oxide beryllium powders. This material possesses superior precision mechanical properties and dimensional stability for specialised applications, principally in the instrumentation field	
	Cast	Fine-grain cast material offering higher purity than that obtainable from beryllium produced by powder metallurgical processing	
Sheet and plate	Standard	The most commonly specified sheet product. Produced by hot cross rolling. Offers good strength and formability characteristics. Forming temperature, 704 °C	Beryllium sheet and plate in cross-rolled panels up to 0·5 m × 9·8 m and in gauges down to 0·5 mm
	High strength	A high-strength beryllium product produced by hot cross rolling. Possesses high mechanical properties and modest formability. Forming temperature, 710 °C	
	High formability	Beryllium sheet hot cross rolled from fine-grained cast ingot. Offers modest strength with highest degree of formability. Forming temperature, 371 °C	
Extrusions	Standard	The most commonly specified extruded product. Produced by hot extruding. Possesses good mechanical properties	Extrusions in the form of rod, bar, tube and other structural configurations
	High strength	A high-strength beryllium product produced by hot extruding. Offers superior mechanical strength characteristics	
Wire		Beryllium wire, drawn to diameters down to 0·05 mm, in the annealed or as-drawn condition. Typical UTS values of 150,000 and 175,000, respectively. Coil lengths up to 825 km. For special requirements, beryllium wire with nickel or special oxide coatings is available	
Foil	—	Beryllium foil in gauges from 0·0013 mm to 0·25 mm is available as-rolled from fine-grained cast ingots. Foil in disk form for X-ray window applications and reasonable requirements for vacuum tightness can be satisfied for all gauges	
Powders	—	A wide selection of beryllium powders of various chemical composition and particle size is available including those that include the starting material. Specialised beryllium powders are produced for high-energy-propellant applications	
Ingot	—	High-purity beryllium cast ingot for re-melt or alloying purposes. Special fine-grain cast ingots for direct conversion to finished products by wrought processing or machining	
Bead	—	Beryllium bead or pebbles for direct conversion into cast products or for alloying purposes	

Wear resistance

Beryllium–copper alloys have shown unusual resistance to wear both with and without lubrication against steel and other materials. Wear resistance, however, varies with operation conditions.

Hardness

Of copper-base alloys, beryllium–copper alloys have the greatest hardness; they can be hardened to Rockwell C-38 to C-45. Thus beryllium–copper parts can be machined in half the time previously required for one part.

The formation of short chips during machining eliminates the need for special cutting fluids or coolants. The function of the fluids is primarily for lubrication between tool and work and for flushing away fine chips. Although the new alloy costs more than phosphor bronze, the saving in labour, tool wear and machine time in many cases result in a lower-cost finished part. Additional economies may be realised by the collection and resale of re-claimed chips and scrap at regular beryllium–copper scrap prices.

By combining the many property advantages of beryllium–copper alloys with low-cost machining, this new alloy is expected to find application in many small parts made on automatic screw machines. Because of the low machining cost, designers will be able to eliminate design comprises formerly required to keep final part costs down. The principal immediate application will be in the manufacture of connectors and other small electrical and electronic parts.

Fabricating beryllium–copper alloys

Beryllium–copper alloys offer an advantage over other copper-base alloys because they are age-hardenable. This allows the consumer to first fabricate or cast them, and then by a simple low-temperature thermal treatment, achieve the physical properties required of the part he is making. These alloys are available as wrought products such as strip, rod, bar and wire for machine fabrication, and as casting ingot for foundry use.

Thermal treatments

Ease of fabrication in the unhardened state, with final properties obtained by a simple low-temperature precipitation hardening treatment, makes beryllium–copper alloy one of the most versatile of the copper-base alloys. With most other copper-base alloys, the part must be formed from material already having the desired properties. For that reason, it is often difficult, or even impossible, to fabricate.

Thermal treatments for beryllium–copper alloy fall into two classifications:

1 Solution heat-treatment.
2 Precipitation hardening, age-hardening, or gang.

Solution heat-treatment indicates a softening operation, precipitation hardening a hardening operation.

Fabricating

Beryllium–copper alloys are fabricated easily by standard production methods. Strip is readily blanked, formed, deep-drawn or spun. Rod and bar respond to hot- or cold-forming methods including forging, machining and swaging.

Joining

Beryllium–copper components can be successfully soft-soldered, silver-brazed, carbon, metal or inert-arc welded, and flash, seam or spot resistance welded. Soft-soldering is done after heat-treatment. Carbon-arc and metal-arc processes are used to repair cracks, defects or worn-surfaces in castings or to join beryllium–copper parts. With various resistance welding techniques, spot seam and flash, some loss of properties can be expected.

Forming

Although age-hardening is the result of a two-step thermal treatment, the fabricator is ordinarily concerned only with the hardening operation; the first step, solution heat-treatment, is generally handled by the supplier. For certain special applications, such as deep-drawing, it may be necessary to provide an intermediate solution heat-treatment, because beryllium–copper alloys, like other alloys, harden with cold work.

It should be noted that the material will shrink approximately 0·508 mm per 25·4 mm on heat-treatment. One should choose the hardest temper from which the part can be formed, and after forming, precipitation harden at the time and temperature required to give the most desirable properties.

Corrosion resistance

Beryllium–copper alloys are frequently specified in applications where corrosion is a problem, not only because of the protection which they afford against many types of attack but for their unique design advantages as well. Resistance to corrosive conditions is all too frequently merely one of the demands made upon critical parts. Accordingly, beryllium copper is often chosen in preference to other materials of comparable corrosion rating, its high strength, good conductivity ease of fabrication and other important features available in numerous combinations make it especially suitable in a large number of applications. The relative corrosion resistance of several materials in salt-spray solution is shown in Fig. 4.6.4.

Tarnishing of beryllium–copper alloys can be expected under certain conditions but the phenomenon does not affect the mechanical properties of the material. Humid or sulphur-bearing atmospheres stain and darken the alloys. Sulphurised atmospheres produce a black surface and marine or salt-spray exposures result in a green film. During hardening or annealing in an

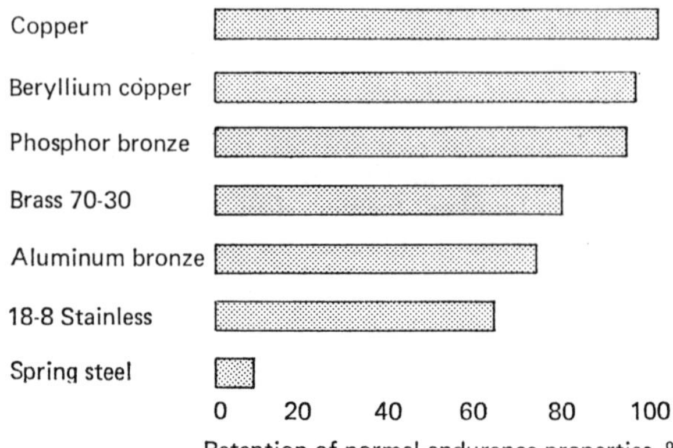

Fig. 4.6.4   Relative corrosion resistance of several metals in salt spray solution

uncontrolled atmosphere, an oxide scale is formed but is easily removed by pickling.

If beryllium–copper parts are to be plated, there is no problem of loss of endurance through hydrogen embrittlement. Plating baths do not produce this condition in the alloys.

The degree of galvanic corrosion to be expected depends upon the type of material used in conjunction with beryllium–copper alloys. It may be safely coupled with other copper alloys and, when used with iron or steel, galvanic action occurs only when there is a relatively large area of beryllium–copper alloy. When connected galvanically with zinc, aluminium or magnesium under corrosive conditions, insulation may be advisable to protect the other material. A plating of tin is sometimes used on beryllium–copper alloys in contact with aluminium alloys. Zinc or cadmium coatings may be employed for use against zinc alloys.

Corrosion rates are generally increased by mechanical action so as to produce stress corrosion cracking or corrosion fatigue. Beryllium–copper alloys resist both phenomena remarkably well.

Beryllium–nickel alloys

Beryllium–nickel alloys offer another alloy form which has many useful properties for design engineers. One such alloy has been designed for usage in fabricated parts that require high-temperature tensile and yield strength, good formability, excellent corrosion resistance in reducing media and good resistance to fatigue.

Current research has shown that the alloy also possesses excellent impact strength and may be used for low-temperature cryostats and pressure vessels. A few of the many applications for beryllium nickel are heat-resistant springs and switches, diaphragms, bellows, retainer clips, feather valves, contact springs and electrical shunts. Although these are common applications for beryllium-nickel alloys, alloys can also be selected for application requiring resistance to abrasion and wear, diffusion bonding, magnetically actuated springs, excellent corrosion resistance, oxidation resistance and excellent secondary emission

Short-time elevated temperature

The room-temperature strength values of the alloy hold up remarkably well up to approximately 426 °C. For instance the yield strength of the 1/2 HT is 6894·8 MN/m² at 426 °C, displaying approximately only a 10 per cent reduction in strength from room temperature.

Flexural fatigue strength

S–N curves for beryllium nickel at room temperature and 426 °C have been determined to ten million cycles. The AT temper material was tested in reserve bending. Mechanical properties were as follows:

Ultimate tensile strength	1574·85 MN/m²
Yield strength (0·2 per cent offset)	1008·04 MN/m²
Elongation	22 per cent

Initial impact property studies indicates that the alloy maintains high impact strength down to 320 °C. The material does not exhibit a brittle transition zero.

Beryllium–aluminium alloys

Another beryllium-containing material contains 62 per cent beryllium and 38 per cent aluminium; it combines a high modulus of elasticity with a low density. It is meant to bridge the gap between magnesium and aluminium alloys and the higher strength titanium alloys and beryllium.

It is not appropriate to expose this alloy to temperatures in excess of 537 °C. This sheet can be bent over radii as tight as 6t at 232–260 °C.

This alloy has been joined by fusion welding using both TIG and electron-beam techniques, brazing, soldering, mechanical fastening, epoxy bonding and diffusion bonding. Fine-grained structures, comparable to or finer than the base microstructure, are developed in the weld zone.

Adhesive bonding requires no special techniques and strengths over 20·684 MN/m² in shear are easily attainable. Mechanical fastening with bolts, screws, pull and squeeze rivets, and other sheet metal techniques are used for joining the alloy in sheet form. Net tensile strengths of 340·474 MN/m² have been attained in bolted lap joints employing titanium fasteners joining the alloy to magnesium plate.

The material is machined using conventional hand or machine tools. Using standard high-speed steel tools, speeds and feeds are conventional although tool wear may be faster. Cold-punching and shearing may be accomplished with the tool materials, clearances and angles used for either aluminium or magnesium. It is also produced as extruded bar, rod, tubing and structural angles and channels and as sheet.

Chapter 4.7 **Clad wires**

Clad wires have been used since the early 1900s, but the explosion into highly
diverse applications has taken place only recently. One of the first 'tonnage'
clad wires was copper-covered steel wire for strength members and communications conductors. Within the last ten to fifteen years a myriad of
other application for clad wires has opened up in several different fields.
With it has come a disciplined engineering approach to their selection —
a 'materials system' approach. This chapter will be limited to the application
of the materials system concept to clad-wire products for the electronics
industry.

One of the first clad wires developed for electrical/electronic use was a
composite of copper clad to a core of nickel–iron alloy. This was designed to
fill a need in glass sealing light bulb leads. The composite wire matches closely
the expansion characteristics of soft glasses and gives better conductivity than
nickel–iron alloy. This alloy is a metallurgically bonded product without any
braze foils or interliners. Bonding is achieved on the atomic level such that
the wire would fail before bond separation occurs. Table 4.7.1 lists typical
properties.

Table 4.7.1 **General physical properties of copper-clad aluminium**

Composition	Resistivity, Ω/cm	Conductivity %IACS	Weight of copper as percentage of total	Tensile strength, kN/m^2	Elongation, %	Specific gravity
Copper	10·27	101·0	100	650	2	8·91
10 per cent Cu/Al	16·08	64·5	26·8	30·0	2	3·34
15 per cent Cu/Al	15·60	66·5	36·8	31·0	2	3·65
20 per cent Cu/Al	15·14	68·5	45	33·0	2	3·98
Aluminium	61·8	61·8	—	19·0	2	8·71

Many new applications for this wire have been found in the electronics
field, e.g., the glass seal in the construction of a diode. In this operation a soft
glass bead is flame-sealed to a 'borated' wire. 'Borated' simply means that a
film of sodium tetraborate has been fused over a controlled amount of copper
oxide. The borate 'fixes' the depth of the oxide to which the glass adheres.
In this case the diode lead uses a small slug of borated wire percussion-
welded to a lead wire of unborated alloy; the glass seal is made to the borated
slug. This type of lead is also used in transistor bases using compression seals

rather than matched seals. This means that the transistor base (or eyelet) is made of steel rather than iron–nickel–cobalt alloy.

Silver-clad copper

Precious metal claddings have found a variety of electronics applications. Silver-clad copper is used for special antennae. Silver-clad nickel is used in high frequency tuners as a fixed plate in the variable capacitor. Gold-clad phosphorus bronze gives connector pins rigidity and long life because of the wrought gold surface. In all cases the clad wire combines the desirable properties of two materials to yield a resultant material that turns out to be functionally superior to either of the individual components.

Another recent application is in high-frequency cables. For cables operating in the range of 10 MHz and higher, copper is needed only on the thin outer skin because this is the location of the signal. Signal depth at 10 MHz is about 0·089 mm. Therefore, the requirements for a conductor that will carry a high-frequency signal are:

1 Minimum copper for high-frequency signal.
2 Good conductivity if other d.c. voltage is to be carried.
3 Good thermal match with shielding.

As a result of these design criteria a new conductor was produced — copper-clad aluminium. Table 4.7.2 presents typical properties. Not only did the copper-clad aluminium conductor meet the design criteria as outlined, thus yielding superior performance, but it also cost much less than solid copper.

A typical use for copper-clad aluminium is in coaxial cables. Other uses for this versatile low-cost conductor include: magnet wire for deflection coils; resistor and capacitor leads; hook-up wire; bus wire and high-frequency trimmer coils. The copper-clad aluminium wire can be processed like solid copper and will accept all common platings normally used on copper.

Cored wires

Another group of clad wires important to the electronics field are called 'cored' wires. Cored wires were developed for a wide range of glass-to-metal seals. Copper-cored wires and solid alloy are closely matched in expansion characteristics. But at the same time there is a significant increase in thermal and electrical conductivity in the cored wire. This means the same size device can handle perhaps as much as eight times the power now being handled.

Lead wires for hermetically sealed reed relays (wet or dry) are another potential application in which the cored-wire concept could simplify manufacturing, improve performance and reduce over-all costs.

Copper-clad steel

Copper-clad steel, one of the oldest clad metals, has a variety of uses in the electronics market. Side rods in radio tubes, capacitor leads, and plastic

T

Table 4.7.2

Composition	Density at 20 °C, g/cm³	Thermal conductivity at 20 °C, W/m. degC	Electrical resistivity at 20 °C, Ω.cm	Conductivity, % 1ACS	Temperature coefficient electrical resistance, 50–80 °C	Linear expansion coefficient, 20–100 °C	Curie temperature, °C
Alloy-clad copper wire	—	0·166	—	16·28	0·00356	0·0000101	427
3:1 ratio	8·4	—	60	—	—	—	—
Solid	8·3	0·032	258	4	0·0029	0·0000098	427
Clad copper wire:							
3:1 ratio	8·2	—	101	12·8	—	$4·5 \times 10^6$	—
solid	8·12	0·029	565	1·8	—	$4·1 \times 10^6$	290
Clad copper wire, 2:1 ratio	7·96	0·271	43·6	26·9	0·00330	$10·8 \times 10^6$	427
Alloy-clad copper wire (F15)	8·42	—	74·0	14·3	—	$10·8 \times 10^6$	430
Solid F15 alloy	8·36	0·04	294	3·5	—	$10·6 \times 10^6$	430

encapsulated transistor leads all employ copper-clad steel. It is also used in a variety of RG/U type coaxial cables.

The unique processing system used by transistor manufacturers yields a clad product which is concentric between steel core and copper-clad. This means uniform signal transmission for high-frequency cables and no 'break-through' in forming.

Although this chapter has emphasised wire, capabilities for improved electronic component performance also exist in strip materials. Some typical examples are:

End product	*Clad product*
Transistor cans	Nickel-clad stainless steel
Lead frames	F15 alloy-clad copper
Battery cases	Copper-clad stainless steel
Battery materials	Magnesium-clad steel and silver-clad magnesium.

Once again these materials are designed to improve performance, simplify manufacturing and reduce cost. The clad- and cored-wire concepts are becoming increasingly important. Many of the products currently available were not in existence five years ago. These concepts free the engineer to design for optimum performance in a way never before available. Each year hundreds of new clad metal combinations are produced for applications that range from the exotic which require a few troy ounces to the commodities which are shipped in car-load quantities. It is expected that this trend will continue.

Section Five

Chapter 5.1 **Chemically deposited thin ferrite films**

There are several applications where chemically deposited thin films have significant advantages over other ferrite materials and techniques. In a resonance isolator, for example, thin slabs of ferrite are cut from bulk material and cemented to ceramic substrates. Cutting thin slabs is a delicate operation which the chemically deposited ferrite film eliminates. Likewise, it is difficult to cut the bulk thin slabs into intricate shapes and forms. With the chemical deposition technique it is possible to cut the substrate into any desired shape or form before or after ferrite film depositions. In a travelling-wave maser, optimum isolation can be achieved by depositing a ferrite film onto the active rutile maser crystal since the RF magnetic field drops rapidly outside this dielectric medium. This chapter reports on the following:

1 Chemical preparations and properties.
2 Crystallographic.
3 Ferrimagnetic resonance on garnet materials.

Chemical preparation and properties

Preparation of films

Individual stock solution of the nitrates of various metals used in ferrite formulation are prepared by heating the compounds in beakers to their melting-points and adding sufficient water to maintain the liquid state on cooling to room temperature. These stock solutions are mixed in the proportions required for stoicheiometry of the ingredients in the desired ferrite compositions. Working solutions for the film-depositing process are prepared by diluting the stoicheiometric mixtures with methanol in the volumes shown in Table 5.1.1. These solutions are usually applied to alumina substrates from a medicine dropper and allowed to spread uniformly over the surface. The coated substrates are then subjected to a preliminary firing between 400 and 700 °C for approximately 1 min to convert the nitrates to oxides and then allowed to cool to room temperature. This coating and firing procedure is repeated until a uniform coat of desired thickness was obtained. Between 0·13 and 0·25 μm of material was deposited during each operation. Finally, the substrates are fired in air, or in appropriate atmospheres in the range of 900 °C for a period of 2 hr to effect the reaction between the oxides for the formation of the ferrite films.

Substrate material

Any substrate material capable of withstanding a temperature of at least

Table 5.1.1   General dilution guide of cationic solutions in preparation of several ferrite films

Stock solutions	Volume of stock* solution, ml	Analysis of stock solution, g/ml	Volume of evaporating solvent, ml	Number of coats†	Film thickness, μm	Related weight, mg
Nickel	6·5	0·2069	35	22	12·0	41·4
				29	17·7	61·6
Iron	20·0	0·1271		42	22·5	77·7
Nickel	7·3	0·2069		23	12·0	41·4
				31	17·7	61·6
Zinc	2·1	0·2613	60	42	22·5	77·7
Iron	30·0	0·1271		41	22·5	28·0
Yttrium	17·6	0·2069		16	12·0	41·4
			82	20	17·7	61·6
Iron	30·0	0·1271		25	22·5	77·7
Yttrium	25·2	0·2069		11	12·0	41·4
Aluminium	12·1	0·0650	100	14	17·7	61·6
Iron	30·0	0·1271		17	22·5	77·7

* Volume of the analysed stock solution containing ratios of the metals necessary for stoicheiometry
† To achieve the desired thickness

1,200 °C may be used. Generally, a 96 per cent alumina substrate material was used, its thickness being 0·508–0·762 mm.

Crystallographic and ferrimagnetic resonance

To determine the crystal structure and guide the preparation procedure of the ferrite films, each film is posed to FE K$_\alpha$ radiation, the standard sample holder is adopted so that the films can be X-rayed without any impairment of the film or its substrates.

The diffraction pattern obtained is quite good, since the films are of sufficient thickness, and thus second phase as well as solid solution studies can be pursued with reasonable reliability. The results showed that zinc and aluminium substitutions can also be made successfully in garnet ferrite films. These crystallographic results are shown in Table 5.1.2 and indicate that a reasonably high degree of purity is obtained, i.e. approximately 5 to 10 per cent of second phase is present. The elimination of these effects could probably be accomplished by films the X-rays show that distinct layers had not been formed; rather, the various layers dissolved in one another to produce a composite ferrite that was a solid solution of the various layers.

Ferrimagnetic resonant properties of various ferrite films have been measured to determine the feasibility of the films for device applications. The plane of the film was located in the transverse plane of the wave-guide, while a d.c. magnetic field applied perpendicular to the RF magnetic field, was varied to obtain a resonance plot. The resonance linewidth (H) obtained from the curves are listed in Table 5.1.2.

Major phase and film thickness as a function of firing temperature

Figure 5.1.1 shows that the maximum per cent of major-phase structure for

Table 5.1.2   Crystallographic results and microwave linewidths of various ferrite films subjected to specific preparation conditions

Films prepared	Firing atmosphere	Firing temperature, °C	Composition attained (crystal phases)	Film thickness, μm	Microwave linewidth, Oe
$MgO.(0.4Al\cdot1.6Fe)O_3$	Air	1,100	100 per cent pure	22·5	750
$(0.75Ni.0.25Zn)O.Fe_2O_3$	Air	1,100	90 per cent pure 10 per cent aFe_2O_3	22·5	580
$(0.9Ni.0.1Zn)O.Fe_2O_3$	Air	1,100	90 per cent pure 5 per cent aFe_2O_3 5 per cent Fe_3O_4	17·7	585
$3Y_2O_3.5Fe_2O_3$	O_2	1,100	90 per cent pure 5 per cent YFe_2O_3 5 per cent $Y(Al_2Fe)O_3$	22·5	875
$3Y_2O_3.1.5Al_2O_3.3.5Fe_2O_3$	O_2	1,100	95 per cent pure 5 per cent $Y(Al_2Fe)O_3$	22·5	125
$(0.5Mg.0.5Zn)O.Fe_2O_3$	Air	1,100	95 per cent pure 5 per cent Fe_3O_4	10·4	540
$NiO.Fe_2O_3$	Air	1,000	90 per cent pure 10 per cent aFe_2O_3	22·5	875
$MnO.Fe_2O_3$	N_2	900	95 per cent pure 5 per cent unknown	22·5	1,000

yttrium iron garnet $3Y_2O_3-_5Fe_3O_3$ occurs from 1,100 to 1,200 °C for constant thickness and are mostly a $Y_2O_3-1·5Al_2O-3·5Fe_2O_3$ film, only temperatures around 1,100 °C were found to be the most useful, since, above 1,100 °C the aluminium tends to come out of the structure and straight YIG begins to form as a second phase. Below 1,100 °C and Fe_{203}, YIG, a Al_{203} appear about in that order. Figure 5.1.2 shows that the thinner the film, the more the second phase increases, and it appears that this method of producing garnet film may be limited for the present to films of about 1·0 μm. A few attempts have been made to improve the 0·5 μm film by using non-

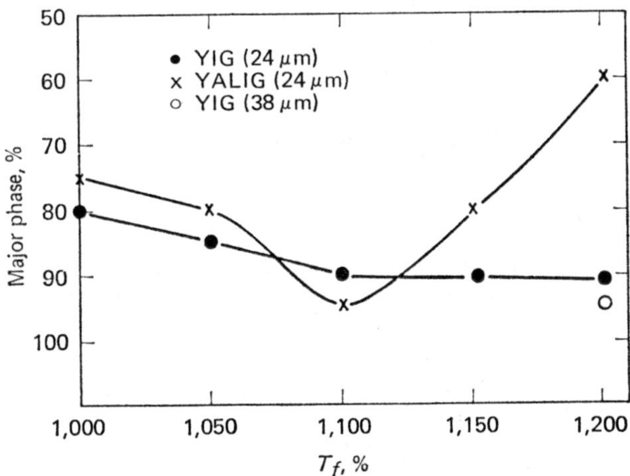

Fig. 5.1.1   Percentage of major phase versus firing temperature soak time with film thickness held constant

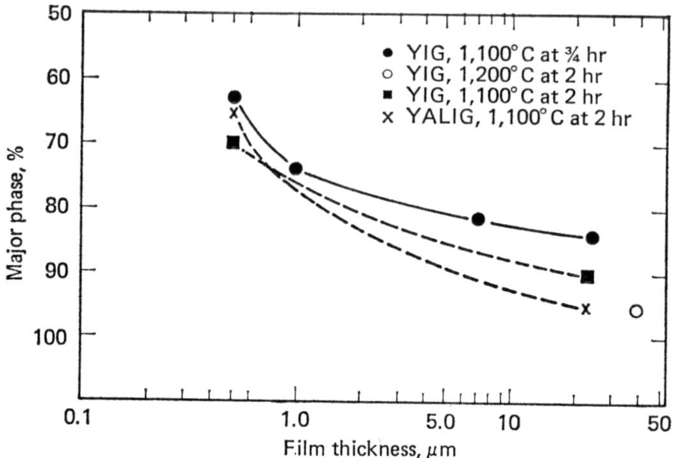

Fig. 5.1.2 **Percentage of major phase versus film thickness for a given time and temperature**

stoicheiometric formulae, changing the concentration of the solution and changing the firing temperature. However, little or no improvement resulted, stronger concentration of solutions yielded some improvement but it apparently was at the expense of the uniformity of the film surface.

Particle size

Investigation of particle size as a function of film thickness, where the firing temperature was held constant at 1,000 °C and soak time held constant for 1 hr, showed only approximately a 0·2 μm particle size (see Fig. 5.1.3). Varying the temperature by 50 degC intervals from 1,000 to 1,200 °C and holding the firing time constant at two hours while maintaining a constant film thickness of 24 μm. showed an increase in particle size per 50 degC rise in temperature. Particle size was measured by a line-broadening technique that was corrected for strain effect. This method, however, becomes inaccurate for particle sizes close to a micrometre.

Ferrimagnetic linewidth

Favourable magnetic properties are dependent on the per cent major phase of the ferrite film in question. The narrowest linewidth, 125 oersteds (see Fig. 5.1.4) was obtained with the aluminium substituted yttrium iron garnet at a temperature of 1,100 °C which appears to be a medium firing temperature and is consistent with the maximum major phase point of 95 per cent and 1,100 °C of Fig. 5.1.1. The yttrium iron garnet curve of Fig. 5.1.2 soak time was held constant for two hours and the film thickness was 24 μm, an exception being 38 μm at 1,200 °C.

A number of different types of ferrite films have been prepared by deposition from chemical solutions, and crystallographic studies have verified

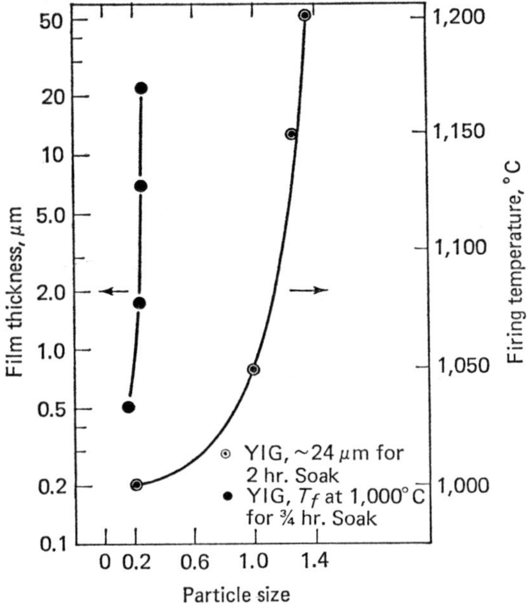

Fig. 5.1.3   Particle size versus film thickness and particle size versus firing temperature, firing time held constant

Table 5.1.3   **Ferrimagnetic linewidth and resonance isolator characteristics**

Ferrite films*	ΔH, Oe, X-band	Maximum insertion loss within the 25 dB isolation points	Maximum reverse to FWD. Loss ratio, dB	Bandwidth percentage at 25 dB isolation points, %
(*Single layer compositions*)				
(1) $(Ni_{2-x})Fe_2O_4$; $x = 1·00$	850	1·8	76 to 1	6·6
(2) $(Ni_{1-z}Zn_z)Fe_2O_4$; $z = 0·25$	610	0·8	124 to 1	6·8
(3) $(Mg_{1-y}Zn_y)Fe_2O_4$; $y = 0·50$	560	0·7	65 to 1	4·9
(*Combined compositions*)				
Single layer compositions (1) and (2)	725	0·7	110 to 1	7·9
Single layer compositions (1) and (3)	750	1·0	65 to 1	7·0
Single layer compositions (1) and (3) and (2)	620	0·8	81 to 1	7·0

* All of the films are 41·5 μm thick deposited on 99·5 per cent Al_2O_3 substrate and fired in air.

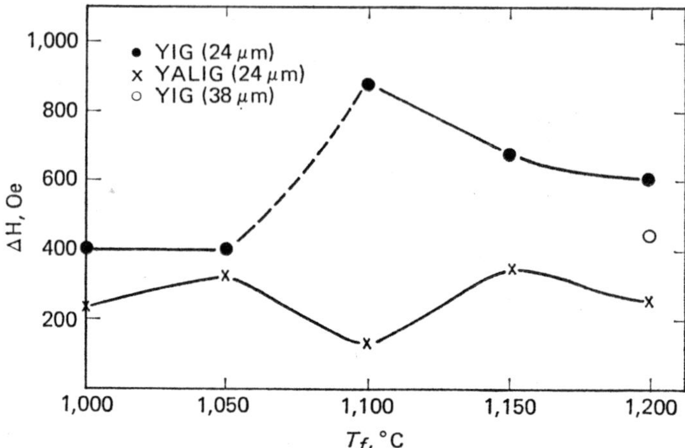

Fig. 5.1.4   Linewidth versus firing temperature soak time and film thickness held constant

the spinel and garnet structure of these ferrite films. Both ferrimagnetic resonant measurements and results obtained with an experimental resonance isolator and millimetre maser prove the feasibility of application of these films in isolators, masers, circulators, phase shifters, switches, and variable attenuators in the millimetre wave region.

Garnet films can be made for a range of thicknesses and therefore may be useful over a range of frequencies for certain applications, although the optimum conditions for applications have not been ascertained. The results of test indicate that the thinner the film, the smaller the particle size and major-phase reduction it appears that they are limited somewhat to thickness greater than $1 \cdot 0\ \mu$m. However, the results are a good approximation for consideration of other materials in that linewidths tend to follow the most favourable structure according to the pore theory. Annealing the sample changes the submicrometre particles to micrometre size; this improves the linewidths. Here the grain growth fills in the pores thereby increasing density. In general a good dense structure is needed to obtain good magnetic properties.

Chapter 5.2 **Magnetic thin films and ferrite core materials**

Planar magnetic thin-film memories have failed by a long way to reach the commercial standing of ferrite core memories despite a considerable development effort over the last decade. To explain this situation and to indicate future commercial trends it is necessary to compare the development of ferrites and thin films for memory systems.

The first storage system embodying the feature of random access was built using toroids of 'square loop' ferrite about thirteen years ago. Developments of ferrites from that time have been directed mainly towards faster operating speeds. It was because of the operating speed limitations of ferrites that, about ten years ago, very great interest was shown in the switching properties of thin ferromagnetic films. It was shown at that time that under ideal conditions a thin film could switch a hundred to a thousand times faster than the fastest ferrite cores and this provided a very great incentive to develop memory systems based on thin films.

Developments of thin-film and ferrite storage systems have, therefore, been carried out in parallel for the last ten years without a significant commercial realisation of a thin-film memory system. Fast, small-capacity thin-film memories have been made for special purpose and military applications, but most general-purpose machines use ferrite cores for their random access store.

The commercial success of a storage system, of course, depends upon performance and cost and these can be related to the magnetic characteristics of the storage element, the method of fabrication of the storage array and the type of storage used. It is instructive to compare ferrite cores and thin films under these headings.

Magnetic characteristics

The use of binary arithmetic in the operation of digital computers requires the storage element to exhibit two distinct stable states which can be used to store the binary digits 'one' or 'zero'. So that the stored information can be selected from one of a number of storage elements the elements should show a distinct threshold level between the stable states. Further, the device should be capable of generating an output signal of sufficient level to identify the stored information, and this operation should take as short a time as possible.

The square-loop ferrite toroids fulfil all the requirements for storage applications. The hysteresis loop shows two stable states of remanent magnetisation and the distinct threshold between them. When a magnetic field is applied which exceeds the threshold field, the magnetisation reverses and generates a voltage which is proportional to the magnetisation and the size of the toroid and inversely proportional to the switching time. The switching time is typically one microsecond (10^{-6} sec) and the output signal greater than 30 mV.

Thin ferromagnetic metal films a few thousand ångstroms ($1 \text{ Å} = 10^{-8}$ cm) thick can also fulfil all the requirements for storage applications. The magnetic properties of the thin films must, however, be distinguished in several important ways from the ferrite toroid. Firstly, the magnetic properties of thin-film films are anisotropic which means that the properties are equal in all directions. Thin films have in fact a 'preferred direction of magnetisation' or 'easy axis' the loop is measured with the field parallel to the easy axis, and the field perpendicular to the easy axici. This direction is the 'hard axis'. It is characteristic of a thin film with a preferred direction of magnetisation that if a field is applied in the hard direction, the magnetisation 'rotates' from the easy towards the hard direction. This is the second property which distinguishes thin films from ferrite toroids in which the magnetisation changes by the movement of 'domain walls'.

Magnetisation rotation is a very much faster process than domain wall motion and hence a typical switching time for a thin film is ten nanoseconds (10^{-8} sec). The third feature that must be distinguished between planar thin films and ferrite toroids is that the output voltage generated is generally very low, less than 1 mV. The low signal results from the geometry of the magnetic element which has a very small volume and does not form a closed magnetic flux path about the voltage sensing conductor.

Method of fabrication

The economics of the method of fabrication of the storage array obviously have considerable bearing upon the commercial success of a storage system. The important factors are the uniformity of the magnetic properties of the storage elements and the ease with which they can be assembled.

Ferrite toroids are individually formed by pressing techniques and batch fired to obtain optimum magnetic properties. Because each element is separate it can be individually tested for quality before assembly into the storage array. In this way the uniformity of properties in the storage array is assured. The assembly process, however, is tedious and in most cases is performed by hand.

Thin films are made by some form of deposition process such as evaporation or electrodeposition over reasonably large flat areas and the individual storage elements formed by selective etching of the continuous film. The many elements, therefore, that go to make up the storage array are made simultaneously. This batch fabrication of the array of storage elements can only be an advantage if all the elements show uniform magnetic properties. An etched array of thin-film elements indicates the assembly process that would appear advantageous compared with ferrite systems since the selection conductors can themselves be batched fabricated.

Development of thin-film memories

Since 1955, when thin films with uniaxial properties were first prepared and shown to have very short switching times, active work has been carried out in many laboratories on the fabrication of thin film storage systems.

It is convenient to discuss the problems and the extent to which they have been overcome in terms of the magnetic element, the method of fabrication and construction, and the system design, although these are more intimately related than with ferrite core systems.

With thin films, the major materials effort over the last ten years has not been directed at obtaining improved switching times (these are sufficiently fast at under 10 nsec) but rather at the problem of threshold fields which are sensitive to disturbing fields. It was pointed out earlier that the successful binary storage element requires two stable states and a distinct threshold between them. It was soon found with thin-film storage elements that successive field pulses of amplitude considerably less than the threshold fields for irreversible switching could cause a gradual magnetisation reversal to take place. This phenomenon of magnetisation 'creep' was found in fact to occur when time varying fields from neighbouring conductors acted on an element which itself was acted upon by the static fields generated by the magnetisation of neighbouring elements.

The first efforts to eliminate the effects of creep were to space the elements sufficiently far apart to make the effects of stray fields negligible. When this was done a low packing density resulted. At the same time considerable effort was spent in arriving at an understanding of the physical mechanism of creep, and observations showed that it was a function of film thickness. In very thin films of less than 300 Å, the effect could be reduced but this approach to the problem presented other difficulties in system design arising from the lower output voltage levels generated by a thinner film. However, by making two such thin films as a pair about the sense conductor, larger output voltages can be achieved and creep effects minimised.

The second important material problem that has presented considerable difficulty has been that of making sufficiently large areas of film with uniform magnetic characteristics. One of the major advantages of the thin-film approach to storage is that the production is by a batch process, but this advantage only holds if high yields can be achieved. To achieve uniform characteristics the film must be homogeneous in composition and structure and uniform in thickness, while the substrate onto which it is deposited must be uniformly smooth. Further, the composition must be held to better than 0·5 per cent to achieve optimum magnetic properties. Technique such as thermal evaporate, electrodeposition and chemical deposition have been developed but in general uniform characteristics are only achieved over limited substrate areas of about 10·16 cm × 10·16 cm at the most and when commercial yields are demanded the areas are considerably reduced from this.

Together with the difficulty of magnetic material preparation goes the problem of applying the address conductors. In the earliest thin-film store these were applied as wires wrapped around the thin-film substrate but more sophisticated techniques have been developed since then. One method compatible with the deposition of the magnetic elements is to evaporate the selection conductors through masks together with the intermediate insulators where necessary. Since this technique takes the batch process one stage further, the final yield will depend upon the success of the combined processes and consequently must be lowered. Alternative techniques in which the conductors are made separately by etching copper-clad circuit boards and the whole sandwiched together can produce higher plane yields but suffer from problems of registration which has to be accurately achieved to produce low-noise systems.

The word-organised system which has to be used with thin films if adequate tolerances on selection currents are to be achieved has presented several limitations to the development of thin films. If the fast switching times of the thin films are to be used, the store must be driven with fast rise-time pulses (10 nsec) and these have presented serious problems from the point of view of noise due to capacitive coupling between word and sense conductors. This is especially important when signal levels are typically less than 1 mV. The drive transistors must have fast rise times and produce several hundred milli-amperes which makes the system costly, especially since each word requires a separate driver.

One outstanding problem with the word-organised system arises from the fact that the digit and sense conductors are common which gives rise to large paralysing pulses on the sense amplifiers during 'write in' of information and dictates a long recovery time before 'read out' can be achieved. The contribution to cycle time from this effect is large compared with element switching times and remains a major problem in the realisation of a system operating with the full potentialities of a fast thin-film switching element.

Trends

It will not be possible to utilise the 3D system in ferrite stores for use at the highest speeds because of the limitations imposed by ferrite material and core size, and already word-organised systems are being designed around 0·304 mm outside diameter toroids. The simpler two-wire address leads to acceptable assembly problems but the systems can only be commercially justified in terms of high speed and the assumption that selection electronics costs will reduce.

Development of multi-million bit mass stores using normal switching time toroids (0·5–1 μsec) is also being carried out. This development relies upon reducing plane assembly costs by using a simpler rectangular three-wire address system and minimising selection electronics with a modified word-organised system.

Since the advantage of the 3D storage system is unlikely to be part of the future development of ferrite core systems and as the problems of manu-facture of smaller and smaller cores grow, the potential competition of thin film storage becomes apparent. Given that the word-organised 2D system, of modification of it, is to be used, the intrinsic properties of the thin films have certain advantages over the ferrite toroid.

The second advantage of a continuous process arises from on-line pulse testing which provides quality control prior to plane assembly. This advantage is shared by ferrite cores but is absent with planar thin films. The storage plane is constructed by passing orthogonal word conductors around the magnetic wire. This relatively simple wiring arrangement presents an advantage over ferrite cores in that effectively large numbers of storage elements (the plated wire) are assembled at a single operation, and is in fact half way between the bit-by-bit assembly of a ferrite core system and the all-bits-at-once assembly of the planar thin film. Besides the advantages in production, several advantages arise from the cylindrical nature of the film. If the easy axis of magnetisation is circumferential the magnetisation forms a closed flux path around the sense conductor which allows, in general, thicker films to be used and higher output voltages to be achieved.

From the system design viewpoint, the larger output voltages mean that slower current rise times can be used to obtain acceptable signal levels and yet still maintain fast switching times (30–50 nsec) compared with ferrite cores, and this can lead to a very much simpler engineering problem.

In conclusion it would appear that the commercial realisation of a thin film storage system could be near at hand. This situation has arisen because the 3D system cannot be used with ferrites at the higher speeds, because the continuous production of cylindrical thin films promises to be economical and because the cylindrical film can be operated at speeds which are competitive with ferrite systems and yet still allow an economic system design.

Telecommunication ferrites

The properties of soft ferrites offer certain advantages over metallic materials and, in addition, the most suitable shape for a particular application may be chosen, as ceramic techniques are used in the manufacture. For example, the material in pot core enables an air gap to be included in the centre of the core, reducing the external coupling to a minimum. Optimisation of core and air gap enables an application requirement to be met at a minimum price, weight and volume.

It is in the telecommunication field that the advantages of ferrite materials are most significant. The communication systems in use today and those systems which are being developed for the future rely in general, on some form of mixing and demixing of a wide range of frequencies, an adequate means of frequency separation, and a means of transferring power from one impedance level to another.

During recent years many have actively participated in the efforts of national and international committees through the International Electrotechnical Commission to establish standard ferrite parts. These have found widespread international acceptance. Although the basic shapes of ferrite parts for transformer and inductor applications have been standardised, the available grades of ferrite material are such that it is possible to meet the range of requirements for an inductor or transformer by choosing a suitable grade of material.

New shapes of core and new ferrite materials are constantly being assessed to improve the performance of high quality inductance coils and transformers.

U

Typical application

A typical application of ferrites in the telecommunication field is in a low pass filter used in the voice frequency circuits of a multiplex telephone equipment. The important characteristics of the filter are that the attenuation response must be stable with time and temperature variation. The loss and size of each inductor used in the filter are normally resonated with capacitors and the frequency of resonance must be stable within $\pm 0 \cdot 25$ per cent with time temperature changes from ambient up to 60 deg C, and the usual mechanical shocks of transportation and installation.

Inductor requirements

The designer of these filter circuits requires inductors that are as small as possible, have a high degree of stability, low loss, low harmonic distortion and are capable of initial adjustment to cover the inevitable tolerances on inductance and capacitance.

In order to achieve the degree of permeability stability required, the permeability variations of the ferrite have to be reduced and this is achieved by introducing an air gap into the magnetic circuit. With pot cores it is an advantage to have this air gap in the centre limb of the cores. The permeability under these conditions is commonly referred to as the gapped or effective permeability. Both the permeability and loss tangents of the magnetic circuit are reduced by this air gap.

To achieve maximum stability, the need for a large air gap is implied, but this would mean that many turns of wire would be required to achieve the inductance value, so the loss due to d.c. resistance would be high. In addition to achieve maximum stability with temperature change the percentage variation in inductance should cancel the capacitance variations when polystyrene capacitors are used.

The capacitance variation with temperature is a fixed value and this implies that a specific value of gapped permeability is necessary to offset it. In practice, a compromise value of permeability is sought which gives good stability, low magnetic losses and a low value of d.c. resistance and eddy current copper loss. It can't be shown that the losses due to d.c. resistance and eddy currents are inversely and the losses in the ferrite material directly, proportional to the gapped permeability. Furthermore the over-all loss of the coil is a minimum when the losses in the ferrite and copper are equal in value. Therefore at any one frequency, the over-all loss can be made a minimum by choosing the value of gapped permeability which makes the ferrite loss and copper loss equal. This is shown in Fig. 5.2.1.

In addition to the effects of an air gap, the ratio of d.c. resistance to inductance R_{dc}/L is dependent on the size and shape of the ferrite core. The IEC range of sizes has been optimised to provide a minimum value of R_{dc}/L for a given volume of core. This has been made possible by the use of ceramic manufacturing techniques which are used in the production of ferrite cores. It can also be shown that the third harmonic distortion is related to the core dimension for a given volume of core.

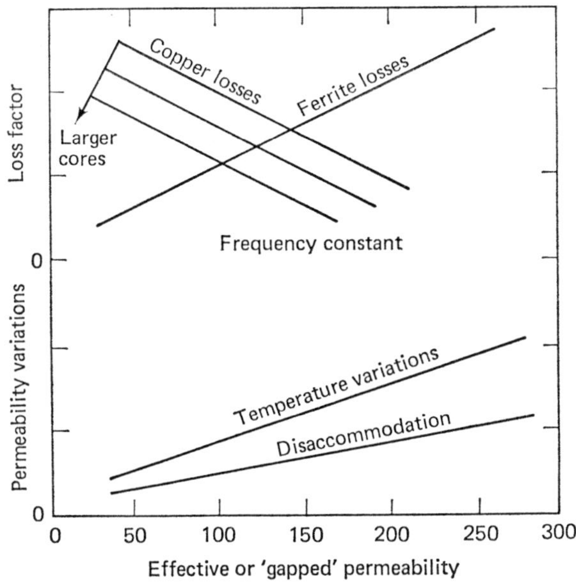

Fig. 5.2.1   Minimisation of over-all loss by selection of correct 'gapped' permeability

The above requirements have been related specifically to the case of inductors for low filters but any application that require precise values of inductance with high stability, low loss, small size and cost will benefit from the use of ferrite material in pot core form.

Transformer requirements

There are numerous applications in the telecommunication field which require low power transformers for impedance matching over a wide band of frequencies. In these cases the stability of the inductance is not of such importance, so the full undiluted values of permeability that are obtainable from modern ferrites can be used to advantage.

The main requirements for these transformers are the permeability and bandwidth be high, and that the copper and ferrite losses, the harmonic distortion, and the size and the cost should be low. The copper losses are proportional to R_{dc}/L and consequently the conditions that led to the minimisation of this value for a range of pot cores for inductors apply with equal importance to transformer use. The harmonic distortion is also a minimum for the same reason. It can be shown that the optimisation of dimensions also gives low figures for the leakage inductance for a fully wound bobbin and this indicates that the band width will usually be more than adequate for most purposes.

Available materials

Material grades enable the designer to choose a ferrite material with optimum properties for each frequency of application from a few hundred cycles up to 20 MHz. The permeability and loss characteristics of these materials are important.

The cores are manufactured using the ceramic techniques, but superior characteristics of the telecommunication grades of ferrite are only obtained by the careful selection of high purity raw materials coupled with specialised firing techniques involving controlled atmosphere conditions and temperature cycle.

A rigid process control is applied to all stages of the operation to ensure a high degree of reproducibility and stability of characteristics. They make the maximum use of this improved stability special attention has been paid to the design of the associated clamping hardware, that takes up the minimum additional space, supplies the necessary clamping pressure under extremes of environment such as temperature, vibration and mechanical tolerances. In addition it has been designed specially for printed circuit use. The variation with permeability when subjected to vibrations and bump testing was negligible.

It is anticipated that future developments in ferrite grades of telecommunication material will involve higher permeabilities, lower losses, better control of temperature coefficient for more exact matching of the temperature coefficient of the associated capacitors and further improvement in disaccommodation which is an indication of permeability stability.

To make the best use of high permeabilities for transformers the inevitable air gap between the mating faces of the two halves of a pot core will have to be reduced to a minimum by improvement of surface finish, also of the adjuster mechanism.

Further attempts to reduce the factor R_{dc}/L has resulted in the introduction of the 'cross core' which has the added advantage of incorporating terminal pins in the bobbin cheeks. These protrude through the core and are suitable for printed circuit application.

Chapter 5.3 **Magnetic switching in materials**

Magnetic materials is a subject poorly understood by the majority of engineers. Recent research, particularly in magnetic films, has provided insights to understand and explain the behaviour of materials that react to applied magnetic fields. Elementary domain theory is central to any significant comprehension. The dynamic behaviour of non-linear magnetic material cannot be analysed entirely from the familiar *B-H* or hysteresis loop. However, the loop shape can be used to confirm behaviour anticipated on the basis of domain theory and observation. Dynamic processes in magnetic material involves a change in the state, which is termed switching.

The fundamental principles pertinent to switching behaviour and experimental data from a typical magnetic core are used to bring together theory and practical observation. Many recent applications of magnetism require a more sophisticated understanding built upon such a knowledge of material behaviour.

Magnetic switching

Switching phenomena are usually associated with non-linear acting devices or at least non-linear operation of a device. Ferromagnetic and ferrimagnetic materials generally exhibit non-linear behaviour to a marked degree. The non-linearity is frequently a nuisance to be swamped out or just ignored in linear circuit analysis. Carrying non-linearity to an extreme provides the so-called square loop behaviour which has been exploited in applications to computer memories, inverters and magnetic amplifiers, to name a few. Much has been learned about basic magnetic switching phenomena. Insights gained are broadly applicable to much of the large variety of commercial magnetic materials available using iron as one of the principal constituent elements whether in oxide or metallic form.

The term 'switching' in this context is used in the most general sense to mean any change in magnetic state as a function of time where the possible final state depends upon the path of approach to that state and where the change in state is a major part of the maximum possible change. The viewpoint is from the electrical terminals of a device containing ferromagnetic or ferrimagnetic material. In these terms, this is simply a conductor passing relatively near to the material, near enough to produce significant interaction. An attempt is made to explain and justify some typical observations from a materials point of view on the basis of magnetic domain theory.

Magnetic materials come in many shapes and sizes for convenience. But often useful terminal properties are obtainable only by careful attention to structural geometry. Basic material fabrication is usually by:

1 Hot: and cold-rolling.
2 Moulding.
3 Direct deposition by vacuum evaporation, electroplating, spraying, or dripping.

Usually some heat treatment is required to develop the desired properties. Even though a considerable variety of fabrication methods is used and the final appearance of the material can be different, the general magnetic behaviour is very similar.

The understanding of detailed behaviour depends upon the partition of energy stored or dissipated in the material. In the natural order a minimisation of stored energy is necessary for stability. Therefore, any study seeks to determine how energy may be minimised within the material and the conditions favourable to achieving such a state. The term 'anisotropy' is used to identify a preference for magnetisation to be orientated in particular directions. Stored energy is minimised by alignment of magnetisation in the preferred direction. Anisotropy can arise from metallurgical considerations such as crystal structure or from the geometrical shape of an entire magnetic device and its constituent microscopic elements. Anisotropy can be induced by a heat treatment in the presence of an external magnetic field with no concomitant metallurgical change apparent. Ferromagnetism and ferrimagnetism depend on the atomic electron spin. A minimum energy state can be shown to obtain from spins being aligned together for large portions of a material.

Magnetostriction

The property of magnetic materials to undergo elongation along or transverse to an applied magnetic field direction, is also an important consideration energetically.

As usual, one attempts to simplify an analytical problem by reducing the number of unknowns and variables to a minimum without inadvertently discarding the problem to be solved. Similarly, magnetic materials become more predictable and consistent by avoiding problems with anisotropy and magnetostriction as much as possible or by coping with these features under well behaved circumstances. An example of the avoidance approach is permalloy-type material (about 80 per cent nickel and 20 per cent iron, with some other constituents such as molybdenum in small quantities). The chemical composition is carefully adjusted to provide approximately zero magnetostriction (plus or minus depending upon the ratio of nickel to iron) and the consumption is fairly close to the minimum of crystalline anisotropy obtained for Ni_3Fe. Permalloy is one of the more widely used square loop materials. Alloys containing approximately 50 per cent nickel and 50 per cent iron also show very desirable magnetic properties for many applications.

This composition is definitely magnetostrictive. It shows a preferred magnetisation direction in three orthogonal directions corresponding to the direction of rolling (for sheets or strip) and each of the other two orthogonal axes. Alloys containing small amounts of silicon (about 3 per cent) are also grain-oriented by the rolling process and show a single preferred direction. For magnetostrictive material, some mechanical freedom must be provided for the dimensional changes occurring as the magnetisation changes. For example, a toroidally shaped core can be placed in a loose fitting box and the magnetisation direction chosen to be one of the preferred axes. Having recognised magnetostriction as a potential problem, one usually avoids it by chemical composition or accommodates it in the physical arrangement.

Shape anisotropy — the tendency for magnetisation to lie along a particular geometrical axis of a material sample (the long axis of a bar), can be circumvented or turned to advantage in most instances by adjusting the length to the cross-sectional area of the magnetic path. In the absence of any externally applied field, a magnetised sample showing some non-zero magnetisation must provide the field energy outside of the material in question. This is called a demagnetising field. It is minimised by making the magnet circuit long in relation to its cross-section. For this reason a toroidal core shape is a favoured form. Essentially its length in the circumferential direction is infinite and the demagnetising field is approximately zero (not exactly zero because of the distributed air gap in a tape would core, for example). In radial or axial directions the demagnetising field is many orders of magnitude greater. In this case, the shape anisotropy tends to constrain (in the absence of any other factors) the magnetisation to be along the circumferential direction. Another technique to accomplish a similar result is the use of thin magnetic film deposited on a flat substrate surface. Typical film thicknesses used are a few micrometres. In this way, magnetisation is constrained by the shape to lie essentially in the plane of the surface. Permanent magnets are intended to supply a relatively large demagnetising field indefinitely. For this reason their magnetisation is intentionally difficult to change. Materials suitable for permanent magnets, in general, are not useful for applications involving magnetic switching as defined above. Whether the material is magnetically soft and thus easily switched or magnetically hard, the physics of the problem remain the same. Alteration of the basic B-H loop shape characteristic of the material is avoided by eliminating a significant demagnetising field in the desired magnetisation direction.

A general discussion would not be complete without at least mentioning the role of eddy currents. From Maxwell's equations an electric field results from a time varying magnetic field.

$$\int E \mathrm{d}l = \int B \mathrm{d}s$$

where E is the electric field intensity along a path of incremental length $\mathrm{d}l$ taken around a closed line c (in this case a path enclosing the cross sectional area of the magnetic material S, with flux density B). This equation may be solved for specific assumed geometries and from it, the power loss determined. In general, such a solution shows a power loss directly proportional to material thickness squared and the time derivative of flux variation squared and inversely proportional to resistivity of the material. Assuming high-frequency switching is required, a high resistivity is an obvious way to reduce eddy-current power loss.

With most metals, resistivity is mainly a function of chemical composition showing on the order of 0·1 μm resistivity, a very low value. The only practical variable remaining is lamination or tape strip thickness. Thickness as shown in Fig. 5.3.1 ranges from about 30 μm for power frequencies down to about

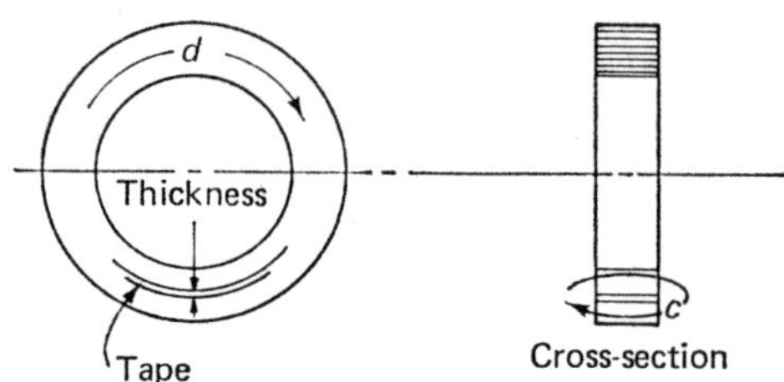

Fig. 5.3.1   Toroidal core

3 μm for very high frequencies. For useful magnetic switching application, eddy-current loss is kept essentially negligible. In fairly high-speed applications a rolled tape thickness of 3 μm, the minimum commercially available, may not be thin enough. Thin metallic magnetic films, 0·1 μm thick, can be switched very rapidly with negligible eddy-current problems. Ferrite materials have a resistivity on the order of 0·1 m or about 10^6 times greater. As cores, their superiority to rolled tape cores for very fast switching as regards eddy currents is quite obvious.

Domains and wall motion switching

A B-H loop, as in Figs. 5.3.2 and 5.3.3, was difficult to understand until the notion of magnetic domains was proposed and developed. Evidently, the material can show a fairly stable value for B anywhere from zero to nearly B_m with just about equal facility. An explanation is that the ferromagnetic

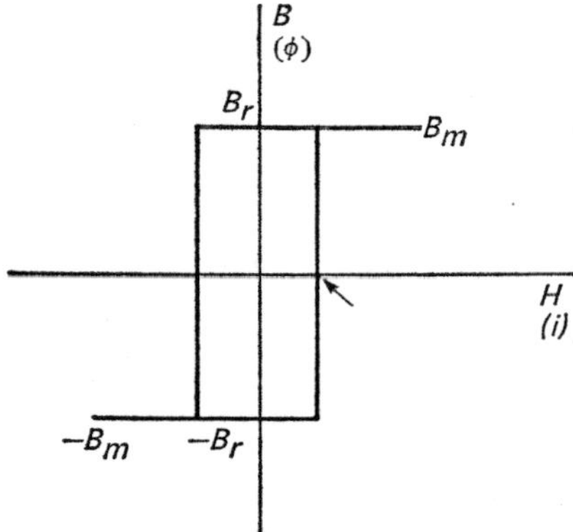

Fig. 5.3.2   Square B-H loop

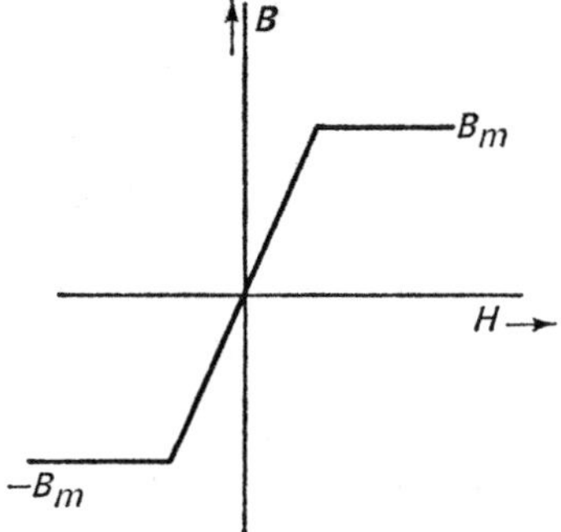

Fig. 5.3.3   Non-square *B-H* loop

and ferrimagnetic materials under consideration are magnetised essentially at saturation levels at all times on a microscopic basis but not all the material volume is magnetised in the same direction. By observing the instantaneous induced voltage (not integrate) across the N-turn winding in Fig. 5.3.4, one usually observes that the flux seems to change discontinuously by steps of variable amounts. This Barkhausen noise evidently results from many atoms realigning their magnetic moments simultaneously rather than a one-by-one process. Nearly all magnetic materials of any commercial interest are poly-crystalline. Although one could believe that each crystal would tend to act independently, the effects observed suggest that changes happen over many crystals simultaneously. Large portions of a material specimen appear to function together. From actual observation one or more contiguous portions of the material volume do behave as a single magnetic entity — these single entities are called domains.

Assume a volume of material has been magnetised to saturation B_m and then left at its residual level B_r. Presumably the entire volume is a single domain with magnetisation in the direction of the applied field H. Practically, such a condition is only realised with a single crystal specimen oriented precisely along one crystalline anisotropy axis. In polycrystalline materials metallurgical perfection is not achieved, with a result that some, hopefully

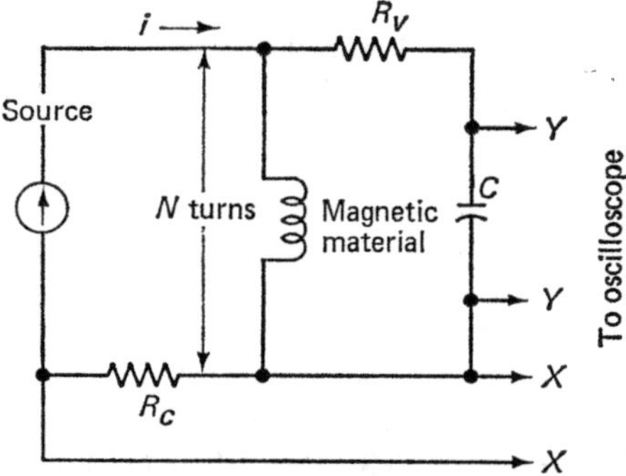

Fig. 5.3.4   *B-H* loop test

small, volume is not magnetised in the same direction as the bulk. In fact, flaws in the structure have quite likely prevented some small portions from being aligned at all. These form nucleation centres of reverse magnetisation, if a field H is applied in opposition to the original direction these centres act as domains favourably aligned in terms of the new H field as shown in Fig. 5.3.5a. The material is switched magnetically by growth of these domains at the expense of the initially large domain aligned antiparallel to the new field. Given some time the picture changes to that shown in Figs. 5.3.5b and 5.3.5c, until essentially the entire volume has been magnetised to the new direction. Of course, some centres of reverse magnetisation relative to the new direction for M can be expected as before, i.e., B_r and B_m.

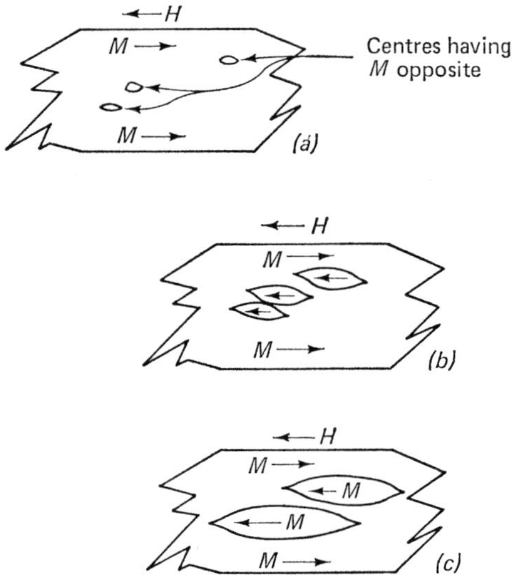

Fig. 5.3.5   Domain growth by wall motion

Between domains having opposite magnetisation directions a discontinuity exists. The exchange energy which is minimised by all atomic spins having the same direction prevents the discontinuity from occurring across one atomic layer because lower total stored energy can be achieved by a more gradual change in magnetisation. A domain wall is found wherein the magnetisation direction changes more or less smoothly from one direction to another typically a change of 90 or 180°.

The most typical wall structure is a Bloch wall where the magnetisation appears to process around an axis perpendicular to the plane of the wall as shown if Fig. 5.3.6. Such a wall requires stored energy to support the field produced by that material in the wall perpendicular to the direction of the two domains it separates. If the wall is bounded on one or two sides by the surface of the magnetic material, a demagnetising field in free space must be supported. The field energy decreases as a function of wall width, as usual, nature seeks an energy condition which is satisfied by a particular wall width

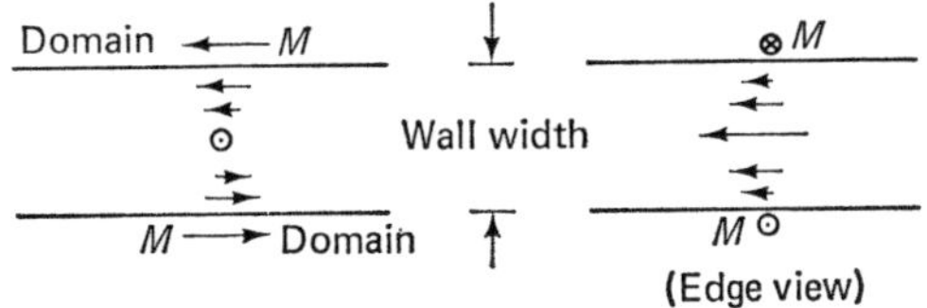

Fig. 5.3.6   Representation of a Bloch wall

wherein the total energy, exchange plus wall, is a minimum. Typical wall widths are of the order of 10^{-6} m.

Magnetic switching most commonly occurs primarily by wall motion. A field is applied to change the magnetisation from one direction to the opposite direction as in a toroidal core. For such a core, Ampere's law leads one to expect that switching (reversal) will begin at the inner diameter of the magnetic core. A given magnetomotive force will produce the greatest field for the smallest path length. Centres of reverse magnetisation begin to grow by movement of the wall surrounding each of the centres. Eventually the centres grow large enough to essentially coalesce into large domains resulting in rather long walls surrounding large domains. Once a domain wall is established, essentially no energy should be required to move because no further change in exchange or wall energy is required. Such behaviour is outside the experience of any observers. In fact, energy is required to keep the wall moving until the material is fully switched. This energy is required to move the wall past metallurgical flaws in the material structure, which tend to cause the wall to stick or hang up. The wall bumps along from one flaw to another (flaws may be voids, impurity sites, grossly misligned crystallites, etc.); it stretches and contracts as it moves. The wall velocity depends upon the magnitude of the applied field. Behaviour is very similar to a viscous damped mechanical system. Techniques have been developed for the study of thin magnetic films to permit visual observation of domain growth.

Recalling the previous discussion of wall motion switching, one might see that a precise determination of the point at which saturation is reached is difficult at best. Saturation certainly cannot be obtained unless all centres of reverse magnetisation are removed. Moreover such centres or perhaps the entire specimen may be somewhat misaligned relative to H_a. In such a case, saturation will only be observed to occur when the magnetisation direction is rotated to correspond with H_a. A rule of thumb is to have H_a about twenty times greater than H_a in the direction that saturation is desired. This is not always convenient and often not possible in the normal operation of a practical device using the magnetic core.

Tests are of use mainly to compare materials for a proposed application and to obtain approximate operating parameters once a material has been selected. The curve is of little use for partially switched material where the driving field is terminated before the natural response time of the core has been accommodated.

Discussion so far has assumed wall motion switching produced by an applied magnetic field of constant strength. In some circuit applications the flux is forced by an applied voltage waveform. For a square B-H loop core, the current tends to approach a squarewave form as a function of time almost without regard to the applied voltage wave form. The actual B-H loop obtained is still essentially square for minor loops approaching the saturation level. However, if the core is driven to saturation, a curious anvil shape begins to develop as shown in Fig. 5.3.7. The peak field to begin reversal may exceed the coercive field H_c appreciably. Also the value for H_c obtained in such a test may be substantially less than H_c obtained with current source drive. Such behaviour is consistent with domain theory.

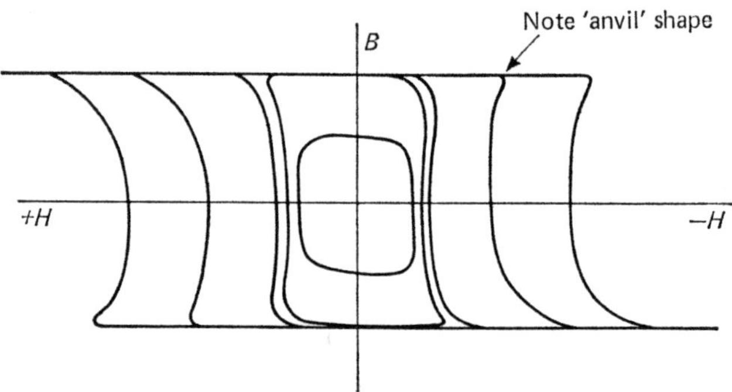

Fig. 5.3.7 B-H loops for several values of sine wave a.c. voltage applied (flux forced)

Relatively much more energy is needed to nucleate a domain than is required to propogate its walls once they are established. Thus, with flux change forced by a voltage source the energy is supplied by a high peak of current during a short interval to nucleate many domains or propogate existing walls at high speed. The wall motion domain switching model accounts for observed behaviour in square B-H loop materials very well. Its basis is the energy required by the material to nucleate and propogate domains.

Rotational switching

Most people are familiar with the tendency of a compass needle to align its magnetisation along the Earth's magnetic field direction. If the needle is displaced, a restoring torque realigns the needle once more. The mechanical motion of the compass needle is quite obvious. Less obvious perhaps is the possibility that the magnetisation of a ferromagnetic or ferrimagnetic material can be turned without any evident mechanical motion. The magnetisation tends to lie along a preferred direction (the preferred direction can come about for several reasons). The anisotropy energy is minimised by alignment in the preferred direction. If the magnetisation is displaced by an applied field a torque must act against the anisotropy field to cause the angular displacement. As the field increases from zero, the magnetisation tends to turn toward the direction of the applied field. When the direction of magnetisation becomes the same as the applied field direction no further change is possible. If H_a is applied perpendicular to the anisotropy axis, the flux density increases as the magnetisation turns from zero to the saturation level, describing the curve shown in Fig. 5.3.3. This curve is linear (just as it would be for an air core inductor) for flux density levels less than saturation. The principle difference is the slope which is steeper than for an air core. An ideal B-H characteristic like this would be most desirable for transformers and indictors. Such a characteristic is very closely approached with thin magnetic film specimens where the anisotropy axis is produced by forming the film in the presence of an intense magnetic field (by vacuum deposition or electroplating in an applied field for example). For practical core materials as stated previously the B-H loop is somewhat between the ideals of Figs. 5.3.2 and 5.3.3: perhaps as shown in Fig. 5.3.7. Such materials are familiar for power transformers where the power loss is the most important consideration. Although the square B-H loop materials usually show somewhat lower losses their use is not justified because of significantly higher cost.

In general one finds experimentally that practical polycrystalline ferromagnetic or ferrimagnetic materials that exhibit electrical behaviour which can be explained in terms of either rotation or wall processes occur as material is switched from one state to another. Since rotational switching is so much faster, a great deal of interest was generated in the computer industry for very fast memory equipment. By using two directions of applied field and thin magnetic film with an induced anisotropy axis, switching times on the order of 10^{-9} sec are attainable. This speed is at least an order of magnitude faster than the fastest ferrite core switching obtained. However, the electronic circuits required to operate a film memory system have not yet been adequate to exploit the higher speed system economically.

Application and testing

Domain theory helps to gain some qualitative insights into magnetic switching. It does not suggest any simple way to test devices for specific applications involving magnetic switching. As discussed, B-H loops show a large variation in a given core depending upon the test conditions. The data obtained from such tests may be of so little value that other tests must be devised to suit the actual use of the core. Equipment such as self-saturating magnetic amplifiers and computer memories as a rule require more than one core. The most critical problem a designer faces is one of matching cores to each other. In memory equipment from 10^4 to 10^7 essentially identical cores must be found to populate a single equipment. Highly sophisticated automatic test machines are required. Basically the machines supply carefully regulated current pulses usually to a single turn winding and evaluate the induced voltage in another single turn winding. The programme of current pulses and the sort of evaluation of voltage response is determined by the application.

Ferromagnetic and ferrimagnetic material behaviour has been discussed on the basis of the material's response to applied magnetic fields. A variety of specialised device applications involve the same basis switching behaviour. The materials discussed operate in a cycle manner which can be described

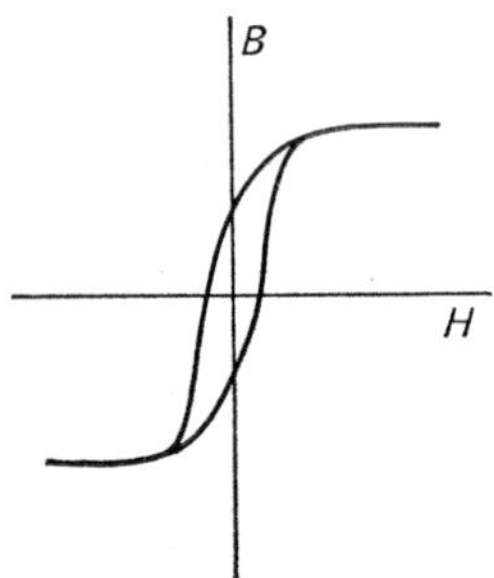

Fig. 5.3.8   Typical *B-H* loop for electrical sheet steel

by a *B-H* loop (see Fig. 5.3.8). A few *B-H* loops have been considered to relate material behaviour to electrical signals. Detailed applications of the magnetic materials have been avoided as beyond the scope of this discussion. Although some sort of *B-H* loop could be determined for any magnetic device in any switching application, the usually design approach is to use data and impose tolerance bands on the basis of a specific test regimen peculiar to the application.

Chapter 5.4 **Testing of electronic materials and parts**

This chapter is written to indicate the purpose of testing electronic materials or parts to determine resistance to deleterious effects of natural elements and conditions surrounding operational, physical and electrical tests.

The purposes described in this chapter give suitable conditions obtainable in a laboratory to give results equivalent to actual working conditions and to make each of the general tests adaptable to a broad range of electronic and electrical materials or parts.

The method of test, procedure, or apparatus required for test is not given, these can be found in the relative 'British Standards'. The purposes are divided into three classes as follows:

1 Environmental.
2 Physical characteristics.
3 Electrical characteristics.

Environmental tests

Salt spray

The salt spray test, in which specimens are subjected to a fine mist of salt solution, has several more or less useful purposes when utilised with full recognition of its deficiencies and limitations. Originally proposed as an accelerated laboratory corrosion test simulating the effects of sea-coast atmospheres on metals, with or without protective coatings, this test has been erroneously considered by many as an all-purpose accelerated corrosion test, which if withstood successfully will guarantee that metals or protective coatings will prove satisfactory under any corrosive condition.

Experience has since shown that there is seldom a direct relationship between resistance to salt-spray corrosion and resistance to corrosion in other media, even in so-called 'marine' atmospheres and ocean water. However, some idea of the relative service life and behaviour of different samples of similar, or closely related, metals or of protective coating-base metal combinations in marine and exposed sea-coast locations can be gained by means of the salt-spray test, provided accumulated data from correlated field surface tests and laboratory salt-spray tests show that such a relationship does exist, as in the case of aluminium. Such correlation tests are also necessary to show the degree of acceleration, if any, produced by the laboratory test.

The salt-spray test is generally considered unreliable for comparing the general corrosion resistance of different kinds of metal or coating-metal combinations, or for predicting their comparative service life. The salt-spray test has received its widest acceptance as a test for evaluating the uniformity — specifically, thickness and degree of porosity — of protective coatings, metallic and non-metallic, and has served this purpose with varying degrees of success. In this connection, the test is useful for evaluating different batches of the same product, once some standard level of performance has been established.

The salt-spray test is especially helpful as a screening test for revealing particularly inferior coatings. When used to check the porosity of metallic coatings, the test is more dependable when applied to coatings which are cathodic rather than anodic toward the basic material. This test can also be used to detect the presence of free iron contaminating the surface of another metal, by inspection of the corrosion products.

Temperature cycling

This test is conducted for the purpose of determining the resistance of a part or material to the shock of repeated surface exposures to extremes of high and low temperatures for comparatively short periods of time, such as would be experienced when materials and parts are transferred to and from heated shelters in cold climates. These conditions may also be encountered in equipment operated intermittently in low-temperature areas. It is not required that the specimen reach thermal stability at the temperature of the test chamber during the short exposure time specified.

Permanent changes in operating characteristics and physical damage produced during temperature cycling result principally from variations in dimensions and other physical properties and from alternate condensation and freezing of atmospheric moisture.

Effects of temperature cycling include:

1 Cracking.
2 Delamination of finishes and embedding compounds.
3 Opening of terminal seals and case seams.
4 Changes in electrical characteristics due to moisture effects or to mechanical displacement of conductors or of insulating materials.

Humidity (steady state)

This test is performed to evaluate the properties of materials used in electronic components as they are influenced by the absorption and diffusion of moisture and moisture vapours. This is an accelerated environmental test, accomplished by the continuous exposure of the specimen to a high relative humidity at an elevated temperature.

These conditions impose a vapour pressure on the material under test which constitutes the force behind the moisture migration and penetration. Hygroscopic materials are sensitive to moisture and deteriorate rapidly under humid conditions. Absorption of moisture by many materials results in swelling, which destroys their functional utility and causes loss of physical strength and changes in other important mechanical properties.

Immersion

This test is performed to determine the effectiveness of the seal component parts. The immersion of the part under evaluation into liquid at widely different temperatures subjects it to thermal and mechanical stresses that will readily detect a defective assembly, or a partially closed seam or moulded enclosure. Defects of this type can result from faulty construction or from mechanical damage such as might be produced during physical or environmental tests. The immersion test is generally performed immediately following such tests because it will tend to aggravate any incipient defects in seals, seams and brushings that might otherwise escape notice.

The test is essentially a laboratory test condition, and the procedure is intended only as a measure of the effectiveness of the seal following test. The choice of fresh or salt water as a test liquid is dependent on the nature of the component part under test.

When measurements are made after immersion cycling to obtain evidence of leakage through seals, the use of salt water solution instead of fresh water will facilitate detection of moisture penetration. The test provides a simple and ready means of detection of the migration of liquids.

Effects noted can include lowered insulation resistance, corrosion of internal parts and appearance of salt crystals. This test is not intended as a thermal-shock or corrosion test, although it may incidentally reveal inadequacies in these respects.

Barometric pressure (reduced)

The barometric pressure test is performed under conditions simulating the low atmospheric pressure encountered in the non-pressurised portions of aircraft and other vehicles in high-altitude flight. The test is intended primarily to determine the ability of component parts and materials to avoid dielectric-withstanding-voltage failures due to the lowered insulating strength of air and other insulating materials, at reduced pressures. Even when low pressures do not produce complete electrical breakdown, coronae and their undesirable effects, including losses and ionisation, are intensified. Low barometric pressures also serve to decrease the life of electrical contacts, since intensity of arcing is increased under these circumstances. For this reason, endurance tests of electromechanical components are sometimes conducted at reduced pressures.

This type of test is also performed to determine the ability of seals in components to withstand rupture due to the considerable pressure differentials that may be developed under these conditions. The simulated high-altitude conditions can also be employed to investigate the influence on components operating characteristics, of other effects of reduced pressure, including changes in dielectric constants of materials: reduced mechanical loading on vibrating elements, such as crystals, and decreased ability of thinner air to transfer heat away from heat-producing components.

x

Moisture resistance

The moisture-resistance test is performed for the purpose of evaluating, in an accelerated manner, the resistance of component parts and constituent materials to the deteriorative effect of moisture vapour and films by vulnerable insulating material and from surface wetting of metals and insulation. These phenomena produce many types of deterioration, including corrosion of metals, physical distortion and decomposition of organic materials, leaching out and spending of constituents of materials, and detrimental changes in electrical properties.

This test differs from the previously described 'steady-state' humidity test and derives its added effectiveness in its employment of temperature cycling which provides alternate periods of condensation and drying essential to the development of the corrosion processes and in addition, produces a 'breathing' action of moisture into partially sealed containers. Increased effectiveness is also obtained by the use of a higher temperature, which intensifies the effects of humidity. The test includes low-temperature and vibration subcycles that act as accelerants to reveal otherwise indiscernible evidence of deterioration since stresses caused by freezing moisture and accentuated by vibration tend to widen cracks and fissures. As a result the deterioration can be detected by the measurement of electrical characteristics, including such tests as dielectric withstanding voltage and insulation resistance, or by performance of a test for sealing.

Provision is made for the application of a polarising voltage across insulation to investigate the possibility of electrolysis, which can promote eventual dielectric breakdown. The test also provides for electrical loading of certain components, if desired, in order to determine the resistance of current carrying components, especially fine wires and contacts, to electrochemical corrosion. Results obtained with this test are reproducible and have been confirmed by investigations of field failures.

Thermal shock

This test is conducted for the purpose of determining the resistance of a part to exposures at extremes of high and low temperatures, and to the shock of alternate exposures to these extremes, such as would be experienced when materials or parts are transferred to and from heated shelters in very cold areas. Although it is preferred that the specimens reach thermal stability at the temperature of a test chamber during the specified exposure, in the interest of saving test time, parts may be tested at the minimum exposure durations specified, which will not insure thermal stability, but only an approach thereto.

Permanent changes in operating characteristics and physical damage produced during thermal shock result principally from variations in dimensions and other physical properties.

Effects of thermal shock include cracking and delamination of finishes, cracking and crazing of embedding and encapsulating compounds, opening of thermal seals, cans, etc., leakage of filling materials and changes in electrical characteristics due to mechanical displacement or rupture of conductors or of insulating materials.

Life — at high ambient temperatures

The purpose of this test is to determine the effects on electrical and mechanical characteristics of a part, resulting from exposure of the part to a high ambient temperature for a specified length of time, while the part is performing its operational function.

Evidence of deterioration resulting from this test can at times be determined by visual examination. However, the effects may be more readily ascertained by measurements or tests prior to, during, or after exposure. Surge current, total resistance, dielectric strength, insulation and capacitance are types of measurements that would show the deleterious effects due to exposure to elevated ambient temperatures.

Flammability (external flame)

This test is performed for the purpose of determining the flammability of a part or a material exposed to an external flame. Flammability is defined as the ability of a part to support combustion. This can be determined by the following:

1 Time taken for a part to become self-extinguishing after application of a flame.
2 Part does not support violent burning.
3 Exposure of a part to a flame does not result in an explosive-type fire.
4 Spreading of surface burning on larger parts is deterred.

The principle factors which affect the results of a flame test are: the heat of the flame; the volume of the part and other heat sink effects; the time of exposure to the flame; the presence of circulating air; variations of the materials and surface of the parts.

Physical characteristics

Vibration

The vibration test is used to determine the effects on component parts of vibration within the predominant frequency ranges and magnitudes that may be encountered during service. Most vibration encountered in field service is not of a simple harmonic nature, but tests based on vibrations of this type have proved satisfactory for determining critical frequencies, modes of vibration and other data necessary for planning protective steps against the effects of undue vibration. Vibration, can produce objectional operating characteristics, noise, wear and physical distortion and often results in fatigue and failures.

Shock — small parts

A shock test is intended to determine the suitability of small parts, with or without auxiliary protection, for use in electronic equipment that may be subjected to moderately severe shocks as a result of suddenly applied forces or abrupt changes in motion produced by rough handling, transportation or field operation.

Shocks of this type may disturb operating characteristics or cause damage similar to that resulting from excessive vibration particularly if the shock pulses are repetitive.

Solderability

The purpose of this test is to determine the solderability of all solid and stranded wires up to 3·2 mm thickness and lugs, tabs, hook heads, turrets, etc., that are normally joined by a soldering operation.

This determination is made on the basis of the ability of these terminations to be wetted by a new coat of solder, or to form a suitable fillet when dip soldered to a specially prepared solderable wire. The procedure will verify that the treatment used in the manufacturing process to facilitate soldering is satisfactory and has been applied to the required portion of the part that is designed to accommodate a solder connection. Accelerated ageing is included in the test method; this simulates a minimum of 6 months' natural ageing under a combination of various storage conditions having different effects.

Radiographic

Radiographic inspection is generally a non-destructive method for detecting internal physical defects in small parts that are not otherwise visible. Radiographic techniques are intended to reveal such flaws as improper positioning of elements, voids in encapsulating or potting compounds, homogeneities in materials, presence of foreign bodies or materials, broken elements, etc.

Resistance to soldering heat

This is performed for the purpose of determining whether component parts can withstand the effects of the heat to which they will be subjected during the soldering of their terminations. The heat can be either conducted heat through the termination into the component part, or radiant heat emanating from the solder dip or soldering iron when in close proximity to the body of the component part, or both. In order to establish a standard procedure for the most nearly reproducible method, the use of soldering irons (with their necessary broad tolerance on temperatures and other variables in the soldering-iron procedure) is not covered by this method. However, the solder-dip method is used because of its more controllable conditions.

The heat of soldering can affect the electrical characteristics of the component part and cause mechanical damage to the materials making up the part, e.g. electrical characteristics of resistive elements, semiconductor devices, soldered connections, plastics, and insulation materials are affected in varying degrees by heat, mechanical damage, such as loosening of terminations or windings, softening of insulation, opening of solder seals, and weakening of mechanical joints, can occur. For this test, some specifications which cover component parts that are fabricated of heat-susceptible materials permit the use of attached heat sinks to reduce conducted heat, or specify some form of shielding to reduce exposure to radiant heat. When allowed or required by the component-part specification, the heat sinks should simulate actual precautions which must be taken during soldering, and the shielding should afford the protection such as would be present when soldering terminations which protrude through a panel or printed-circuit board.

Terminal strength

This test is performed to determine whether the design of the terminals and their method of attachment can withstand one or more of the applicable mechanical stresses to which they will be subjected during installation or disassembly in equipment. These stresses must be withstood by the component part without sustaining damage which would affect either the utility of the terminals or the operation of the component part itself. Evidence of damage caused by this test may not become evident until subsequent environmental tests are performed, such as seal, moisture resistance, or life. Procedures are established in this method for testing wire-lead terminals, flexible-flat-strip or tab-lead terminals, and rigid-type terminals which are threaded or have other arrangements for attaching conductors. The forces applied consist of direct axial, radial or tension pulls, twist, bending torsion, and the torque exerted by the application of nuts or screws on threaded terminals. These applied stresses will disclose poor workmanship, faulty designs, and inadequate methods of attaching terminals to the body of the part. Other evidence of damage may be disclosed by mechanical distortion of the part, breaking of seals, cracking of materials surrounding the terminals, or changes in electrical characteristics, such as shorted or interrupted circuits and change in resistance values.

Shock components and subassemblies

This test is conducted for the purpose of determining the suitability of component parts and subassemblies of electrical and electronic components when subjected to shocks such as those which may be expected as a result of rough handling, transportation and military operations. It differs from other shock tests in that the design of the shock machine is not specified, but the half-sine and saw-tooth shock pulse waveforms are specified with tolerances. The frequency response of the measuring system is also specified with tolerances.

Electrical characteristics

Resistance to solvents

The purpose of this test is to verify that the markings or colour-coding will not become illegible or discoloured on the component parts (including

X2

printed-wiring boards) when subjected to solvents (normally used to clean solder-flux, fingerprints, and other contaminants from printed-wiring and terminal-board assemblies, etc.). The solvents should not cause deleterious, mechanical or electrical damage, or deterioration of the materials or finishes.

Dielectric withstanding voltage

Dielectric withstanding voltage (also called high-potential, over-potential, voltage, voltage-breakdown, or dielectric-strength test) consists of the application of a voltage for a specific time between mutually insulated portions and ground. This is used to prove that the component part can operate safely at its rated voltage and withstand momentary over-potentials due to switching, surges, and other similar phenomena.

Although this test is often called a voltage breakdown or dielectric strength, it is not intended to cause insulation breakdown or that it be used for detecting coronae, rather it serves to determine whether insulating materials and spacings in the component part are adequate. When a component part is faulty in these respects, application of the test voltage will result in either disruptive discharge or deterioration. Disruptive discharge is evidenced by flashover (surface discharge), sparkover (air discharge), or breakdown (puncture discharge). Deterioration due to leakage currents may change electrical parameters or physical characteristics.

Insulation resistance

The intended purpose of this test is to measure the resistance offered by the insulating members of a component part to an impressed direct voltage tending to produce a leakage of current through or on the surface of these members. A knowledge of insulation resistance is important, even when the values are comparatively high, as these values may be limiting factors in the design of high-impedance circuits. Low insulation resistances, by permitting the flow of large leakage currents, can disturb the operation of circuits intended to be isolated, for example, by forming feedback loops.

Excessive leakage currents can eventually lead to deterioration of the insulation by heating or by direct-current electrolysis. Insulation-resistance measurements should not be considered the equivalent of dielectric withstanding voltage or electric breakdown tests. A clean, dry insulation may have a high insulation resistance, and yet possess a mechanical fault that would cause failure in the dielectric withstanding voltage test.

Conversely, a dirty deteriorated insulation with a low insulation resistance might not break down under a high potential. Since insulating members are composed of different insulation resistances, the numerical value of measured insulation resistance cannot properly be taken as a direct measure of the degree of cleanliness or absence of deterioration.

The test is especially helpful in determining the extent to which insulating properties are affected by deteriorative influences, such as heat, moisture, dirt, oxidation, or loss of volatile materials.

D.C. resistance

The purpose of this test is to measure the direct-current (d.c.) resistance of resistors, electromagnetic windings of components and conductors. It is not intended that this should apply to the measurement of contact resistance.

Resistance — temperature characteristics

This test determines the percentage change in direct-current (d.c.) ohmic resistance from the d.c. ohmic resistance at the reference temperature, per unit temperature difference between the test temperature and the reference temperature.

The equation used to calculate this characteristic, commonly called the 'temperature coefficient of resistance', is based on the assumed straight-line relationship between resistance and temperature over a range of specified test temperatures.

Capacitance

The purpose of this test is to measure the capacitance of component parts. Preferred test frequencies for this measurement are 60 Hz, 120 Hz, 1 kHz and 1 MHz.

Quality factor (Q)

The purpose of this test is to measure the quality factor, commonly called Q, of electronic parts such as capacitors and inductors. By definition, the factor Q expresses the ratio of reactance to effective resistance of a circuit element. This numerical ratio is considered a 'figure of merit' for a reactive component (or a resonant circuit utilising such components) as it is a measure of the ability of the component (or circuit) to store energy compared to the energy it wastes. For this reason Q is called 'storage factor'. Q is thus equal to the inverse of the dissipation factor. A relationship exists between Q and the properties of a tuned circuit, such as the resonant rise in voltage phenomena.

Contact resistance

The contact-resistance test is to determine the resistance offered to a flow of current during its passage between the electrical-contacting surfaces of connecting components, such as plugs, jacks, connectors and sockets, or between the electrical contacts of current-controlling components, such as switches, relays and circuit breakers. For practical reasons lead and terminal resistances may be included in the actual measurement, as well as the contact resistance proper. In many applications it is required that the contact resistance be low and stable, so that the voltage drop across the contacts does not affect the accuracy of the general circuit conditions. If large currents are passed through high-resistance contacts, excessive energy losses and dangerous over-heating of the contacts may occur.

Current noise test for fixed resistors

This resistor noise test is performed for the purpose of establishing the 'noisiness' or 'noise quality' of a resistor in order to determine its suitability for use in electronic circuits having critical noise requirements. This method is intended as a standard reference for the determination of current noise present in a resistor, for use in an application with specific current-noise requirements.

It is not intended as a general specification requirement. Interference caused by the generation of spurious noise signals in parts tends to mask the desired output signal, thus resulting in loss of information. For low-level audiofrequency and other low-frequency circuits, where low-noise parts are used, resistors may become an important source of interfering noise.

One source of noise in a resistor is molecular thermal motion which generates a fluctuation voltage termed 'thermal noise'. It is not necessary to determine the magnitude of thermal noise by measurement since the mean-square value of the fluctuation voltage is predictable from Nyquist's equation, which shows the mean-square value to be proportional to the product of resistance, temperature, and the pass band of the measuring system.

Generally, an increase in fluctuation voltage appears when direct current (d.c.) is passed through resistive circuit elements. The increase in fluctuation voltage is termed 'excessive noise'. The magnitude of current noise is dependent upon many inherent properties of the resistor such as resistive material and other factors such as processing, fabrication, size and shape of resistive element.

Since there is no apparent functional relationship between current noise and many of these factors, current noise generally cannot be predicted from physical constants. Therefore, it is necessary to measure current noise to determine its magnitude. In this test a method has been designed to evaluate accurately the 'noisiness' or 'noise quality' of individual resistors in terms of a noise-quality index. The noise-quality index, expressed in decibels (dB), is a measure of the ratio of the root-mean-square (r.m.s.) value of current-noise voltage, in microvolts (μV), to the applied d.c. voltage (V). The pass band associated with the noise-quality index is one frequency decade, geometrically centred at 1 kHz. This index is termed the microvolts-per-volt-in-a-decade 'index'. In the design of circuits, an added advantage accrues from the definitiveness of the index which allows the estimation of interference attributable to current noise. Conversely, for a given limit of current-noise interference in a particular circuit design, a maximum acceptable value of the index may be established. Ordinarily, it is not necessary to duplicate the operating conditions of the particular circuit design when measuring the current noise.

The noise quality of populations of resistors may be reasonably estimated by measurement of the index of representative groups of resistors using suitable sampling procedures. Measurements on sample groups tend to have a normal distribution and once representative parameter values for the distribution have been established (the mean and standard deviation), such parameter values would serve as norms in judging 'noisiness' and product uniformity insofar as noise is concerned.

Voltage coefficient of resistance determination

Certain types of resistors exhibit a variation of resistance with changes in voltage across the resistor. This is measurable characteristic; a test to determine the magnitude of such a characteristic is the voltage coefficient of resistance determination procedure.

Contact-chatter monitoring

This test is conducted for the purpose of detecting contact-chatter in electrical and electronic component parts having moveable electrical contacts, such as relays, switches and circuit breakers, where it is required that the contacts do not open or close momentarily, as applicable, for longer than a specified time under environmental test conditions, such as vibration, shock, or acceleration. This test provides standard procedures for monitoring such 'opening of closed contacts' or 'closing of open contacts'.

Annotated bibliography

This bibliography has been arranged in subjects order as below:

1 Bonding.
2 Dielectrics.
3 Encapsulation.
4 Ferrites.
5 Flat cables.
6 Insulating materials.
7 Microelectronics.
8 Tubing and sleeving.
9 Thick and thin films.

1 Bonding

AMLINGER, P. R., and ROGERS, S. L., *Microbonds for hybrid microcircuits*, 5th Quarterly Report, Hamilton Standard Division, United Aircraft Corporation (April 1965).

RIBEN, A. R., and SHERMAN, S. L., *Microbonds for hybrid microcircuits*, Final Report, Hamilton Standard Division, United Aircraft Corporation (January 1967).

SAWYER, H. F., OLDAKER, D. R., and BATEMAN, V. G., *Research and development for three-dimensional welded circuit packaging design requirements*, Final Report, General Dynamics, Pomona, California (June 1963).

BELSER, R. D., 'Electrical resistance of thin-metal films before and after artificial ageing by heating', *J. appl. Phys.*, V28(1) (January 1957).

OWEN, D. B., *Factors for one-sided tolerance limits and for variables sampling plans*, Technical Sandia Corporation (March 1963).

2 Dielectrics

VON HIPPEL, A., *Dielectrics and waves*, John Wiley, New York (1954).

CLARK, F. M., *Insulating materials for design and engineering practice*, John Wiley, New York (1956).

MILLER, A. N., *Non-destructive high potential testing*, Haydon Book Co., New York (1964).

VAN BLADEL, J., *Electromagnetic fields*, McGraw-Hill, New York (1964).

KRAUSS, J. D., *Electromagnetics*, McGraw-Hill, New York (1953).

HARPER, C. A., *Electronic packaging with resins*, McGraw-Hill, New York (1961).

HEWLETT PACKARD COMPANY, *Microwave theory and measurements*, Prentice-Hall, Englewood Cliffs, New Jersey (1963).

VOLK, –., LEFFORAGE, –., and STETSON, –., *Electrical encapsulation*, Reinhold, New York (1962).

KIP, F., *Electricity and magnetism*, McGraw-Hill, New York (1962).

BAKER, W. P., *Electrical insulation measurements*, Chemical Publishing Co., New York.

3 Encapsulation

JOHNSTON, C. A., and MARTIN, R. E., 'Glass resins', IEEE Paper 32 C3. 52 1965, Electrical Insulation Conference (1965).

BATTAGLINI, T. E., *Current trends and future needs for plastics in computers*, Society of Plastic Industry, National Conference New York (6–7 June 1966).

DORMAN, E. N., 'Cycloaliphatic epoxy resins. Premium properties for electrical insulation', Paper XXI, 1965 Conference, Society of Plastic Engineers.

BAKER, E. C., 'Electronic encapsulation thermal conductivities extended on order of magnitude', *Electronic Packaging and Production*, p. 80 (June 1966).

WOODLAND, J. H., 'Rigid urethane foam-substitute polyester casting compound', Paper XXVI, 1964 Conference, Society of Plastic Engineers.

4 Ferrites

BRINKER, H., and HELLAND, A., 'An evaluation of magnetic logic', *Electro-Technology*, p. 94 (March 1966).

FREEMAN, D., 'Iron powder cores from a users point of view', *Magnetic Core Conference Proceedings*, Volume 7 (16–17 June 1965).

ROBINSON, J., 'Fundamentals of EMI shielding', *Electro-Technology*, p. 36 (June 1966).

JORGENSEN, C. M., 'Electromagnetic-interference', *ibid.*, p. 95 (May 1966).

ATHEY, S. N., *Magnetic tape recording*, NASA Publication S.P. 5038 (January 1966).

MANTELL, C., *Engineering materials handbook*, McGraw-Hill, New York (1958).

GREENWOOD, D., *Manual of electromechanical devices*, McGraw-Hill, New York (1965).

BOZORTH, R., *Magnetic properties of metals and alloys*, American Society for Metals, Cleveland (1959).

5 Flat cables

ANGELE, W., *Flat cable manufacture and installation techniques*, NASA.

PRISE, W., 'A-state-of-the-art review', Lockhead Missile and Space Company, Sunnyvale, California, *Electronic Packaging and Production* (December 1966).

GODWIN, E. F., 'Flat cables of lightweight interconnections', USAECOM, Fort Monmouth, NJ, 14th Annual Wire and Cable Symposium, Atlantic City (December 1965).

LANGE, R. O., 'Flat cables — test before flying', Lockhead Missiles and Space Company, Sunnyvale, California, *Packaging and Materials EDN Magazine* (February 1967).

ALEXKAIS, N. G. (Consultant), 'Microelectronics', *EDN Magazine*, Los Angeles, California (December 1965).

THE INSTITUTE PRINTED CIRCUITS, *Multilayer printed circuit boards technical manual* (March 1966).

SANDERS ASSOCIATES, *Flexprint circuit design handbook*, Bulletin FT–169, Nashua, NH (1965).

FIDERER, L., Hughes Aircraft Company, Culver City, California, 'Shielding module interconnections', *Electronic Packaging and Production* (August 1964).

6 Insulating materials

HARPER, C. A., *Electronic packaging with resins*, McGraw-Hill, New York (1961).

JOHNSON, L. M., *Epoxy resin stress behaviour at low temperatures. Electrical shorts*, 3M Company (July 1964).

DOSSER, D. M., 'Defects in epoxy resin castings', *Insulation Magazine* (November 1962).

OLYMPHANT, M., 'Thermal shock tests for casting resins', Conference Electrical Insulating Materials, Cleveland, Ohio (1958).

HANSON, W. M., *Electrolytic corrosion*, 3M Company (May 1963).

JAVITZ, A. E., 'Epoxy resin system for embedded circuits and components', *Electrical Manufacturing* (April 1955).

HARPER, C. A., 'Electronic package designs', *Electro-Technology*, p. 89 (November 1964).

7 Microelectronics

FRÖHLICH, H., *Theory of dielectrics*, Clarendon Press, Oxford (1949).

YOUNG, L., *Anodic oxide films*, Academic Press, New York (1961).

VERMILYEA, D. A., *J. appl. Phys.*, V36, p. 3663 (1965).

YOUNG, L., *Trans. Faraday Soc.*, V55, p. 842 (1959).

GEEL, VAN, CH. W., *Halbleiterprobleme*, VI, p. 291 (Braunschweig 1955).

SASKI, Y., *J. Phys. Chem. Solids*, V13, p. 177 (1960).

MCLEAN, D. A., and POWER, F. S., *Proc. IRE*, V44, p. 872 (1956).

BERRY, R. W., and SLOAN, D. J., *ibid.*, V47, p. 1070 (1959).

VROMEN, B. H., and KLERER, J., *IEEE Trans. Comp. Parts*, PMP–1, S–1 (1965).

MILES, J. S., and SMITH, P. H., *J. Electrochem. Soc.*, V110, p. 1240 (1963).

JOHNSON, M. C., *IEEE Trans. Comp. Parts*, CP–11, 1 (1964).

WHITMORE, R. E., and VOSSEN, J. L., *ibid.*, PMP–1, S–10 (1965).

MEAD, C. A., *Phys. Rev. Letters*, V6, p. 545 (1961).

SIDDALL, G., *Vacuum*, V9, p. 274 (1959).

HIROSE, H., and WADA, Y., *Japan J. Appl. Phys.*, V3, p. 179 (1964).

LEWIS, B., *Microelectronics and Reliability*, V3, p. 109 (1964).

FEUERSANGER, A. E., HAGENLOCHER, A. K., and SOLOMON, A. L., *J. Electrochem. Soc.*, V111, p. 1378 (1964).

VRATNY, F., and HARRINGTON, D. J., *ibid.*, V112, p. 484 (1965).

DAVIDSE, P. D., and MAISSEL, L. J., *J. appl. Phys.*, V37, p. 574 (1966).

CLARK, R. S., *IEEE Trans. Comp. Parts*, PMP–1, S–31 (1965).

JACKSON, N. F., HOLANDS, E. J., and CAMPBELL, D. S., 'IERE-IEE Conference on "Applications of Thin Films in Electronic Engineering" ' (1966).

LAKAHMANAN, T. K., WYSOCKI, C. A., and SLEGESKY, W. J., *IEEE Trans. Comp. Parts*, CP–11, 14 (1964).

FRISER, R. G., *J. Electrochem. Soc.*, V113, p. 357 (1966).

CLARK, R. S., *Trans. Metal Soc.*, *AIMEE*, V233, p. 592 (1965).

SINCLAIR, W. R., and PETERS, F. G., *Rev. scient. Instrum.*, V33, p. 744 (1962).

STERLING, H. F., and SWANN, R. C., *Solid state Electronics*, V8, p. 59 (1965).

STUART, M., *Nature*, V199, p. 59 (1963).

WHITE, P., *Insulation*, p. 57 (1963).

HALLER, I., and WHITE, P., *J. Phys. Chem.*, V67, p. 1784 (1963).

CARIOU, F. E., and VALLEY, D. J., *IEEE Trans. Comp. Parts*, PMP–1, S–54 (1965).

LIGENZA, J. R., *J. appl. Phys.*, V36, p. 2703 (1965).

CALLINAN, T. D., and ROMANS, J. B., *Elect. Manufacturing*, p. 146 (May 1957).

MEDOWSKI, J., *Prace Inst. Tele-i Radiotechnicznego*, V8, p. 61 (1964).

HENNIGER, P., *Frequenz*, V17 (September 1963).

HERBERT, J. M., IEE Conference on Dielectric and Insulating Materials, London (1964).

SKANAVI, G. I., and MATVEEVA, E. N., *Zh. eksp. teor. Fiz.*, V30, p. 1047 (1956).

8 Tubing and sleeving

Staffs Reports. 'Flexible Tubings, Sleevings and Heat-Shrinkable Special Devices.' Insulation Directory/Encyclopedia, Lake Publishing Company (1967).

Technical Data Bulletin. A. 7000–2 Rayclad Tubes Incorporated (1967).

NEMA. Publication No. VS 1–1962. Coated Electrical Sleeving.

ASTM. D 922–65, 'Standard Specification for Non-Rigid P.V.C. Tubing.'

WEBB, J. A., 'Heat Shrinkable Tubings — Material and Applications.' Electronic Packaging and Production (April 1966).

9 Thick films

COLLINS, F. M., and PARKS, C. F., 'Thallium oxide glaze resistors', Electronic Components Conference Proceedings, pp. 432–7 (1967).

SPIGARELLI, D. J., 'Selection of furnace equipment for thick film firing', Proceedings of First Technical Thick-film Symposium, pp. 37–43.

STEIN, S. J., and EBLING, W. F., 'Some practical considerations in the fabrication of printed glazed resistors and circuits', Proceedings of the 1966 Electronic Components Conference, pp. 8–16.

HOFFMAN, L. C., BUCCHETTA, V. L., and FREDRICK, K. W., 'Adhesion of platinum–gold glaze conductors', Proceedings of the 1963 Electronic Components Conference, pp. 381–6.

DRIEAR, J. R., 'Observations on formation of "cleavage" at interface between glaze resistors and terminations', Proceedings of the 1965 Electronic Components Conference, pp. 387–92.

O'CONNELL, J. A., 'Thick film technology', 1966 Proceedings of NEP/CON, pp. 112–29.

HOFFMANN, L. C., 'Precision glaze resistors', *The American Ceramic Society Bulletin*, pp. 490–3 (September 1963).

SHORT, A. O., *Silver migration in electric circuits*, Tele-Tech and Electronic Industries (February 1956).

MELAN, E. H., and MONES, A. H., 'The glaze resistor — its structure and reliability', Proceedings of the 1964 Components Conference, p. 76.

SCHNEIDER, M., and AUDA, D. K., 'Problems in establishing a thick-film microcircuit facility', Proceedings of First Technical Thick-film Symposium 1967, pp. 91–101.

BURKS, D. P., GREENSTEIN, B., and MAHER, J. P., 'Screened thick-film resistors', Proceedings of 1967 Components Conference, pp. 217–28.

ILGENFRITZ, R. W., 'Controlled processing for precision thick film resistors', Proceedings of the 1967 Components Conference, pp. 224–37.

SAILER, E., and KENNEDY, A., 'The use of Silicons in a low cost high reliability microcircuit package', 1966 Proceedings of NEP/CON, pp. 396–405.

STEIN, S. J., 'Recent Advances in Thick-Film Symposium 1957', pp. 133–43.

HUGHES, D. C., *Screen printing of microcircuits*, Dan Mar Publishing Company, Somerville, NJ (1967).

HOLLAND, L., *Thin film microelectronics*, Chapman and Hall (1965).

LLOYD, P., 'Review of thin film techniques for microelectronics', *Microelectronics and Reliability*, V6, p. 177 (1967).

ANDERSON, D., 'Summary of thin film deposition techniques', *SCP and Solid-State Technology*, V9, p. 27 (December 1966).

ILGENFRITZ, R., 'Thick film hybrid microelectronic circuit technology', *ibid.*, V9, p. 35 (June 1966).

ROBINSON, P. H., and MULLER, C. W., 'Deposition of silicon upon sapphire substrates', *Trans Met. Soc. AIME*, V236, p. 268 (March 1966).

ANDERSON, J. C., and STIRLAND, D. J., 'Electron microscopy of very thin film', p. 163 et seq of *The use of thin films in physical investigations*, Academic Press, London (1966).

WARNER, R. M., and FORDEMWALT, J. N., *Integrated circuits*, McGraw-Hill, New York (1965).

MICHAEL, P. C., 'Some applications of thin film circuits in electronic equipment', *Electronic Equipment News* (November 1966).

GALPIN, R. K., TARBIN, F. G., HAWKES, P. L., and SAROGA, W., 'Practical tantalum thin-film single sideband demodulator', *ibid.* (Microelectronics in Equipment — Conference, paper A5) (1966).

MCLEAN, –., and ORR, W. H., 'Tantalum integrated circuits', *Bell Lab. Record*, V44, p. 304 (October/November 1966).

WHITE, D., and HOPPS, R. E., 'Tantalum thin-film circuits', *AEI Telecommunications Journal*, V1, p. 21 (1966).

PITT, K. G., POOL, W. J., and WALTON, B., 'Comparison of the use of thin and thick resistors and conductors in hybrid integrated circuits', *IEE Convention Publication*, Nos. 30 and 91 (1967).

DU PONT COMPANY LEAFLET, *Screen printable dielectric* (28 February 1967).

HAWKES, P. L., and HOBBS, R. E., 'Specification and design of film and hybrid integrated circuits', Paper delivered at a Symposium, Northern Polytechnic, London (31 May to 2 June 1967).

BINGHAM, K. C., CUTTEL, W. R., and GILLINGHAM, A., *IEE Conference Publication*, No. 30, p. 65 (1967).

COOMBE, R. A., *Electrical properties and applications of thin-films*, Pitman (1967).

Index